AF751331

Methods in Molecular Biology

Series Editor
John M. Walker
School of Life and Medical Sciences
University of Hertfordshire
Hatfield, Hertfordshire, AL10 9AB, UK

For further volumes:
http://www.springer.com/series/7651

Morpholino Oligomers

Methods and Protocols

Edited by

Hong M. Moulton

Department of Biomedical Sciences, College of Veterinary Medicine, Oregon State University, Corvallis, OR, USA

Jon D. Moulton

Gene Tools, LLC, Philomath, OR, USA

Humana Press

Editors
Hong M. Moulton
Department of Biomedical Sciences
College of Veterinary Medicine
Oregon State University
Corvallis, OR, USA

Jon D. Moulton
Gene Tools, LLC
Philomath, OR, USA

ISSN 1064-3745 ISSN 1940-6029 (electronic)
Methods in Molecular Biology
ISBN 978-1-4939-6815-2 ISBN 978-1-4939-6817-6 (eBook)
DOI 10.1007/978-1-4939-6817-6

Library of Congress Control Number: 2017937324

© Springer Science+Business Media LLC 2017
This work is subject to copyright. All rights are reserved by the Publisher, whether the whole or part of the material is concerned, specifically the rights of translation, reprinting, reuse of illustrations, recitation, broadcasting, reproduction on microfilms or in any other physical way, and transmission or information storage and retrieval, electronic adaptation, computer software, or by similar or dissimilar methodology now known or hereafter developed.
The use of general descriptive names, registered names, trademarks, service marks, etc. in this publication does not imply, even in the absence of a specific statement, that such names are exempt from the relevant protective laws and regulations and therefore free for general use.
The publisher, the authors and the editors are safe to assume that the advice and information in this book are believed to be true and accurate at the date of publication. Neither the publisher nor the authors or the editors give a warranty, express or implied, with respect to the material contained herein or for any errors or omissions that may have been made. The publisher remains neutral with regard to jurisdictional claims in published maps and institutional affiliations.

Printed on acid-free paper

This Humana Press imprint is published by Springer Nature
The registered company is Springer Science+Business Media LLC
The registered company address is: 233 Spring Street, New York, NY 10013, U.S.A.

Preface

Morpholino phosphorodiamidate antisense oligos (Morpholinos) were originally developed for therapeutic applications but have found widespread use as research reagents. They are readily available, easy-to-handle, and stable oligos that bind to complementary RNA and prevent other large molecules from accessing their bound target, like sequence-specific masking tape for RNA. Their function depends on the usual biological function of their target sequence. They have been used in animals, bacteria, protists, plants, and fungi. They don't rely on the cells for enzymatic activity such as RNase-H or RISC. Instead the oligos simply bind to and block their complementary targets—and sometimes other partially complementary RNA sequences, so appropriate controls are important. Clinical trials with Morpholino oligos have had excellent safety profiles, and have led to the first approved Morpholino drug, eteplirsen (ExonDys51), for treatment of some mutations causing Duchenne muscular dystrophy. Meanwhile, their efficacy, specificity, and great versatility as antisense effector and diagnostic molecules in a broad range of biological systems bode well for their long-term place in the toolbox of molecular biology.

This book begins with a historical perspective on Morpholinos by Jim Summerton, the biochemist who originally conceived of these oligos. The following three chapters present an overview of good Morpholino practices, the first publication of studies of aggregation of Morpholinos in solution, and a description of techniques for chemically modifying Morpholinos to add additional functions. Chapter 5 reviews and presents techniques for controlling Morpholino activity with light. Techniques for modulating microRNA activity in zebrafish embryos are presented in Chapter 6. Chapter 7, on probing genes during fin regeneration, presents local Morpholino delivery techniques based on injection then electroporation for zebrafish. Next methods for determining the structure of gene networks during development are discussed in the context of sea urchin embryos. Electroporation again appears, this time for knockdowns in chick embryos in Chapter 9. In Chapter 10 bacterial knockdowns, a potential alternative to antibiotics, are introduced using Morpholinos conjugated with cell-penetrating peptides (PPMO). Electroporation has proven effective to introduce Morpholinos into the protist *Giardia lamblia*, discussed in Chapter 11. A few less-commonly used RNA targets are discussed next, polyadenylation sequences in Chapter 12 and noncoding RNAs in Chapter 13. Pretargeting, an application that uses a pair of complementary Morpholinos to stick an effector (such as a radioisotope) onto a targeting moiety (such as an antibody) in vivo, is discussed in Chapter 14. Chapter 15 reviews diagnostic applications of Morpholinos and presents protocols for capacitance-based detection of nucleic acid analytes bound to complementary surface-attached Morpholinos. A PPMO is effective to inhibit a host factor that is involved in influenza infection in the lung of mice, as discussed for influenza in Chapter 16. Multiple-exon-skipping cocktails of Morpholinos have been delivered to dogs carrying a genetic model of Duchenne muscular dystrophy; protocols from these experiments are presented in Chapter 17. The retina is amenable to knockdown with Morpholinos in several organisms; protocols for knockdowns in developing mouse retina are given in Chapter 18. When Morpholinos are introduced to cerebrospinal fluid, they enter cells more effectively than from blood;

protocols for intracerebroventricular delivery in mice are discussed in Chapter 19. Techniques for delivering Morpholinos in utero to developing embryos have been established in sheep and are presented in Chapter 20. Pharmacokinetic, pharmacodynamic, and biodistribution studies can benefit from (1) a surface plasmon resonance-based assay that is label-free, antibody-free, and a walk-away assay for Morpholino quantification, as described in Chapter 21, and (2) in Chapter 22 an ELISA method offering rapid and inexpensive assays of Morpholino concentrations in biological extracts.

Thanks to all of the authors who have contributed to this volume, creating a broad survey of the diverse applications of Morpholinos along with protocols that will assist new labs in moving the frontier. We hope that Morpholino oligos will continue to be key tools in new biological discoveries, advancing our understanding and the development of drugs and devices to better the health of humans, our associated animals, and our environment.

Corvallis, OR, USA *Hong M. Moulton*
Philomath, OR, USA *Jon D. Moulton*

Contents

Contributors

YOSHITSUGU AOKI • *Department of Molecular Therapy, National Institute of Neuroscience, National Center of Neurology and Psychiatry, Kodaira, Japan*

JORDAN BOUTILIER • *Department of Biomedical Sciences, College of Veterinary Medicine, Oregon State University, Corvallis, OR, USA*

UMAR BURKI • *The John Walton Muscular Dystrophy Research Centre, The MRC Centre for Neuromuscular Diseases, Central Parkway, International Centre for Life, Institute of Genetic Medicine, Newcastle University, Newcastle upon Tyne, UK*

OANA CALUSERIU • *Department of Medical Genetics, University of Alberta Faculty of Medicine and Dentistry, Newcastle University, Edmonton, AB, Canada*

GARRICK CHOW • *Department of Biomedical Sciences, College of Veterinary Medicine, Oregon State University, Corvallis, OR, USA*

MADZIA P. CROSSLEY • *Department of Zoology, University of Cambridge, Cambridge, UK; Department of Chemical and Systems Biology, Stanford University School of Medicine, Stanford, CA, USA*

SETH M. DALY • *Departments of Internal Medicine, University of Texas Southwestern Medical Center, Dallas, TX, USA*

KATHRIN A. DUNLAP • *Department of Animal Science, Texas A&M University, College Station, TX, USA*

YUSUKE ECHIGOYA • *Department of Medical Genetics, University of Alberta Faculty of Medicine and Dentistry, Edmonton, AB, Canada*

ALEX SUTTON FLYNT • *Department of Biological Sciences, University of Southern Mississippi, Hattiesburg, MS, USA*

QIUMING GONG • *Knight Cardiovascular Institute, Oregon Health & Science University, Portland, OR, USA*

DAVID E. GREENBERG • *Department of Internal Medicine, University of Texas Southwestern Medical Center, Dallas, TX, USA; Department of Microbiology, University of Texas Southwestern Medical Center, Dallas, TX, USA*

M. KATHRYN IOVINE • *Department of Biological Sciences, Lehigh University, Bethlehem, PA, USA*

RAHUL KANADIA • *Department of Physiology and Neurobiology, University of Connecticut, Storrs, CT, USA*

DEVI KRISHNA PRIYA KARUNAKARAN • *Department of Physiology and Neurobiology, University of Connecticut, Storrs, CT, USA*

URSULA KONIGES • *Department of Chemical & Biomolecular Engineering, Tandon School of Engineering, New York University, Brooklyn, NY, USA*

JANA KRTKOVÁ • *Department of Biology, University of Washington, Seattle, WA, USA; Department of Experimental Plant Biology, Faculty of Science, Charles University in Prague, Prague, Czech Republic*

TORSTEN KRUDE • *Department of Zoology, University of Cambridge, Cambridge, UK*

RASTISLAV LEVICKY • *Department of Chemical & Biomolecular Engineering, Tandon School of Engineering, New York University, Brooklyn, NY, USA*

YONG-FU LI • *Gene Tools, LLC, Philomath, OR, USA*

GUOZHENG LIU • *Department of Radiology, University of Massachusetts Medical School, Worcester, MA, USA*
RIKA MARUYAMA • *Department of Medical Genetics, University of Alberta Faculty of Medicine and Dentistry, Edmonton, AB, Canada*
STEFAN C. MATERNA • *School of Natural Sciences, University of California Merced, CA, USA*
PAUL A. MORCOS • *OligoExpert, Corvallis, OR, USA*
JON D. MOULTON • *Gene Tools, LLC, Philomath, OR, USA*
HONG M. MOULTON • *Department of Biomedical Sciences, College of Veterinary Medicine, Oregon State University, Corvallis, OR, USA*
M. NIZZARDO • *Dino Ferrari Centre, Neuroscience Section, Department of Pathophysiology and Transplantation (DEPT), University of Milan, and Neurology Unit, IRCCS Foundation Ca' Granda, Milan, Italy*
ALEXANDER R. PAREDEZ • *Department of Biology, University of Washington, Seattle, WA, USA*
JAMES G. PATTON • *Department of Biological Sciences, Vanderbilt University, Nashville, TN, USA*
RICARDO RAJSBAUM • *Department of Microbiology and Immunology, University of Texas Medical Branch, Galveston, TX, USA*
MAHESH RAO • *Department of Biological Sciences, Vanderbilt University, Nashville, TN, USA*
M. RIZZUTI • *Dino Ferrari Centre, Neuroscience Section, Department of Pathophysiology and Transplantation (DEPT), University of Milan, and Neurology Unit, IRCCS Foundation Ca' Granda, Milan, Italy*
CLAUDIO D. STERN • *Department of Cell and Developmental Biology, University College London, London, UK*
VOLKER STRAUB • *The John Walton Muscular Dystrophy Research Centre, The MRC Centre for Neuromuscular Diseases, Central Parkway, International Centre for Life, Institute of Genetic Medicine, Newcastle University, Newcastle upon Tyne, UK*
CAROLYN R. STURGE • *Departments of Internal Medicine, University of Texas Southwestern Medical Center, Dallas, TX, USA*
SAULIUS SUMANAS • *Division of Developmental Biology, Cincinnati Children's Hospital Medical Center, Cincinnati, OH, USA*
JAMES E. SUMMERTON • *Gene Tools, LLC, Philomath, OR, USA*
SHIN'ICHI TAKEDA • *Department of Molecular Therapy, National Institute of Neuroscience, National Center of Neurology and Psychiatry, Kodaira, Japan*
NAPOLEON TERCERO • *Cytec Solvay Group, Stamford, CT, USA*
RYAN THUMMEL • *Department of Anatomy and Cell Biology and Opthamology, Wayne State University School of Medicine, Wayne State University, Detroit, MI, USA*
OCTAVIAN VOICULESCU • *Department of Physiology, Development and Neuroscience, Cambridge University, Cambridge, UK*
XIAOQUI WANG • *Department of Animal Science, Texas A&M University, College Station, TX, USA*
TOSHIFUMI YOKOTA • *Department of Medical Genetics, University of Alberta Faculty of Medicine and Dentistry, Edmonton, AB, Canada; The Friends of Garrett Cumming Research & Muscular Dystrophy, Canada HM Toupin Neurological Science Endowed Research Chair, Edmonton, AB, Canada*
ZHENGFENG ZHOU • *Knight Cardiovascular Institute, Oregon Health & Science University, Portland, OR, USA*

Chapter 1

Invention and Early History of Morpholinos: From Pipe Dream to Practical Products

James E. Summerton

Abstract

Beginning with my concept in 1969 to treat disease at the nucleic acid level using antisense nucleic acids, antisense has evolved to the current Morpholino oligos. Morpholinos have been the dominant gene knockdown system in developmental biology. Lack of delivery technologies has limited their use in adult animals (including humans), though alteration in muscles in Duchenne muscular dystrophy (DMD) allows delivery into adult muscle. Morpholinos are currently in Phase 3 clinical trials for DMD and a Morpholino oligo for skipping dystrophin exon 51 has been approved by the US FDA. With improved delivery techniques, such as those in development at Gene Tools, therapeutic Morpholinos for many difficult-to-treat diseases will be possible. Initial applications are expected to be custom cocktails of delivery-enabled Morpholinos for treating cancers.

Key words Morpholino oligomers, Antisense, Anti-gene, History, Delivery

1 Introduction

1.1 Conception

Evolution of Morpholino antisense oligos began in 1969 with my conception of a drug design strategy expected to provide treatments for most or all viral diseases, and perhaps a host of other diseases as well. That strategy is now referred to as “antisense” because it entails developing and using an antisense sequence complementary to the sense genetic sequence which one wishes to block or alter.

1.2 Research Applications

The first gene blocking agents evolved through many stages to our current Morpholino antisense oligos, which were launched in the year 1999 as custom research reagents that have been used mainly as gene modulating agents for studying the intricate cascades of gene activations and deactivations that are precisely controlled with respect to both time and position in rapidly maturing embryos, particularly zebrafish embryos, a preferred model organism in the

Hong M. Moulton and Jon D. Moulton (eds.), *Morpholino Oligomers: Methods and Protocols*, Methods in Molecular Biology, vol. 1565, DOI 10.1007/978-1-4939-6817-6_1, © Springer Science+Business Media LLC 2017

developmental biology field. Morpholinos, micro-injected into fertilized eggs, have dominated antisense applications in the developmental biology field [1, 2, 3] because only Morpholinos: (1) provide adequate sequence specificity; (2) are completely stable in biological systems; and (3) are generally free of the off-target effects that commonly plague other antisense structural types, including S-DNAs, siRNAs, and shRNAs.

1.3 Therapeutics: The Low-Hanging Fruit

The lack of a technology for safely delivering Morpholinos into the proper subcellular compartments in animals (including humans) has long deterred broad therapeutic applications of Morpholinos. However, for the special case of Duchenne muscular dystrophy (DMD) the underlying mechanism of the disease itself facilitates delivery of therapeutic levels of Morpholinos into just those cells (muscle cells) that need the therapeutic treatment. To exploit this special case, the biotech company, Sarepta Therapeutics (previously named: AVI Biopharma and Antivirals Inc.), has developed a safe and effective Morpholino drug for treating DMD, and has carried that drug through successful Phase 1 and 2 clinical trials (completed in late 2015) through Phase 3 to approval of the Morpholino drug eteplirsen in September 2016.

1.4 Therapeutics: The Final Component?

A safe and effective delivery technology has long been needed in order for Morpholinos to fulfill their promise of therapeutics for a broad range of diseases. As of March 2016 it appears researchers at GENE TOOLS have finally developed a delivery technology capable of providing truly safe and effective delivery of Morpholinos in vivo. While this new delivery technology is still being optimized, preliminary results suggest that Morpholinos will now finally be able to provide the long-promised therapeutics for a broad range of difficult-to-treat diseases. Barring unexpected problems, our first application of these new delivery-enabled Morpholinos will comprise custom cocktails for treating cancers.

2 A difficult Start Toward Antisense: 1969–1980

2.1 A Key Seminar

In 1969, I was a graduate student who had recently joined the lab of Dr. Christopher Mathews in the Biochemistry Department of the new medical school at the University of Arizona in Tucson. Dr. Mathews' lab was focused on the biochemistry of nucleic acids. The department's weekly seminar was being given by B. R. Baker, a famous scientist in the field of drug design. His topic was "active-site-directed irreversible-enzyme-inhibitors for viral diseases." His development strategy started with a known substrate for both the viral enzyme and the corresponding human enzyme. A large number of simple chemical derivatives of that substrate were made and tested for increased binding to the viral enzyme, but not to the human enzyme. Any derivatives that afforded greater binding to the viral enzyme, but not to the human enzyme, were carried through one or more additional cycles of derivatization and testing until several compounds were produced which had a strong preference for binding to the viral enzyme, but not to the human enzyme. The best compounds were then derivatized with each of a wide variety of reactive groups until a compound was generated that irreversibly bound to the viral enzyme, but not to the human enzyme. Such a compound was expected to constitute a near-final therapeutic for that virus.

In essence Baker was: (1) identifying a target viral structure; then (2) laboriously generating a complementary structure; and finally, (3) adding a reactive group to the complementary structure such that upon binding of the complementary structure to the target viral structure the reactive group covalently linked the two structures and thereby inactivated the target viral structure.

2.2 Initial Conception

During Baker's seminar it occurred to me that if one switched from targeting viral proteins to instead targeting viral nucleic acids, then as much as 99% of the development effort could be avoided, and one should be able to develop therapeutics for a hundred different viruses almost as easily as for one virus. In this drug design strategy one strand of the viral nucleic acid would comprise the viral target. A complementary "carrier" genetic strand would then be prepared (using a polymerase), which would serve to carry multiple crosslinking agents. When that carrier strand paired to its complementary target strand, the crosslinking agents would irreversibly inactivate the target strand. The principal challenge would be to develop the crosslinking agent. As a graduate student focused on the biochemistry of nucleic acids, it appeared that designing a suitable crosslinking agent should be simple because even in 1969 the structures of duplex nucleic acids were known at subatomic resolution, and the relative reactivities of the various sites on both RNA and DNA were well known. A particular advantage of such a new drug design strategy was that once the required crosslinking agent was in hand, the strategy could be quickly and easily used for selectively targeting virtually any single-stranded genetic sequence—affording therapeutics for most or all viruses, plus a host of other diseases.

2.3 An Indelicate Suggestion

After Baker's seminar I went up to him and enthusiastically made the rather indelicate suggestion that he should abandon his current strategy for targeting viral proteins and instead switch to a far more efficient strategy of targeting viral genetic sequences, and I laid out the compelling advantages such a switch would provide, and explained why designing the key crosslinking agent should be easy in light of the precise structural information and chemical reactivities then known about nucleic acids.

My brash suggestions were not well received by the world-famous Baker. Perhaps my red beard, shorts, and sandals, and my excessive enthusiasm for and confidence in an idea that was only minutes old, may have influenced his reception of my new drug design strategy. While I am sure that my brash suggestions to Baker were hugely embarrassing to Dr. Chris Mathews , my dissertation advisor who was standing beside Baker, nonetheless, even then, and throughout the nearly five decades since, Chris has been very supportive and helpful in my efforts to develop this genetic blocking strategy.

While that initial concept probably sounded wildly speculative, it should be appreciated that each month my company, GENE TOOLS, currently designs, synthesizes, and ships to labs around the world Morpholino antisense oligos targeted against about 500–1000 different RNA sequences specified by our research customers, and these Morpholinos, which have revolutionized the developmental biology field over the past 15 years, function by the same mechanism underlying that 1969 drug design strategy.

But as will be described later herein, there has been much evolution of that initial idea. The crosslinking agents are no longer used. And the antisense "carrier" strand has been massively redesigned relative to natural nucleic acids to obviate the need for crosslinking agents, and to provide a number of superior properties not possible with natural genetic material. Such changes also simplify synthesis and reduce costs. Also, genetic sequencing and computers now play a major role in selecting optimal targets for maximal biological activity, and achieving very high specificity for the targeted genetic sequences. Finally, a great deal of time and effort has gone into, and continues to go into, developing technologies for safely and efficiently delivering such compounds into the proper subcellular compartments in cultured cells and animals (including humans).

2.3.1 "Pipe-Dream" Publication

On nearing completion of my unrelated doctoral dissertation work, for my postdoctoral project I decided to pursue my 1969 idea on targeting genetic sequences for therapeutic purposes. To this end, in 1973, I wrote a paper describing that genetic targeting strategy, and submitted it to the Journal of Theoretical Biology. It came back with a rejection notice, but no comments. Five years later in 1978, after having carried out proof-of-principal work as a postdoc at Berkeley, I resubmitted that paper, now accompanied by a companion paper providing supporting experimental results generated while at Berkeley. This time the paper was accepted and finally published in 1979 [4], carrying the original submission date of 1973 and its companion paper was also published [5]. The letter of acceptance came with a kind note from the same reviewer who had rejected the earlier version submitted in 1973. In that note the reviewer stated that the 1973 version had been rejected because it appeared to be just a "pipe dream" with no real chance of success. The reviewer also complemented my perseverance and progress in pursuing that "pipe dream."

2.3.2 Experimental Work Begins as a Postdoctoral Fellow at Berkeley

In order to be able to pursue one's own project as a postdoc, it is generally necessary to obtain a postdoctoral fellowship. To start this process, I applied for and was accepted to work on my proposed genetic blocking project at the Chester Beatty Research Institute in London—contingent on obtaining a postdoctoral fellowship to pay my salary and expenses. With that acceptance in hand, I next applied to National Institutes of Health (NIH) for the requisite postdoctoral fellowship.

Early in 1973 I was thrilled to receive notification from NIH awarding the requested fellowship to work on the genetic blocking project at the Chester Beatty Research Institute. However, just a few weeks later another notification came from NIH regretfully informing me that the fellowship was being withdrawn. The reason was that to reduce government spending President Nixon had unexpectedly sequestered a significant portion of NIH's approved budget. That depressing news put an end to plans to go to England to begin developing my proposed sequence-specific genetic blocking agents.

2.3.3 A Year of Delay

Suddenly at loose ends because of this devastating news, and needing some income to live on, I embarked on a collaborative project with an Electrical Engineer at the Univ. of Arizona who had a small grant from the US Department of Occupational Safety and Health Administration (OSHA) to study the mechanisms underlying silicosis in miners and foundry workers. That very-low-budget project (around \$10,000) was carried out in an autoclave room with a borrowed spectrophotometer and centrifuge, plus a fair amount of my own red blood cells. The project was quite successful in identifying the probable mechanism underlying silicosis, and the information generated in that project led to an important publication [6] that improved OSHA's strategies for reducing silicosis. Results from that project also led to our development of a simple, inexpensive, and efficient technology for selectively removing the pathogenic fraction of air-borne silica dust from foundries and mines, including even mile-wide open pit mines.

2.3.4 Progress Resumes

Early in 1974, as that silicosis project was nearing completion, I unexpectedly received still another notification from NIH informing me that the postdoctoral fellowship was again available for pursuing my genetic blocking agents. The reason that fellowship had been reinstated was that by 1974 President Nixon was embroiled in his Watergate scandal—and so was not inclined to mess with

NIH's appropriated funds. Since the position at the Chester Beatty Research Institute in England had fallen through the previous year due to Nixon, I applied to several labs in the USA which appeared to have programs somewhat relevant to my genetic blocking project. Bringing postdoctoral fellowship funds helped favor a welcome at several labs, and I selected the lab of Dr. H. Fraenkel-Conrat and B. Singer in the Department of Molecular Biology at Berkeley as being most suitable for the genetic blocking project.

On arriving in Berkeley in the summer of 1974 it was clear that more chemical synthesis expertise and equipment would be needed than was available in the Molecular Biology Department. Inquires led to Dr. Tinoco in the Chemistry Department, which was just across the street from the Molecular Biology Department. Dr. Tinoco, in turn, suggested the younger Dr. Paul Bartlett (so called because there were two Dr. Paul Bartletts in the Chemistry Department at that time) might be interested. Bartlett turned out to be the perfect mentor and collaborator, providing the practicalities of organic synthesis needed for designing and synthesizing the crosslinking agent for the genetic blocking strategy. Because my graduate training was as a nucleic acid biochemist, advice and assistance from Bartlett in organic synthesis was invaluable for the project.

2.3.5 Research Associate at National Jewish Hospital and Research Center

When my 2 year postdoctoral fellowship expired at Berkeley I obtained another fellowship from NIH for a research associate position for the purpose of continuing the genetic blocking project. With that funding in hand, I lined up a position at the National Jewish Hospital and Research Center in Denver, Colorado. Building on the work carried out at Berkeley, the focus now shifted from chemical synthesis of the crosslinking agent, to its attachment to a suitable viral genetic sequence, and then assessment of its capability to selectively crosslink and thereby inactivate the complementary viral genetic sequence. That phase of the project went well and completed the work we (Paul Bartlett at Berkeley and myself at National Jewish) needed for a publication describing the design and implementation of this genetic blocking strategy. The resulting paper was submitted to and published in the Journal of Molecular Biology in 1978 [7].

2.3.6 Assistant Professor-Senior Research at Oregon State University (OSU)

After 18 months at National Jewish in Denver, I was offered an assistant professor–senior research position in the Biochemistry-Biophysics Department at Oregon State University (OSU) by Dr. Christopher Mathews, who had been my dissertation advisor at the University of Arizona, and who had recently been appointed Chairman of the Biochemistry-Biophysics Department at OSU. On my arrival at OSU in Feb. of 1978, Chris made available one of his labs suitable for continued development of the genetic blocking agents. Soon after arriving I began a fruitful collaboration with Dr. Dwight Weller, a professor of organic synthesis in the Chemistry

Department just across a parking lot from my lab in the Biochemistry-Biophysics. Department at OSU.

In early work at OSU it was learned that after attaching the crosslinking agent to a carrier strand of nucleic acid, that agent slowly underwent substantial intramolecular reaction which short circuited the genetic blocking strategy. To avoid that problem, cross-complexing agents were designed to stabilize the carrier/target duplex much like the crosslinking agents, but without the covalent crosslink. However, after synthesizing these cross-complexing agents, they were also found to suffer serious limitations due to intramolecular interactions that blocked pairing to their target genetic sequences.

These limitations provided the motivation in 1979 to devise a radically different design strategy for blocking specific gene sequences. The idea was to design a set of four subunits, each of which could hydrogen bond to three polar sites facing the major groove of one of the four oriented base pairs in duplex DNA (those being: A:T, T:A, G:C, and C:G). By linking these four subunits in a selected order, the resultant oligomer was predicted to bind and block the functioning of any selected gene. Such oligos were named "anti-genes," because unlike antisense agents, which block the sense RNA transcripts, anti-genes were designed to block specific genes in their duplex-DNA form. Another version, with a different backbone structure, was designed to block A-form DNA/RNA and RNA/RNA duplexes.

2.3.7 1978–1979: Birth of the Antisense Field

The antisense field is considered by many to have started in 1978 with two publications by Paul Zamecnik and Mary Stephenson at Harvard. They reported the use of a 13-mer antisense DNA oligo (purchased from a commercial source) to inhibit translation of the sense Rous sarcoma viral RNA in a cell-free translation system [8]. In a companion paper they also reported results from experiments in cultured cells wherein that same 13-mer antisense DNA was added at the same time as an infecting Rous sarcoma virus—resulting in inhibition of virus production [9]. Those two papers played a major role in bringing the antisense therapeutics strategy to the attention of researchers studying gene function, and others interested in new approaches to drug design. Being published in a high-impact journal (Proc Natl Acad Sci), coming out of Harvard, including results from a live virus in cultured cells, and utilizing an antisense oligo that could be purchased from a commercial source, all contributed to a very high impact for those two papers.

Also in 1978 Paul Bartlett and I published an extensive antisense paper in the Journal of Molecular Biology which demonstrated the functioning of our sequence-specific crosslinking agent for specifically linking a viral RNA to its complementary DNA [7]. Also in 1978 the first patent was issued in the antisense field (US

Patent 4,123,610). This was issued to me and Paul Bartlett and assigned to National Institutes of Health.

The following year (1979) two of my antisense papers were published in the Journal of Theoretical Biology. The first paper, which described my basic antisense drug design strategy, was initially submitted in 1973, but rejected, and then a more extensive version was resubmitted in 1978 and published in 1979 [4]. A companion paper focused on design of the crosslinking agent used in that genetic blocking process [5].

Also in 1979, Paul Miller, Paul Ts'o, and coworkers at Johns Hopkins University published a paper on a new uncharged linkage structure for DNA [10]. While that methylphosphonate linkage was not usefully incorporated into an antisense oligo until the mid-1980s [11, 12], it, and an earlier phosphotriester linkage type developed by the same group [13], suggested an effective means for achieving good stability of antisense oligos in biological systems. This was significant because achieving good stability in biological systems is a major challenge in developing antisense agents.

From that modest beginning in 1978 and 1979, between the mid-1980s through the end of the 1990s the antisense field ballooned into a huge research and development effort that led to dozens of different antisense structural types, involved many hundreds of researchers, was funded by hundreds of millions of dollars from NIH and other research funding agencies, and was pursued by most of the major pharmaceutical companies.

2.3.8 1980–Leaving OSU and Starting the First Antisense Company

My newly conceived anti-genes were an exciting conceptual breakthrough (detailed years later in US Patent 5,166,315) and appeared to have great promise for therapeutics. Therefore, before moving forward with their reduction to practice I started looking into the terms of my grants and other sources of funding to make sure there would be no impediments to taking anti-genes all the way to therapeutic products that could benefit humanity. In the course of looking through the fine print of my grant from the US National Institutes of Health, I was appalled to find a clause that specified that for grants to *nonprofit and governmental entities* (such as OSU), for any patents issued on inventions made with those grant funds, said patents could not be licensed *exclusively* for more than 5 years to any nongovernmental organization, such as a major drug company. The consequences of that clause were that if my new anti-gene design strategy worked out as hoped, then it would be virtually impossible to ever get any of the resulting anti-gene therapeutic products to patients. This is because if no patent application was filed then no drug company would be willing to put up the hundreds of millions of dollars required for development and regulatory approvals of each of the therapeutic products—because any such company would lack market exclusivity that is essential to recoup the huge development and regulatory expenses required for new drugs. If instead a patent application was filed and a patent issued, again no drug company would be willing to fund the development and regulatory approvals because their maximum of 5 years of license exclusivity under that patent would have expired long before they could ever get any of the products into the market. Further, the US Government, and virtually all other governments, did not fund late-stage clinical development and regulatory approvals—they left that extremely expensive component of drug development to be paid for by commercial entities. Thus, I concluded that continuing on at OSU with funding from government grants would be a complete waste of my time, and a waste of government funds, because even if the technology was successful, no therapeutic products were likely to make it to patients.

In sharp contrast, at that time if a *for-profit* company was awarded a grant from NIH, that absurd clause (limiting license exclusivity to no more than 5 years) would not apply.

At that point it appeared I had two choices: (1) I could continue working at OSU on my newly conceived anti-genes, but with the likelihood that such anti-genes would *never* be used as therapeutics for human patients; or, (2) I could abandon my position and lab at OSU, give up my funding from NIH (and several other sources), and take the risk of starting up a private company to develop my genetic blocking agents.

The folly of developing a therapeutics technology that could never serve patients was unacceptable. This, and other factors, motivated me to start ANTIVIRALS Inc. (AVI) on 1 January 1980, the first company focused on developing antisense therapeutics. Note that in 1997 GENE TOOLS, LLC was spun off to focus on custom Morpholinos for the research community. In 2002 ANTIVIRALS Inc. was renamed AVI BioPharma Inc., and in 2012 again renamed Sarepta Therapeutics Inc.

2.3.9 Major Contributions to Developing Morpholino Antisense Oligos

Founding a new biotechnology company was a huge risk, and without key contributions from my wife it would not have been feasible. Her contributions were as follows. In 1977 while working at National Jewish in Denver, I met a young teacher, Patricia Rusnak (Pat), on a hike up a 13,000 ft high mountain near Denver, and not long thereafter we became engaged. Late in 1977, I received an offer of a research faculty position at Oregon State University in Corvallis, and in Jan. of 1978 Pat and I married and a week later packed up and moved to the beautiful city of Corvallis, Oregon. Since my position at OSU was not tenure track, and since I was pursuing a then-quite-speculative line of research, Pat prudently elected to enroll at the local community college to earn degrees in computer programming and in accounting to add to her Masters in Education. Her foresight in getting that added practical training would soon make a major contribution to the future development of Morpholino antisense oligos—by providing us with an income while I focused on starting up my new 1-person company. Further, during her years of teaching Pat had frugally saved a significant portion of her income—which allowed her to make the down payment on our first home in Corvallis. That home was to also play a key role in the future development of Morpholinos. Specifically, the basement served as my company's first very-low-cost laboratory. In addition, several years later that basement laboratory served as the location for a site visit from NIH. The site visit team was there to judge whether or not my proposal for a Small Business Innovation Research grant could likely be successful in my rather unconventional lab facilities. Happily, they concluded the project had a reasonable chance of success and so the grant was approved. But approval was contingent on moving the lab from the basement of our house (by then with three small children living upstairs) to a new facility, and NIH added $15,000 to the requested grant to cover the cost of the move and the increased operating expenses. Soon thereafter I set up a small well-vented organic synthesis lab in a 400 square foot abandoned tombstone dealership.

3 Antisense Comes of Age

3.1 Development of Simple Prototype Anti-gene

Research in 1980 was spent designing, synthesizing, and testing a very simple prototype anti-gene to test the feasibility of gene blocking via major-groove-binding agents. While results appeared promising, it also became clear that developing effective anti-genes would likely require a great deal of time and resources—far more than my fledgling 1-person company could muster.

3.2 Return to Antisense

Therefore, in 1981, I redirected my focus to antisense agents targetable against single-stranded RNA sequences. In my renewed focus on antisense agents my designs were strongly influenced by the Miller and Ts'o strategy of utilizing a nonionic backbone to prevent degradation in biological systems [10]. I also postulated that a properly designed antisense oligo with a suitable nonionic backbone might bind its target sense strand with sufficient affinity that there would be no need for a cross-linking agent—thereby avoiding the intramolecular troubles I'd previously encountered with antisense sequences carrying cross-linking and cross-complexing agents.

By the mid-1980s, the antisense drug design strategy was rapidly gaining popularity; ANTIVIRALS was no longer just a 1-person operation; and, we were making very good progress toward increasingly effective antisense structural types. As a consequence, it was becoming easier to raise funds for our research. This allowed several moves to progressively larger facilities, and hiring an increasing number of personnel.

A	B	C
Carbamate DNA	Carbamate Morpholino	Phosphorodiamidate Morpholino
1984 ***(14)***	1987 ***(15)***	1989 ***(16)***

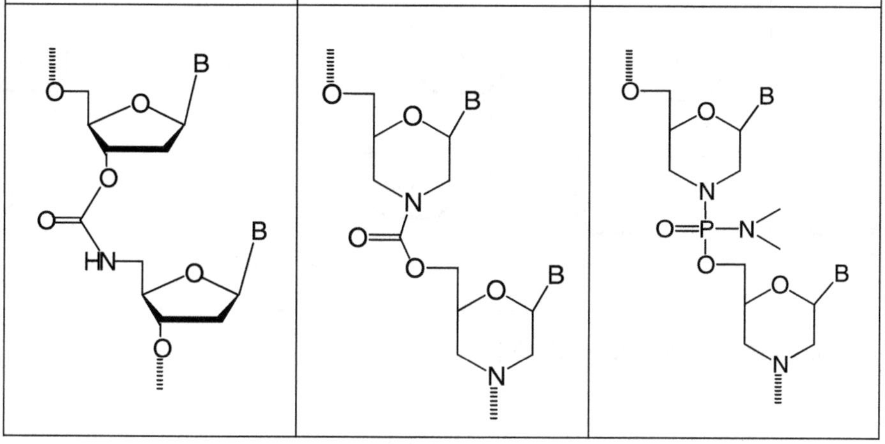

Fig. 1 ANTIVIRALS' antisense structural types [14–16]

During the 1980s at ANTIVIRALS Inc. we developed and tested a substantial number of antisense structural types. Of those many structural types, Fig. 1 shows the three most significant types in our long progression of structures with increasingly desirable properties. Between structure A and B there were about 6 other structures made and tested, but judged inadequate, and between structure B and C there were a dozen structural variations made and tested, but again judged less than optimal. Structure C in Fig. 1 is our current Morpholino type, first synthesized and tested in 1989.

By the year 2000 antisense oligos of the structure C-type were fully developed and had been carefully tested at the biophysical, biochemical, and cell culture levels, and testing had begun in living animals. Extensive experimental evidence at all of these experimental levels clearly indicated that our Morpholino structural type (structure C of Fig. 1) had an unmatched combination of properties that, in complex biological systems, led to their greatly outperforming all of the popular competing structural types, particularly in regard to the key properties of: sequence specificity; general absence of off-target effects; long-term stability; and predictable targeting [17, 18, 19, 20, 21, 22].

Aside from the new structural types being devised and developed at ANTIVIRALS Inc., during the 1980s and into the 1990s most new antisense structural types were being developed at universities and national laboratories (NIH and the FDA). Figure 2 shows the most popular of those competing antisense structural types reported by other groups in the years between 1978 and 2001.

D	E	F	G	H
DNA	Methyl Phosphonate DNA	Phosphorothioate DNA	PNA	siRNA
1978 ***(8,9)***	1983 ***(11)***	1987 ***(23,24)***	1993 ***(25,26,27)***	2001 ***(28,29,30)***

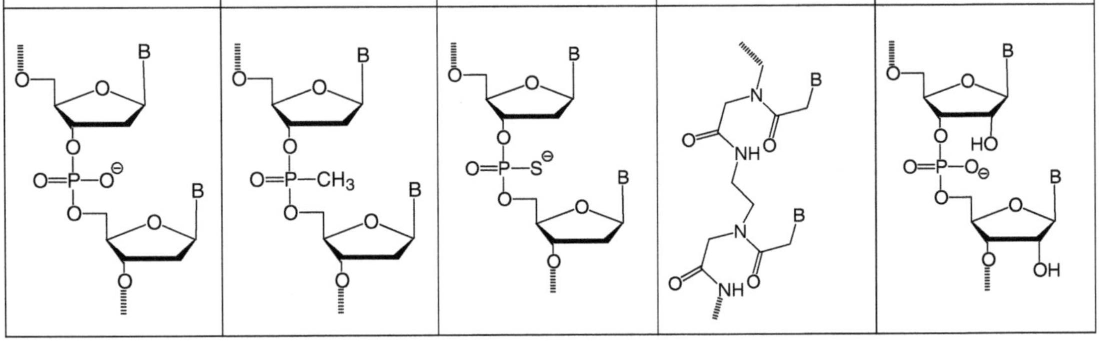

Fig. 2 Popular antisense structural types [23–30]

Table 1
Qualitative comparison of properties of antisense structural types

Desirable properties	Structural types: C Morpholino	D DNA	E Me-phosphonate DNA	F S-DNA	G PNA	H siRNA
1 Sequence specificity	++	+	+	−	+	+
2 Resistance to degradation	++	−	++	+	++	−
3 General lack of off-target effects	++	+	++	−	++	−
4 Lack of effect on innate immune system	++	−	++	−	++	−
5 Not toxic in embryos	++	+	++	−	+	−
6 Predictable targeting	++	+	++	−	++	−
7 Effective for splice modification	++	−	++	−	++	−
8 Good water solubility	++	++	−	++	−	++
9 Does not require assist from cell factors	++	−	++	−	++	−
10 Publication data base	++	−	−	−	−	−

Table 1 provides a qualitative comparison, for the most used antisense structural types, of the key properties required for effective antisense activity [17, 18, 19, 20, 21, 22].

While ANTIVIRALS Inc. was the sole antisense company from 1980 until the late 1980s, beginning in 1987 four more antisense companies and one anti-gene company were founded. These were: Gilead Sciences Inc. in 1987; Genta Inc. in 1988; Hybridon Inc.

in 1989; ISIS Pharmaceuticals Inc. in 1989; and the anti-gene company, Triplex Pharmaceutical Corp. in 1989.

While ANTIVIRALS Inc. devised and developed multiple new antisense structural types, the principal business strategy of the newer companies founded in the late 1980s (and generally funded by venture capital) was primarily to use established antisense structural types to target particular genetic sequences associated with known diseases, and then take the resultant antisense oligos into clinical trials. Because of the multiple serious limitations of structural types D, E, F, G, and H (*see* Fig. 2 and Table 1), and virtually all of the other less popular structural types, the business strategy of taking antisense oligos of those flawed structural types into clinical trials has been pretty much a disaster—leading to many failures to obtain regulatory approvals and wastage of multiple billions of dollars. And even in those rare cases where FDA approval was obtained (e.g., Vitravene, approved in 1998), at best the antisense products are of only marginal value to patients.

Not surprisingly, the big pharmaceutical companies have almost all abandoned the antisense therapeutics strategy—with the exception of structure H of Fig. 2 (siRNA and its precursor, shRNA), which is currently the latest fad in the antisense field. However, I believe the popularity of this structural type is destined to soon fade when researchers begin to do rigorous sequence specificity studies. This is because of a serious limitation of siRNAs—there are potentially hundreds of RNA targets for each siRNA [21, 22]—a consequence of the seed sequence, which determines the sequence specificity of the siRNA, recognizing too little sequence information to assure an adequate level of specificity for the targeted RNAs.

4 Key Requirements for Antisense Drugs

I contend that for antisense drugs to finally begin to provide substantial benefits to patients, the drug developer needs to:

1. Develop an antisense structural type (such as the Morpholino type) which provides all of the key properties listed in Table 1.
2. Develop a safe and effective in vivo delivery technology sufficient to get adequate amounts of the antisense drug into cells of the tissues/organs to be treated.
3. Only after **steps 1** and **2** are achieved should clinical trials proceed on a specific drug.

I believe that the final frontier in the antisense field entails achieving safe and effective delivery of antisense oligos into the cytosol/nuclear compartment of a broad range of cell types in living animals, particularly humans. We have been working toward this objective for years, and it appears that the crucial breakthrough was finally achieved in March of 2016.

If this March 2016 advance in delivery technology continues to be as promising as it now appears, then I suspect the long-promised flood of antisense therapeutics may finally arrive, perhaps beginning as soon as 2017. Our first objective is to develop custom cocktails of Morpholinos for curing any patient's cancer. Each component of the cocktail for a given patient will be targeted against one of the RNA transcripts absent from the patient's normal cells, but found (by sequencing a biopsy sample of the patient's cancer) to be present in and essential for the viability of that patient's cancer. Such cocktails are explicitly designed to destroy the cancer without harming the patient.

References

1. Ekker S (2000) Morphants: a new systematic vertebrate functional genomics approach. Yeast 17:302–306
2. Heasman J (2002) Morpholino oligos: making sense of antisense? Dev Biol 243:209–214
3. SPECIAL ISSUE (2001) Morpholino gene knockdowns – all 27 research reports in the July issue of the journal *Genesis* 30:89–200
4. Summerton J (1979) Intracellular inactivation of specific nucleotide sequences: a general approach to the treatment of viral diseases and viral-mediated cancers. J Theor Biol 78:77–99
5. Summerton J (1979) Sequence-specific cross-linking agents for nucleic acids: design and functional group testing. J Theor Biol 78:61–75
6. Summerton J, Hoenig S, Butler C, Chvapil M (1977) The mechanism of hemolysis by silica and its bearing on silicosis. Exp Mol Pathol 26:113–128
7. Summerton J, Bartlett P (1978) Sequence-specific crosslinking agents for nucleic acids. Useof6-bromo-5,5dimethoxyhexanohydrazide for crosslinking cytidine to guanosine and crosslinking RNA to complementary sequences of DNA. J Mol Biol 122:145–162
8. Stephenson M, Zamecnik P (1978) Inhibition of Rous sarcoma viral RNA translation by a specific oligodeoxyribonucleotide. Proc Natl Acad Sci U S A 75:285–288
9. Zamecnik P, Stephenson M (1978) Inhibition of Rous sarcoma virus replication and cell transformation by a specific oligodeoxynucleotide. Proc Natl Acad Sci U S A 75:280–284
10. Miller PS, Yano J, Yano E, Carroll C, Jayaraman K, Ts'o POP (1979) Nonionic nucleic acid analogues. Synthesis and characterization of didexoyribonucleoside methylphosphonates. Biochemistry 18:5134–5143
11. Miller PS, Agris CH, Blake KR, Murakami A, Spitz SA, Reddy MP, Ts'o POP (1983) Nonionic oligonucleoside analogs as new tools for studies on the structure and function of nucleic acids inside living cells. In: Pullman B, Jortner J (eds) Nucleic acids: the vectors of life. D. Reidel Publishing, Dordrecht, p 521
12. Blake KR, Murakami A, Spitz SA, Glave SA, Reddy MP, Ts'o POP, Miller PS (1985) Hybridization arrest of globin synthesis in rabbit reticulocyte cells by oligodeoxyribonucleoside methylphosphonates. Biochemistry 24:6139–6145
13. Miller P, Barrett J, Ts'o P (1974) Synthesis of oligodeoxyribonucleotide ethylphosphotriesters and their specific complex formation with transfer ribonucleic acid. Biochemistry 13:4887–4986
14. Stirchak E, Summerton J, Weller D (1987) Uncharged stereoregular nucleic acid analogues. 1. Synthesis of a cytosine-containing oligomer with carbamate internucleoside linkages. J Org Chem 52:4202–4206
15. Stirchak E, Summerton J, Weller D (1989) Uncharged stereoregular nucleic acid analogues. 2. Morpholino nucleoside oligomers with carbamate internucleoside linkages. Nucleic Acids Res 17:6129–6141
16. Summerton J., and Weller D. (1993) Uncharged Morpholino-based polymers having phosphorous containing chiral intersubunit linkages. US Patent 5,185,444.
17. Summerton J, Weller D (1997) Morpholino antisense oligomers: design, preparation, and properties. Antisense Nucleic Acid Drug Dev 7:187–195
18. Summerton J, Stein D, Huang S, Matthews P, Weller D, Partridge M (1997) Morpholino and phosphorothioate antisense oligomers compared in cell-free and in-cell systems. Antisense Nucleic Acid Drug Dev 7:63–70
19. Summerton J (1999) Morpholino antisense oligomers: the case for an RNase H-independent structural type. Biochim Biophys Acta 1489:141–158

20. Summerton J (2006) Chapter 6: Morpholinos and PNAs compared. In: Christopher J, Matthew D (eds) Peptide nucleic acids, morpholinos and related antisense biomolecules. Springer, New York, NY, pp 89–113
21. Summerton J (2007) Morpholinos, siRNA, and S-DNA compared: impact of structure and mechanism of action on off-target effects and sequence specificity. Curr Top Med Chem 7:651–660
22. Wilczynska A, Bushell M (2015) The complexity of miRNA-mediated repression. Cell Death Differ 22:22–33
23. Matsukura M, Shinozuka K, Zon G, Mitsuya H, Reitz M, Cohen JS, Broder S (1987) Phosphorothioate analogs of oligodeoxynucleotides: inhibitors of replication and cytopathic effects of human immunodeficiency virus. Proc Natl Acad Sci U S A 84:7706–7710
24. Agrawal S, Goodchild J, Civeira MP, Thornton AH, Sarin PS, Zamecnik PC (1988) Oligodeoxynucleoside phosphoramidates and phosphorothioates as inhibitors of human immunodeficiency virus. Proc Natl Acad Sci U S A 85:7079–7083
25. Egholm M, Buchardt O, Christensen L, Behrens C, Freier SM, Driver DA, Berg RH, Kim SK, Norden B, Nielsen PE (1993) PNA hybridizes to complementary oligonucleotides obeying the Watson-Crick hydrogen bonding rules. Nature 365:566–568
26. Nielsen P, Egholm M (1999) Peptide nucleic acids protocols and applications. Horizon Scientific Press, Norfolk
27. Nielsen P (2006) Chapter 1: the many faces of PNA. In: Christopher J, Matthew D (eds) Peptide nucleic acids, morpholinos and related antisense biomolecules. Springer, New York, NY, pp 3–17
28. Elbashir S, Harborth J, Lendeckel W, Yalcin A, Weber K, Tuschl T (2001) Duplexes of 21-nucleotide RNAs mediate RNA interference in cultured mammalian cells. Nature 411:494–498
29. Scherer L, Rossi J (2006) Chapter 8: recent applications of RNA interference (RNAi) in mammalian systems. In: Christopher J, Matthew D (eds) Peptide nucleic acids, morpholinos and related antisense biomolecules. Springer, New York, NY, pp 133–147
30. Gruber J, Manninga H, Tuschl T, Osborn M, Weber K (2005) Specific RNAi mediated gene knockdown in zebrafish cell lines. RNA Biol 2:101–105
31. Summerton J (2016) Custom cocktail for curing any cancer: a strategy for destroying any cancer without harming the patient. J Drug Discov Dev Deliv 3(1): id 1020
32. Summerton J (2003) Custom cancer therapies: safe and effective treatments for most or all cancers. Ann NY Acad Sci 1002:189–196

Chapter 2

Making a Morpholino Experiment Work: Controls, Favoring Specificity, Improving Efficacy, Storage, and Dose

Jon D. Moulton

Abstract

A good Morpholino experiment starts with oligos that have been carefully designed to minimize off-target RNA binding. Performing a successful, reproducible, and well-controlled Morpholino experiment requires oligos that are single stranded and in solution at a known concentration. The outcome of treatment with the oligo needs to be checked for specificity, that is, that the observed outcome is due to interaction with the intended RNA and not an interaction with an unexpected RNA. In this chapter, I will discuss Morpholino use mostly in the context of embryonic microinjection experiments, though many techniques and warnings will be applicable to cell culture or adult animal experiments as well. Controls are critical to a good experiment, but good techniques in designing, preparing, storing, and using the oligos can improve the strength and specificity of the knockdown. Finally, it is important to know the solution concentration of the oligo to ensure that the results are reproducible.

Key words Aggregation, BLAST, Compensation, Coinjection, Concentration, Controls, Dose, Efficacy, Hypochromic effect, Humidor, Lyophilization, MALDI-TOF, Microinjection, Mispair, Morpholino, Nonsense-Mediated Decay, Phenocopy, p53, Rescue, Specificity, Storage, Synergy, Tm, Turnover, Vehicle-only, Vivo-Morpholino

1 Introduction

This chapter is written from my perspective working in product development and technical support for Gene Tools LLC. Over 17 years I have had contacts with many of the labs working with Morpholinos, selecting Morpholino targets and assisting in designing and troubleshooting experiments. I've tracked the new literature and maintained the database of Morpholino publications. I've attended many conferences, speaking with investigators about their experiences with Morpholinos and introducing scientists to the nature and applications of Morpholinos. I've learned that some approaches to using Morpholinos have met with success, others have led to struggle and frustration; here I will suggest some techniques that have worked well and some warnings about the rougher

Hong M. Moulton and Jon D. Moulton (eds.), *Morpholino Oligomers: Methods and Protocols*, Methods in Molecular Biology, vol. 1565, DOI 10.1007/978-1-4939-6817-6_2, © Springer Science+Business Media LLC 2017

roads. If we have never met at a conference, I hope we'll have opportunity to chat in person; I'll be at the Gene Tools table.

2 Controls

Typical controls used in Morpholino experiments include a specificity control to show that the effect of oligo activity is caused by interaction with the intended RNA and a negative control to show that the Morpholino is not inducing a non-sequence-specific effect [1]. For cell cultures, the effect of the delivery system (e.g., Endo-Porter, electroporation) should be tested without the oligo to determine the effect of the delivery method alone on the cells. For injections a vehicle-only control is used to show that there is not a non-Morpholino component of the injection solution causing an effect. For zebrafish injections, once a phenotype has been associated with a transcript-targeting oligo then a p53 oligo is co-injected with the targeting oligo to determine whether loss of the targeted expression is causing p53-mediated changes in phenotype.

2.1 Specificity Control

Whenever an oligo is used to produce a measurable/observable outcome by knocking down gene expression, an important question to address is whether the outcome is a result of knocking down the intended RNA or is due to unexpected interaction with a different RNA. Catastrophic off-target effects have been observed with Morpholinos, even with pairs of Morpholinos targeting the same RNA [2] but these are relatively rare; other knockdown types that have shorter minimum recognition sequences on their target RNA, such as phosphorothioate oligos, locked nucleic acids or siRNA, are far more prone to off-target RNA interactions [3]. Two methods for determining if an outcome is caused by loss of a particular protein are the RNA rescue experiment and the two nonoverlapping oligo experiment. A less ideal specificity control is the five-mispair oligo experiment.

Comparison of a Morpholino phenotype with a mutant phenotype is not a reliable specificity control because mutant phenotypes can be affected by changes in expression of other genes to homeostatically compensate for loss of the mutant gene [4]. It can also be difficult to ascertain whether a mutant is a true null or a hypomorph [5, 6]. Oligo-based knockdowns and genetic knockouts alter gene expression at different levels (RNA vs. DNA) and on different time scales (days versus generations), so the different outcomes seen with some knockdowns versus knockouts are not surprising. With less compensation observed with knockdowns than knockouts, Morpholinos can reveal gene functions a mutant would conceal. Injecting a Morpholino into a mutant null for the Morpholino's target is an emerging technique for assessing off-target RNA interaction of the oligo; maternal RNA effects or excessive oligo dose can be confounding factors.

2.2 Second Nonoverlapping Oligo

The purpose of the second nonoverlapping oligo experiment is to link an oligo's outcome to a knockdown of a particular target. In this experiment, two oligos are used which target the same RNA. Each individual oligo is used in its own experiment and the outcomes are compared. If each of the oligos produces the same outcome, for example a structural change in the organism, this supports the hypothesis that the outcome is a result of binding oligos to the target RNA and not to an unexpected RNA. However, phenotypic agreement between two oligos targeting the same RNA does not provide certainty that the phenotype is an effect of knocking down the target [7].

The second nonoverlapping oligo experiment is a commonly-used technique for supporting oligo specificity, but there has been one well-known case in which phenocopying with a pair of non-overlapping oligos has been shown to be a phenotype caused by off-target interactions of each of the oligos interacting individually with transcripts of two different histones [2]. There are certainly circumstances where more evidence of specificity should be sought, and in which even use of additional nonoverlapping oligos might produce additional phenocopying by off-target RNA interaction. Imagine a case where a gene has been recently duplicated to form genes A and B, such that each of the paralogs has very similar sequence, but the function has diverged, perhaps though mutations occurring soon after the duplication event at sequence encoding a protein active site. In this case numerous oligos designed to target gene A might have sufficient sequence similarity with gene B to alter its expression and the phenotype observed from using the oligos might be mistakenly associated with unintentional knockdown of gene B. This is a case where, if ectopic expression does not present a problem (discussed below), an RNA rescue experiment could show that restoring expression of gene A does not rescue the phenotype caused by the unintentional knockdown of gene B.

Another experiment proposed to support specificity of an oligo is the dose synergy experiment [8]. When two translation-blocking oligos target translation of a particular mRNA, they exhibit dose synergy when used in combination. To use this as a specificity test, the individual oligos are tested at several doses until doses have been found that result in roughly the same observable outcome that is distinct from wild-type; call these the individual test doses. Next the oligos are coinjected, each at half of its individual test dose and at a few smaller fractional doses, such as 1/3, 1/4, and 1/6 of the individual test dose. If coinjection of the oligos at 1/2 of their individual test doses most closely resembles the outcome of a single oligo at its original test dose, then the oligo's combined effects are additive. If the outcome of coinjecting smaller fractional doses (less than the 1/2 doses) more closely resembles the outcome of a single oligo at its individual test dose, then the coinjected oligo efficacy is greater-than-additive and the oligos are showing dose synergy. If dose synergy is observed this supports specificity of the oligos, linking the outcome to the targeted RNA.

While dose synergy has been shown for translation-blocking oligos, splice-modifying oligos present more complex possibilities. Consider a situation where a splice-modifying oligo induces a frameshift by skipping an exon. If another frameshifting splice-modifier is targeted to the adjacent exon, there is a chance that this second frameshift could restore the reading frame of the transcript, conceivably resulting in a stable internally truncated protein that retains some or all activity and rescues the phenotype. Furthermore, no study exploring dose synergy of Morpholinos targeting to trigger a range of various splicing outcomes has been published. Until this has been more rigorously explored, a prudent choice would be to use translation-blocking oligos for dose synergy experiments.

2.3 RNA Rescue

In an RNA rescue experiment, an RNA is injected along with the targeting antisense oligo. This "rescue" RNA does not contain the target sequence for the antisense oligo. A good method is to target a translation-blocking oligo to the 5′-UTR sequence and to clone the target RNA without most of its 5′-UTR, and in particular without the oligo target site. When the RNA and oligo are injected together, the oligo knocks down expression of the target RNA and the rescue RNA is translated, substituting its product for the endogenous transcript's product. In many cases, the RNA can rescue the phenotype, that is, the phenotype is returned to its wild-type state.

Some groups have used Morpholinos targeting the start of the coding sequence and engineered a rescue RNA with mispairs at the wobble-base positions. Based on unpublished discussions with groups attempting to rescue with a mutated coding sequence and groups replacing a target in the 5′-UTR, it is the latter method, where the Morpholino target is completely removed, that has more easily produced strong RNA rescues. Struggles to produce a mutated start-of-coding sequence that rescues effectively can be circumvented by targeting 5′-UTR with the Morpholino and rescuing with sequence lacking the Morpholino target. The lesson is, if you are going to rescue, the easiest path is probably to start with a 5′-UTR-targeting Morpholino.

Unfortunately, if only a very specific and difficult-to-duplicate RNA dose range can recapitulate the wild-type phenotype then, for some genes, overexpression from a rescue RNA may present a problem. For other genes the rescue RNA cannot rescue the wild-type phenotype because the rescue RNA causes ectopic expression; the RNA is translated at inappropriate times and places. When it works, the RNA rescue experiment is a good proof of specificity, showing that a particular oligo sequence causes a particular outcome through the oligo's action on a particular target RNA [7].

2.4 The Five-Mispair Experiment and Its Disadvantages

The five-mispair oligo has long been used as a specificity control for antisense experiments. If an oligo with five mispairs to an RNA target does not have the same outcome as a fully complementary oligo, this suggests that regions of the RNA which contain nearly complementary sequence to the targeting oligo but also bear five mispaired bases should not be affected by the targeting oligo. Ideally, a specificity experiment should test the hypothesis that the outcome of using a targeting oligo is a result of binding the oligo to the target RNA and not an unexpected RNA. While the second nonoverlapping oligo experiment tests this directly, the five-mispair experiment is at best indirect support for specificity.

The distribution and nature of mispairs affects the binding affinity of a mispaired oligo for RNA compared to the completely complementary oligo. If the mispairs are clustered, especially if they are clustered toward one or both ends of the oligo, they will decrease binding affinity less than if they are distributed fairly evenly throughout the oligo sequence [9]. If a CG pair is disrupted, three hydrogen bonds are lost while if an AT is disrupted, two hydrogen bonds are lost. If a base is changed so that a GT pair is created, this forms a weak noncanonical base pair and can weakly stabilize binding (though less than a canonical base pair would).

The five-mispair oligo has been used to help determine the effective-and-specific dose range for a targeting oligo. This experiment uses the interaction of mispaired oligos with RNA as a function of dose; the higher the oligo dose, the more interaction it will have with a mispaired target. The targeting oligo, a negative control oligo, and the five-mispair oligo are separately administered at a range of doses and their outcome on the biological system is scored. The lowest dose where the targeting oligo has an effect on the system relative to the negative control is considered the bottom of the effective dose range while the highest dose where the targeting oligo produces its phenotype while the five-mispair does not produce the same phenotype is considered the top of the specific range. Subsequent experiments with the targeting oligo would be performed within the effective-and-specific dose range.

Practically, there are problems with this experiment. The targeting oligo might be altering expression of unexpected transcripts at concentrations below that concentration where the five-mispair starts to phenocopy the targeting oligo. In some cases, a dose was not found where the targeting oligo has an effect and the five-mispair does not (unpublished reports from various groups). Finally, an easier approach is to test the targeting oligo at a range of doses and, in subsequent experiments, use the dose just sufficient to elicit the phenotype associated with the targeting oligo.

There are usually better options for a specificity control than the five-mispair oligo. In cases where there is no good additional nonoverlapping oligo and an RNA rescue cannot work, the five-mispair oligo can offer indirect support for oligo specificity.

2.5 Negative Controls, Standard and 25-N

The Standard Control Morpholino is a widely reported sequence that has been used in many organisms as a negative control. This oligo was originally designed against a mutant site within an intron of human beta-globin that causes beta-thalassemia and was used in engineering a luciferase-based reporter system for splice-modifying oligos [10]. Its only known fully complementary target is in humans with a particular form of beta-thalassemia.

Because any definite oligo sequence will have some interaction with off-target RNA if the dose is sufficiently high, an oligo mixture called 25-N was developed as a negative control. To manufacture this oligo, the four different activated Morpholino subunits (A, C, G, and T) are mixed together in a single container and loaded onto a Morpholino synthesizer. At each base addition, the mixture is poured over the synthesis resin so that each individual growing oligo receives a random base. After 25 subunit-addition cycles, the oligos are removed from the synthesizer, cleaved from the synthesis resin, purified, quantitated, lyophilized, and sterilized. The result is a mixture of different sequences of 25-base oligos. Used as a negative control, no one sequence is present at more than trace quantities in an oligo solution but the concentration based on the 25-base Morpholino backbone can be made the same as the concentration of a targeting oligo. When RNA from cultured cells treated with the Standard Control oligo were compared on a OneArray (Phalanx Biotech, San Diego CA) with cells treated with the 25-N oligo and with cells not treated with oligo, some changes in transcript levels versus untreated cells were measured in the Standard Control cells while changes of smaller magnitude were produced in the 25-N cells.

2.6 Vivo-Morpholino Negative Control to Control for Vivo-Morpholino Toxicity

The Vivo-Morpholino standard control oligo provides a way to control for the toxicity associated with the delivery moiety of a Vivo-Morpholino. If a Vivo-Morpholino is used toward the top of its practical dose range, toxicity may contribute to the observed phenotype. Using the Vivo-Morpholino Standard Control can reveal components of a phenotype that are caused by delivery moiety toxicity.

2.7 Culture Delivery System Controls

When Morpholinos are delivered into cultured cells, cells should also be subjected to the delivery procedure without the targeting oligo to determine whether the delivery system is causing changes in the biological system. Chemical delivery methods, such as Endo-Porter, can alter expression of some genes. Physical delivery methods, such as electroporation, can also alter expression. A delivery system control can be done using a negative control oligo, but if changes are observed compared to untreated cells then the delivery system alone should also be tested without an oligo to determine whether the negative control oligo contributes to the observed outcome.

2.8 Injection Vehicle Control

The physical process of microinjecting fluid into a fertilized egg can result in developmental delay. Issues with apparent toxicity of a Morpholino oligo have sometimes been traced to the vehicle fluid used to dilute the Morpholino for injection. It is prudent to run a few vehicle-only injections to control for effects of the physical injection process and for possible contamination of the vehicle. If these are done prior to injecting the oligo, the vehicle-only controls can be done with the same needle used for subsequent Morpholino injections, eliminating another variable. If an aliquot of the fluid used to initially dissolve the lyophilized Morpholino is saved along with the oligo stock solution, this can be mixed with the injection diluent to prepare injections missing only the Morpholino.

2.9 p53 Oligos for Zebrafish

A set of changes in zebrafish embryos were sometimes observed resulting from knockdown of some apparently unrelated transcripts. This came to be called the "Morpholino phenotype" and included shortened body axis, spinal curvature, heart edema, craniofacial alterations, and specific neurodegeneration. Appearance of the Morpholino phenotype was correlated with expression of an N-terminal truncated form of p53. Concurrent knockdown of p53 can often make the Morpholino phenotype disappear, revealing the underlying phenotypes associated only with knockdown of the protein activity associated with the target RNA. It was demonstrated that this p53-associated phenotype occurred independent of the kind of gene knockdown by replicating a Morpholino knockdown effect with a different oligo type, a gripNA [11]. Similar p53-mediated effects have been reported associated with mutants [12].

The concurrent p53 knockdown works well for zebrafish embryos because they can complete development without functional p53. There are other organisms, such as *Xenopus laevis*, that require p53 activity to complete development [13].

3 Getting a Stronger Knockdown

A Morpholino oligo sequence is only as good as the sequence used for design. Most Morpholinos are designed against sequences deposited in public databases. As the quality of public database sequence improves, knockdowns based on those sequences have become more reliable. However, the strain of an organism used by an investigator will contain some differences from the database sequence. Sequencing the target in the particular organism and strain used for the Morpholino knockdown will give the best probability of a strong knockdown.

3.1 Oligo's RNA Affinity (Tm)

In a collaboration between Gene Tools LLC and DNA Software (www.dnasoftware.com), the melting temperatures of over a hundred Morpholino-RNA heteroduplexes were measured in an iterative hypothesis-testing process as an equation was developed to predict the Tm of a Morpholino-RNA heteroduplex based on the oligo sequence. This proprietary equation is built into some of the oligo analysis software offered by DNA Software. Gene Tools customer support has found that oligos designed with Tm between 80 and 100 °C (calculated for 10 μM oligo) usually have good activity. Oligos with lower Tm generally have poorer to no activity. Higher Tm oligos have subsequences with Tm in the 80–100 °C range and so are more prone to off-target RNA interactions.

3.2 Timing of Protein Degradation

Physiological effects of knocking down a gene's expression generally do not appear until the preexisting protein has had time to degrade substantially away. This may not be an issue in developing embryos, in which the onset of expression of a transcript might occur without preexisting protein present and a Morpholino can prevent expression of new protein with no preexisting background. In cultured cells, the protein product of a target transcript is typically already present and so there is a lag between the suppression of target gene expression and the disappearance of a protein signal or appearance of a knockdown phenotype. If splice-modifying oligos are used, care must be taken not to interpret alteration of RNA splicing (typically revealed by a PCR-based method) as onset of the physiological result of a knockdown. The RNA signal changes quickly, while the alteration of protein concentration will depend on the turnover time of the protein and the efficacy of the knockdown. Sometimes, seeing a stronger knockdown is as simple as waiting longer (you might need to re-dose with Morpholino to maintain oligo activity over a long wait).

3.3 Exploiting Synergy

The dose synergy discussed in the section on the nonoverlapping oligo experiment (above) can be used to achieve stronger knockdowns. Often a pair of oligos targeting the same RNA will be available from a nonoverlapping oligo control experiment. Be sure to try co-dosing with both oligos to check for dose synergy and test the effect of a stronger knockdown. For experimental systems that use lots of oligo quickly, such as knockdowns in cell cultures in volumes sufficient for Western blots or Vivo-Morpholino experiments in rodents, using a pair of oligos at reduced dose can be an effective way to get more experiments from your oligos.

3.4 Splice Targeting for Nonsense-Mediated Decay

If a splice-modifying oligo causes excision of an upstream exon with a number of bases not evenly divisible by three, the resulting frameshift usually brings a premature termination codon in-frame and triggers the nonsense-mediated decay (NMD) system to degrade the transcript. This is why, when using splice-modifying oligos for knockdowns, upstream frameshifting exons are the preferred targets. Note that if a splice-modified transcript is rapidly degraded by NMD, when PCR products are run on an electrophoretic gel there might be no visible band corresponding to the splice-modified RNA. In this circumstance, look for dimming of the wild-spliced band as an indication of splice-modifying activity. Take care to load the wells lightly with RNA, because dimming is not readily observed from a saturated band.

4 Favoring Specificity in Oligo Design and Use

With a few considerations during the oligo design process you can improve the specificity of oligos. These include avoiding common sequence motifs during design and checking your designs against a transcriptome database to look for possible unexpected RNA interactions. Once you have an oligo in-hand, using the lowest practical dose for your experiment will help achieve knockdowns with good RNA specificity.

4.1 BLAST Your Sequence Designs

Always BLAST the target of a proposed Morpholino oligo against the genome of your experimental organism to see if there are mostly-complementary sequences where the Morpholino might cause unexpected effects. Potential off-target effects can occur if partially complementary sequences are located in 5′-UTR sequence or in the first 25 bases of the coding sequence. Partially complementary sites in the intron abutting splice junctions, or even approaching to within about 30 bases of the splice junction in the intron, might alter splicing. There are other sites where an oligo might have an effect, such as at poly-A tailing signals, on micro-RNA response elements, or on splice regulatory binding sites. Generally if an oligo is binding exonic sequence more than 30 bases downstream of the start codon it will have no effect, though if it binds an exonic splice enhancer it could alter splicing. If the partially complementary sequence has less than a run of about 14 contiguous bases of complementarity, you can probably safely ignore it as the affinity of the interaction will not be very high. However, flanking complementary sequence or a high CG content can stabilize the Morpholino-RNA interaction.

4.2 Avoid Simple Sequences (Single Base Runs, Dinucleotide Repeats)

Single base runs and short repeat sequences (dinucleotide, trinucleotide) occur commonly in the transcriptome and should be minimized in oligos where possible.

4.3 Keep the Dose Low

As the concentration of an oligo increases, it will interact more strongly with partially complementary sequence. Ideally, your targeting oligo should be at a concentration where its entire length provides just enough stability to the Morpholino-RNA interaction that the oligo is active against its RNA target. That way, subsequences of the oligo that might interact with unexpected RNAs will not have enough binding affinity to have activity. As the dose increases, the knockdown of the intended target may become stronger, but the risk of subsequences becoming active against other targets also grows.

4.4 Avoid Excessively High Tm Sequences or at Least Use Less

If an oligo has very high Tm, subsequences within the oligo are also likely to have high Tm. If used at typical experimental concentrations, such oligos are at increased risk of off-target RNA interactions compared with oligos having lower Tm. Using a high Tm oligo at reduced concentration can help to avoid off-target RNA interactions, but the best strategy is to use a 25mer with mid-range Tm.

5 Physical State and Storage

Pure water is the best solvent for Morpholino oligos. There are two reasons for this: (1) if the oligos must be concentrated or put into long-term storage, they can readily be lyophilized from pure water and (2) if the oligo mass needs to be characterized, they can be measured in a MALDI-TOF mass spectrometer from a sinipinic acid matrix; the presence of salts produces a messy and harder-to-interpret MALDI-TOF spectrum.

Aggregation of Morpholinos in solution can decrease their activity over time as single-stranded active antisense is clumped into soluble inactive complexes. A chapter in this volume reports assessing the formation of these particles over time and their disruption by autoclaving. Morpholinos can tolerate some trips through an autoclave. If you have an oligo stock solution that has lost some biological activity, autoclave the stock using liquid cycle and, to avoid evaporation, remove the oligo solution from the autoclave chamber as soon as the autoclave has returned to room pressure.

Morpholinos can also leave solution by associating with the walls of their storage container. This occurs more often if the oligos are stored cold or especially if they are subjected to freeze-thaw cycles. It can be difficult or impossible to remove the oligos from the container walls in a useful form. An indication that this has likely happened is decreasing UV activity (265 nm in 0.1 N HCl) over time in the stock solution. The wall-associated oligo can be detected by removing the liquid from the container and pipetting in 0.1 N HCl; the activity missing from the stock solution appears in the wall-wash solution as low-pH protonation of the oligos makes them cationic and they enter the 0.1 N HCl solution. Unfortunately, such acidic conditions can damage the oligo backbone and in such acidic solution they are no longer useful in biological systems. To avoid this situation, room-temperature storage of Morpholino stock solutions is recommended; at room temperature, the oligos are more likely to form aggregates in solution where they can usually be recovered with autoclaving but less likely to associate with container walls where they are probably lost.

If you wish to save Morpholinos long-term, the best method is to lyophilize the oligos. Dissolve a new Morpholino, autoclave it once to ensure it is in homogeneous solution, aliquot it, freeze it, and lyophilize it. A method that has worked well is to put the stock solution into a small glass vial and freeze the vials on an angle so that part of the bottom of the vial is not covered by ice. That way, no vapor can form below the ice and blow the pellet from the vial when it is under vacuum in the lyophilizer. Freezing the samples at −40 to −80 °C or on dry ice gives you more time to get them into the lyophilizer and get the pressure down below the triple point of water before the first trace of liquid water can form.

To store oligo solutions at room temperature, use a well-sealed container. However, it is prudent to put those sealed containers into a humidor, a larger sealed container such as a bell jar or empty desiccator that contains an open beaker of water. With the air in the humidor saturated with water vapor, there is no driving force for evaporation from the oligo stock so even if the oligo is in a small container with a defective seal, you are much less likely to see the oligo dry out during storage.

6 Determining Concentration

When nucleic acid bases stack with their pi-electron clouds associated, they decrease their UV absorbance near 260 nm; this is called the hypochromic effect and can cause a decrease of around 10% in the absorbance of an oligo solution. This effect is also observed for the 265 nm absorbance of Morpholino oligos. The molar extinction coefficients for Morpholinos are calculated by summing the absorbance of individual bases. For the bases in an oligo to have the same absorption as individual bases, the stacking of the bases in the oligo must be disrupted. At very low pH, the A, C, and G bases become protonated and so cationic, repelling other cationic bases and disrupting the long stack of pi-pi interactions along the oligo bases. This increases the absorption of the oligo, approaching the per-base absorption of a solution of the individual bases. Dilution of an oligo solution in 0.1 N HCl for spectrometry is a good method for disrupting base stacking and the hypochromic effect; accurate concentrations can be calculated from UV measurements of Morpholinos using these acidic samples.

When injecting Morpholinos into embryos, both concentration and delivered volume determine the oligo dose. Both accuracy and precision of injection volumes affect outcomes of Morpholino injections. Delivered dose correlates with oligo activity [14]. Variations in volume injected can account for much of the difference in oligo activity among a group of injected embryos; when fluorescence of a co-injected lissaminated Morpholino was correlated with phenotype from a targeting Morpholino administered at low dose, variation in fluorescence correlated with intensity of phenotype and protein concentration within groups of zebrafish embryos that were injected with the same microinjector settings [15]. Injection volume can be measured by microinjecting water into oil using a calibrated optical micrometer ([16] Suppl. Info.).

7 Conclusion

Over a decade and a half of commercial research use of Morpholinos has taught us that each new oligo sequence has its own characteristics of specificity, biological response, solubility, and aggregation. A Morpholino used at known concentration in single-stranded form, with an appropriate specificity control and using techniques to achieve a strong knockdown, is a powerful and reproducible tool for learning about the function of a particular gene product. The lessons learned from research use of Morpholinos will be invaluable for informing development and application of future therapeutic Morpholinos.

Acknowledgment

Thank you Alex Vincent for editing and discussing this chapter.

References

1. Corey DR, Abrams JM (2001) Morpholino antisense oligonucleotides: tools for investigating vertebrate development. Genome Biol 2(5(reviews)):1015.1–1015.3
2. Coffman JA, Dickey-Sims C, Haug JS, McCarthy JJ, Robertson AJ (2004) Evaluation of developmental phenotypes produced by morpholino antisense targeting of a sea Urchin Runx3 gene. BMC Biol 2:6
3. Summerton JE (1999) Morpholino antisense oligomers: the case for an RNase-H independent structural type. Biochim Biophys Acta 1489:141–158
4. Rossi A, Kontarakis Z, Gerri C, Nolte H, Hölper S, Krüger M, Stainier DYR (2015) Genetic compensation induced by deleterious mutations but not gene knockdowns. Nature 524(7564):230–233. doi:10.1038/nature14580
5. Novodvorsky P, Watson O, Gray C, Wilkinson RN, Reeve S, Smythe C, Beniston R, Plant K, Maguire R, KA RM, Elworthy S, FJ v E, TJ C (2015) klf2ash317 mutant Zebrafish do not recapitulate morpholino-induced vascular and haematopoietic phenotypes. PLoS One 10(10):e0141611. doi:10.1371/journal.pone.0141611
6. Blum M, De Robertis EM, Wallingford JB, Niehrs C (2015) Morpholinos: antisense and sensibility. Dev Cell 35(2):145–149. doi:10.1016/j.devcel.2015.09.017
7. Eisen JS, Smith JC (2008) Controlling morpholino experiments: don't stop making antisense. Development 135(10):1735–1743
8. Bill BR, Petzold AM, Clark KJ, Schimmenti LA, Ekker SC (2009) A primer for morpholino use in zebrafish. Zebrafish 6(1):69–77
9. Kamachi Y, Okuda Y, Kondoh H (2008) Quantitative assessment of the knockdown efficiency of morpholino antisense oligonucleotides in zebrafish embryos using a luciferase assay. Genesis 46(1):1–7
10. Kang SH, Cho MJ, Kole R (1998) Up-regulation of luciferase gene expression with antisense oligonucleotides: implications and applications in functional assay development. Biochemistry 37(18):6235–6239
11. Robu ME, Larson JD, Nasevicius A, Beiraghi S, Brenner C, Farber SA, Ekker SC (2007) p53 activation by knockdown technologies. PLoS Genet 3(5):e78
12. Danilova N, Kumagai A, Lin J (2010) p53 Upregulation Is a Frequent Response to

Deficiency of Cell-Essential Genes. PLoS One 5(12):e15938
13. Wallingford JB, Seufert DW, Virta VC, Vize PD (1997) p53 activity is essential for normal development in Xenopus. Curr Biol 7(10):747–757
14. McKeon J, Cho MJ, Khaledi MG (2001) Quantitation of intracellular concentration of a delivered morpholino oligomer by capillary electrophoresis-laser- induced fluorescence: correlation with upregulation of luciferase gene expression. Anal Biochem 293(1):1–7
15. Umemoto N, Nishimura Y, Shimada Y, Yamanaka Y, Kishi S, Ito S, Okamori K, Nakamura Y, Kuroyanagi J, Zhang Z, Zang L, Wang Z, Nishimura N, Tanaka T (2013) Fluorescent-based methods for gene knockdown and functional cardiac imaging in zebrafish. Mol Biotechnol 55:131. doi:10.1007/s12033-013-9664-6
16. Riedel-Kruse IH, Muller C, Oates AC (2007) Synchrony dynamics during initiation, failure, and rescue of the segmentation clock. Science 317(5846):1911–1915

Chapter 3

Aggregation and Disaggregation of Morpholino Oligomers in Solution

Garrick Chow, Paul A. Morcos, and Hong M. Moulton

Abstract

Morpholino oligomers are effective antisense molecules to regulate gene expression and the US FDA has approved a Morpholino drug for the treatment of Duchenne muscular dystrophy. However, it has been observed that the antisense activities of aqueous solutions of some Morpholinos decrease over time. We hypothesize that the decreased activity is caused by the formation of soluble aggregates of the Morpholinos. Here, we analyzed three Morpholino sequences by size exclusion chromatography and found two of them have over time formed soluble aggregates in water. The degree of aggregation is sequence-, temperature-, and time-dependent. We describe a simple procedure for detecting and breaking down the aggregates to return the Morpholinos to their monomeric forms.

Key words Morpholino, PMO, Antisense oligonucleotides, Aggregation, Soluble aggregates

1 Introduction

Phosphorodiamidate morpholino oligomers (also known as PMO, Morpholinos, or MO) are antisense oligos that are widely used in developmental biology to knock down gene expression. Recently, peptide-conjugated PMO (PPMO) have become valuable tools to investigate the roles of viral and bacterial genes as well as host genes involved in pathogenesis of various viruses and bacteria [1–7]. In their development as therapeutics, a Morpholino drug, eteplirsen, has been approved by the US FDA to skip dystrophin exon 51 for treatment of Duchenne muscular dystrophy (DMD) and Morpholinos targeting other dystrophin exons are currently in ongoing clinical trials for the treatment of DMD (www.sarepta.com). The structure of a Morpholino is superficially similar to DNA as they bear A, C, T, and G bases but, instead of deoxyribose rings of DNA, Morpholinos contain morpholine rings that are coupled together by methylene phosphordiamidate linkages. Therefore, in contrast to the anionic phosphate linkage of natural nucleic acids [8], Morpholinos are not charged at physiological pH.

Hong M. Moulton and Jon D. Moulton (eds.), *Morpholino Oligomers: Methods and Protocols*, Methods in Molecular Biology, vol. 1565, DOI 10.1007/978-1-4939-6817-6_3, © Springer Science+Business Media LLC 2017

Researchers have noticed some Morpholinos losing their antisense activities over time when stored in aqueous solutions at neutral pH [9]. The decreased activity of Morpholinos cannot be caused by oligo degradation because Morpholinos have shown to be completely biological stable and cannot be degraded by any of the many enzymes tested or in cells, tissues, and serum [10–12]. We hypothesize that some Morpholinos can form soluble polymeric aggregates in aqueous solution, possibly due to the nonionic nature of the Morpholinos and promoted by higher G base content. We investigated the solubility and aggregation of the aqueous solutions of three Morpholinos (S3, A3, and E3) with the same length but varied sequences and G contents (Table 1). S3 is the standard control MO, offered by Gene Tools. S3-C is the complementary sequence of S3. A3 and E3 have biological targets and their biological activities have been found to decrease over storage.

We found that all MO samples stayed in solution over a 6 month period stored at 4 or 22 °C as 1 or 2 mM aqueous solutions, indicated by little change in UV absorbance at 260 nm (data not shown). To separate aggregates from monomers, we used a size exclusion chromatography column; because the aggregates are larger in size than their monomers, they are eluted off the column before the monomers (that is, at shorter retention times). As demonstrated by size exclusion high pressure liquid chromatograph (HPLC) analysis, the S3 solutions did not form any detectable soluble aggregates over the 6 month period under various storage conditions (Fig. 1a–c). The size exclusion column was capable of distinguishing the heteroduplex of S3 and S3-C from the monomers (Fig. 1d). Some soluble aggregates were detected in A3 samples stored at 22 °C but very little at 4 °C (Fig. 2) in a time-dependent manner, more pronounced at 2 mM solution (Fig. 2c). When the E3 solution was stored at 22 °C compared to the one at 4 °C (Fig. 3a, b) and at higher storage concentration (2 mM) compared to 1 mM (Fig. 3c), there were more aggregates as shown by UV detection of a greater area under the curve of the HPLC trace formed in the E3 solution over time and the complexity of the

Table 1
Morpholino names, sequences, % guanine (G) content, and length

Name	Sequences (5′–3′)	G%	Length
S3	CCTCTTACCTCAGTTACAATTTATA	4	25
S3-C	TATAAATTGTAACTGAGGTAAGAGG	28	25
A3	AGAGGGAGACCGTGGAAGGAGACTG	48	25
E3	GAGGCAGGAGAATCAGGCAGGGAGG	52	25

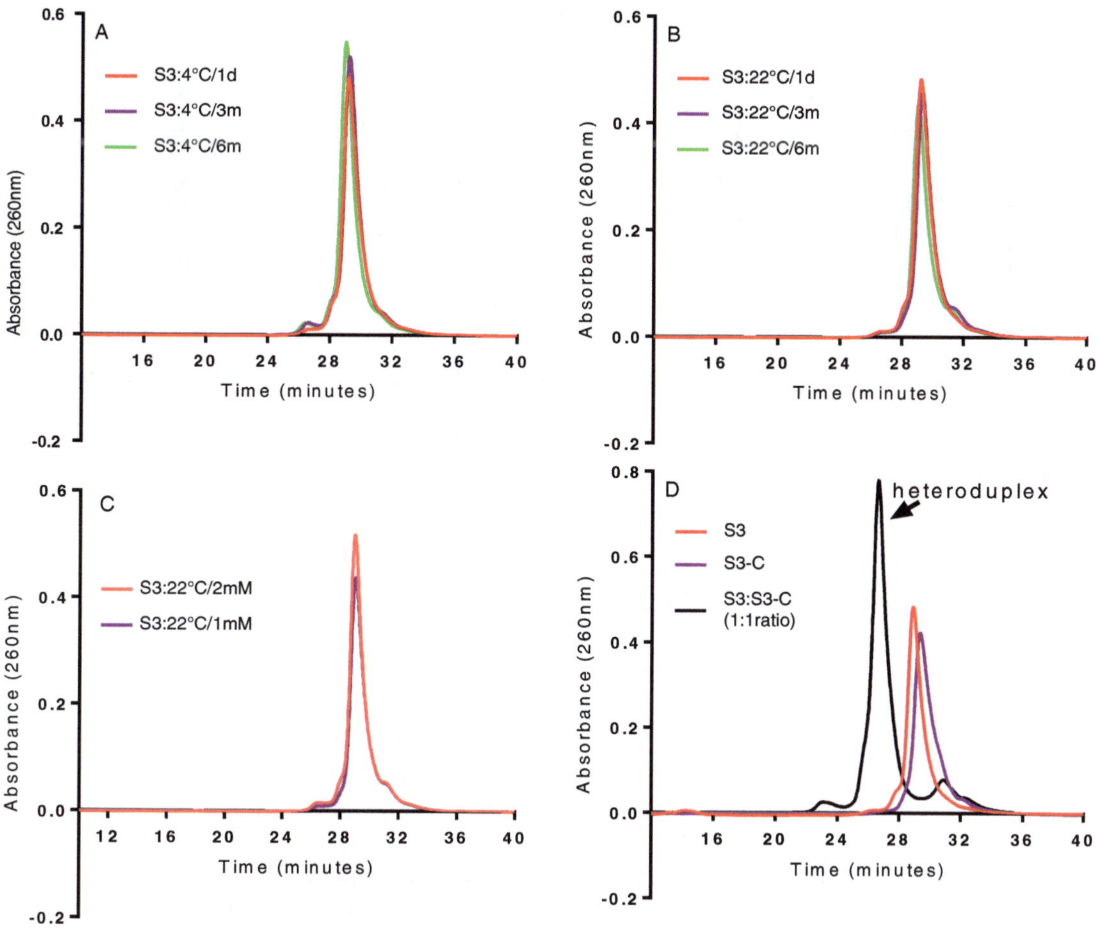

Fig. 1 Size exclusion HPLC profiles of S3 solutions stored under various conditions. The S3 MO was dissolved in sterile water at concentrations of 1 and 2 mM. The solutions were stored either in a humidified chamber at 22 or at 4 °C for up to 6 months. At 1 day (1d), 3 month (3m), and 6 month (6m) time points, aliquots of each MO solution were taken for HPLC analysis. Over the 6 month study period, there was very little change in the peak distribution of S3 for the 1 mM MO solutions stored at 4 °C (**a**) or 22 °C (**b**) or at higher 2 mM concentration (**c**). The size exclusion column was able to resolve monomeric MO (S3 or S3-C) at 29 min from the heteroduplex of the S3/S3-C (**d**) at 26 min. The heteroduplex peak has shorter retention time with a molecular size roughly twice that of S3 or S3-C monomer

aggregation is more pronounced with a broadened plateau and a greater range in retention time, indicating formation of larger aggregates in the E3 samples. The % aggregates under various conditions are shown in Table 2. The A3 and E3 aggregates can be dissociated to their monomeric forms by autoclaving the solutions at 121 °C, while heating the solutions at 65 °C was much less effective (Fig. 4).

Below, we describe a simple method for the detection of Morpholino aggregation and for Morpholino disaggregation.

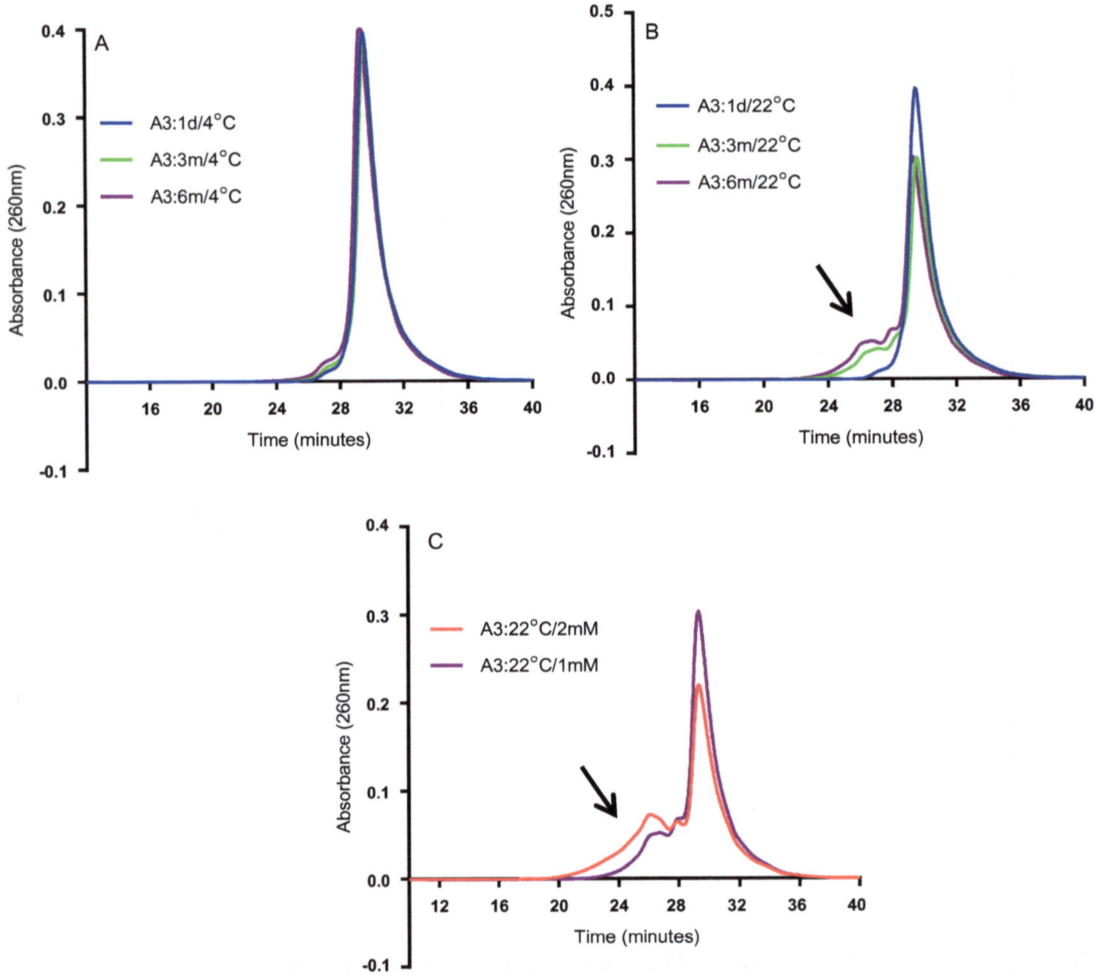

Fig. 2 Size exclusion HPLC profiles of A3 solutions stored under various conditions. The A3 solutions were prepared and stored under conditions identical to the S3 solutions. Over the 6 month study period (**a** and **b**), there was a time-dependent change in peak distribution with decreased main peak intensity accompanied by an increase in a broad peak at shorter retention time (indicated by the *arrow*). The broad peak at shorter retention time corresponds to larger sizes of A3 aggregates. This was much more prominent with the A3 solution stored at 22 °C (**b**) compared to the one stored at 4 °C (**a**). At the 6 month time point, there was a concentration-dependent change in peak distribution for the A3 solutions stored at 22 °C with a higher amount of aggregates in the 2 mM A3 solution (39%) as indicated by decreased main peak intensity at 29 min and increased intensity and width of the peak at retention time spanning 22–28 min (**c**)

2 Materials

2.1 Detection of Morpholino Aggregates in Solutions

1. Morpholinos to be analyzed. Stock solution: 1 or 2 mM concentration dissolved in sterile water or sodium phosphate-buffered saline.
2. A HPLC instrument with a UV detector.
3. Mobile phase: 100 mM sodium phosphate (pH 7.3) filtered through a 0.2 μm filter flask.

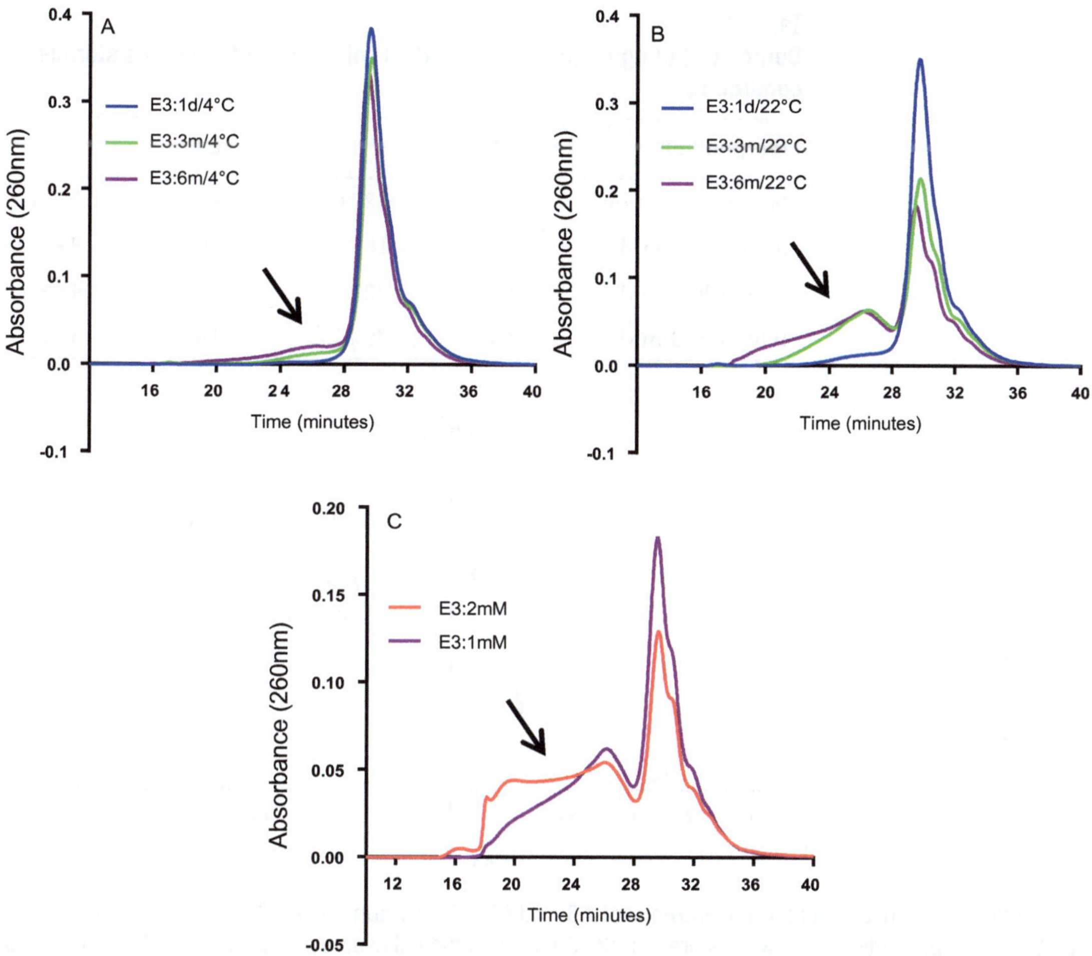

Fig. 3 Size exclusion HPLC profiles of E3 MO solutions stored under various conditions. The E3 solutions were prepared and stored under the same conditions as for the S3 solutions. Over the 6 month study period (**a** and **b**), there is a time-dependent change in peak distribution with a decrease in the main peak intensity at 29 min accompanied by an increase in a broad peak at shorter retention time (shown by the *arrow*), indicating the formation of larger sizes of molecules. This is much more prominent with the E3 solution stored at 22 °C (**b**) compared to the one at 4 °C (**a**). At the 6 month time point, the 2 mM E3 solutions stored at 22 °C developed more and larger sized soluble aggregates (57%) than the 1 mM E3 solution, as indicated by decreased main peak intensity at 29 min and increased intensity and width of the peak at retention time spanning 15–28 min (**c**)

4. Stationary phase: a size exclusion HPLC column (WTC-015S5, Wyatt Technology).

2.2 Disaggregation of Morpholino Solutions

1. Morpholino stock solution, normally at 1 or 2 mM.
2. Boil-Proof 1.5 mL microcentrifuge tubes (Axygen Scientific).
3. A vortexer.
4. A centrifuge.
5. A microcentrifuge tube rack.

Table 2
Degree (%) of aggregates of A3 and E3 solutions under various storage conditions

Morpholinos	A3		E3	
Storage conditions	4 °C	22 °C	4 °C	22 °C
3 month, 1 mM	1%	15%	5%	34%
6 month, 1 mM	4%	22%	10%	43%
6 month, 2 mM	8%	39%	18%	57%

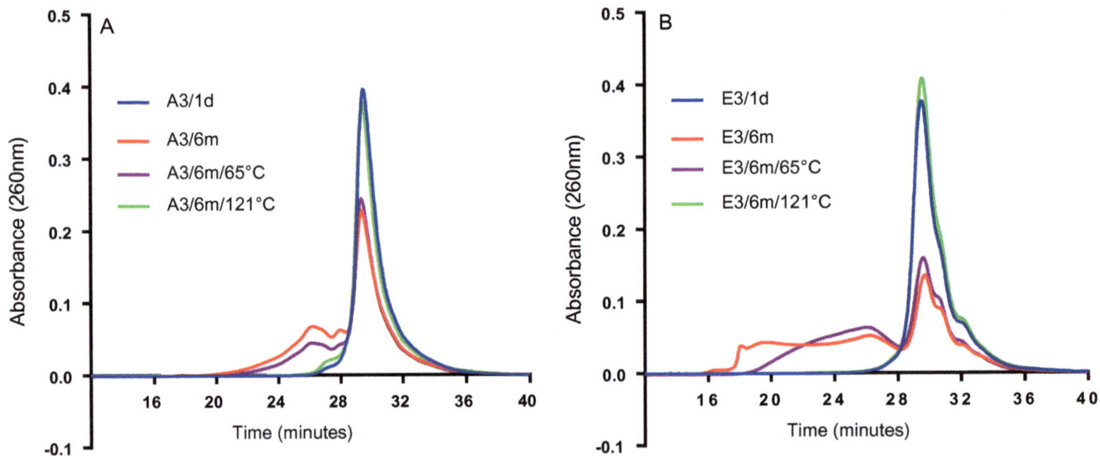

Fig. 4 Effect of high-temperature treatment of A3 and E3 MO solutions. The HPLC profiles of 2 mM A3 (**a**) and 2 mM E3 (**b**) solutions that were stored at 22 °C for 6 months followed by incubation at 65 °C for 30 min (*purple curve*) or in an autoclave at 121 °C for 15 min (*green curve*) compared to the original solutions at 1 day (*blue curve*) and 6 month storage (*red curve*). The data show that autoclaving treatment effectively dis-aggregated both A3 and E3 MO solutions, indicating by the disappearance of the broad peak spanning the time range of 20–26 min for A3 and 17–26 min for E3 and increased the area under the curve of the monomer peak

6. An autoclave (e.g., Consolidated Sterilizer system model SSR-BA).
7. A UV-Vis spectrometer or Nanodrop.
8. 0.1 N HCl.

3 Methods

3.1 Detection of Morpholino Aggregates in Solutions

1. Set up a HPLC system with an attached UV detector according to the instruction given by the HPLC manufacturer.
2. Equilibrate the stationary phase (the HPLC column) with the mobile phase at 0.3 mL/min for 60 min.
3. Dilute the Morpholino stock solution into the mobile phase to make 40 μL of 0.1 mM of analysis solution.

4. Load/inject 30 μL of 0.1 mM of the analysis solution onto the column.
5. Start the HPLC run at a flow rate of 0.2 mL/min for 45 min.
6. Analyze the data.

3.2 Disaggregation of Morpholino Solutions

1. Vortex the tube containing Morpholino stock solution for 30 s (*see* **Note 1**).
2. Transfer an aliquot of the stock solution to 0.1 N HCl to determine the concentration of the Morpholino by measuring its absorbance at 265 nm in 0.1 N HCl.
3. Carefully pipet desired volume of the Morpholino stock solution into a boil-proof microcentrifuge tube.
4. Measure out equivalent volume of deionized water into a separate microcentrifuge tube to serve as a counter balance.
5. Centrifuge the Morpholino solution for 30 s at 1000 × *g* (*see* **Note 2**).
6. Carefully seal the Morpholino tube with autoclave tape (*see* **Note 3**).
7. Place the tube onto a microcentrifuge tube rack.
8. Place the rack in an autoclave.
9. Autoclave the rack using a liquid sterilization cycle for 15 min of sterilization time at 121 °C and chamber gauge pressure of 1 atm (15 PSI). Exhaust should last 15 min followed by 10 min of cooldown (*see* **Note 4**).
10. After the autoclave cycle has completed, carefully remove microcentrifuge rack from the autoclave and allow rack to cool at room temperature for 10 min. After cool down period, vortex oligomer tube for 10 s followed by centrifuging for 30 s at 1000 × *g* (*see* **Note 5**).
11. Unseal the Morpholino tube.
12. Transfer an aliquot of the autoclaved solution to 0.1 N HCl to determine the concentration of the Morpholino after autoclaving by measuring its absorbance at 265 nm in 0.1 N HCl (*see* **Note 6**).

4 Notes

1. Visually inspect tube to ensure solution is homogenous after vortex. If solution is not homogenous, continue vortexing and/or add deionized water until solution is homogenous.
2. Visually inspect tube to see if any oligomer solution is still on the sides or top of the tube. If so, centrifuge tube for another

30 s and repeat inspection until the oligomer is pooled into bottom of tube.

3. When sealing tube with autoclave tape, be careful that the pooled oligomer on bottom does not splash back onto the upper sides of the tube.
4. If using a Consolidated Sterilizer system model SSR-BA, after setting autoclave for liquid cycle and 15 min of sterilize time, the rest of the parameters should be the default settings and no changes should be needed.
5. After centrifugation, inspect tube for any murkiness. If seen, allow solution to cool down for another 5 min and repeat vortex and centrifuge.
6. We have found there is very little change in Morpholino concentration after autoclaving if the tube is sealed correctly.

References

1. Burrer R, Neuman BW, Ting JP, Stein DA, Moulton HM, Iversen PL, Kuhn P, Buchmeier MJ (2007) Antiviral effects of antisense morpholino oligomers in murine coronavirus infection models. J Virol 81:5637–5648
2. Rajsbaum R, Versteeg GA, Schmid S, Maestre AM, Belicha-Villanueva A, Martinez Romero C, Patel JR, Morrison J, Pisanelli G, Miorin L et al (2014) Unanchored K48-linked polyubiquitin synthesized by the E3-ubiquitin ligase TRIM6 stimulates the interferon-IKKepsilon kinase-mediated antiviral response. Immunity 40:880–895
3. Tripathi S, Pohl MO, Zhou Y, Rodriguez-Frandsen A, Wang G, Stein DA, Moulton HM, DeJesus P, Che J, Mulder LC et al (2015) Meta- and orthogonal integration of influenza "OMICs" data defines a role for UBR4 in virus budding. Cell Host Microbe 18:723–735
4. Geller BL, Marshall-Batty K, Schnell FJ, McKnight MM, Iversen PL, Greenberg DE (2013) Gene-silencing antisense oligomers inhibit acinetobacter growth in vitro and in vivo. J Infect Dis 208:1553–1560
5. Greenberg DE, Marshall-Batty KR, Brinster LR, Zarember KA, Shaw PA, Mellbye BL, Iversen PL, Holland SM, Geller BL (2010) Antisense phosphorodiamidate morpholino oligomers targeted to an essential gene inhibit Burkholderia cepacia complex. J Infect Dis 201:1822–1830
6. Shen N, Ko JH, Xiao G, Wesolowski D, Shan G, Geller B, Izadjoo M, Altman S (2009) Inactivation of expression of several genes in a variety of bacterial species by EGS technology. Proc Natl Acad Sci U S A 106:8163–8168
7. Xiao G, Wesolowski D, Izadjoo M, Altman S (2010) Morpholino oligonucleotides do not participate perfectly in standard Watson-Crick complexes with RNA. RNA 16:2218–2225
8. Summerton J, Weller D (1997) Morpholino antisense oligomers: design, preparation, and properties. Antisense Nucleic Acid Drug Dev 7:187–195
9. Lambert JD, Johnson AB, Hudson CN, Chan A (2016) Dpp/BMP2-4 mediates signaling from the d-quadrant organizer in a spiralian embryo. Curr Biol 26(15):2003–2010
10. Amantana A, Moulton HM, Cate ML, Reddy MT, Whitehead T, Hassinger JN, Youngblood DS, Iversen PL (2007) Pharmacokinetics, biodistribution, stability and toxicity of a cell-penetrating peptide-morpholino oligomer conjugate. Bioconjug Chem 18: 1325–1331
11. Hudziak RM, Barofsky E, Barofsky DF, Weller DL, Huang SB, Weller DD (1996) Resistance of morpholino phosphorodiamidate oligomers to enzymatic degradation. Antisense Nucleic Acid Drug Dev 6:267–272
12. Youngblood DS, Hatlevig SA, Hassinger JN, Iversen PL, Moulton HM (2007) Stability of cell-penetrating peptide-morpholino oligomer conjugates in human serum and in cells. Bioconjug Chem 18:50–60

Chapter 4

End-Modifications on Morpholino Oligos

Yong-Fu Li

Abstract

Modifications at either end, both ends, or in-between the ends of a Morpholino oligo provide functional groups for further conjugation. Amino groups are the most useful and efficient reactive entities for chemical bonding with other molecules. The combination of modifications at both ends, especially with double functionalization at the 3′-end, yields myriad opportunities for diverse applications. An orthogonally protected diamine for advanced 3′-end double modification on the solid phase synthesis support allows the convenient assembly of a vast variety of custom-designed molecules. A particular application is the assembly of a class of Vis-Vivo-Morpholino where at the 3′-end an optically visible fluorophore is installed at one side for fluorescent detection and an in vivo delivery moiety is attached at the other side for intracellular activity studies.

Key words Morpholino oligo, Functionalization, End-modification, Conjugation, Amino groups, Vis-Vivo-Morpholino

1 Introduction

The dream of modern drug research to develop a therapeutic technology that can act specifically only on the target responsible for the disease has led to the development of drugs that can turn off defective mRNAs when aberrant proteins are made in the body, or usefully alter the mRNAs to treat the genetic diseases in which a required protein cannot be produced [1]. In the quest for effective antisense drug candidates, various chemical modifications of the natural oligonucleotides (ONs) have been studied, such as modifications in the phosphodiester backbone, heterocyclic nucleobase and sugar moiety, some conferring high affinity or specificity for their target nucleic acid sequences [2, 3]. Among those antisense oligomers, including siRNA, PNA, S-DNA, and LNA, Morpholino oligos have two remarkable chemistry features [4]: (a) The internucleoside linkage contains a phosphorodiamidate group that confers a neutral (uncharged) backbone; and (b) The ribose sugar is replaced by a six-membered morpholino ring

Hong M. Moulton and Jon D. Moulton (eds.), *Morpholino Oligomers: Methods and Protocols*, Methods in Molecular Biology, vol. 1565, DOI 10.1007/978-1-4939-6817-6_4, © Springer Science+Business Media LLC 2017

that provides excellent nuclease stability [5] in comparison to that of the unmodified ONs. These two modifications together make the Morpholino oligos superior as a steric blocker for specific inhibition of gene expression. Because their uncharged backbone does not interact electrostatically with proteins, they are free of non-antisense effects and because they are resistant to cleavage by nucleases, they are completely stable in biological systems. Since demonstrating efficacy, specificity, and non-toxicity, Morpholino antisense technology has entered into drug research and development.

However, like other types of antisense oligomers, Morpholino oligos themselves face two major problems: (a) They are invisible by optical microscopy in biological assay systems, and (b) They cannot easily pass through the cell membrane to reach their intracellular targets. In order to solve these problems and open doors for other applications such as disease diagnosis and high-throughput screening, novel methods for a mild and site-specific bioconjugation of Morpholino oligos have been developed. A comprehensive review [6] has portrayed functionalizing Morpholino oligos including the reactive group installation, fluorophore attachment, cell-permeable moiety conjugation, and photo-switch assembly. These powerful methods owe their existence to the discovery of chemoselective reactions that enable modifications for specific purposes.

The success of conjugation schemes depends on the presence of the correct chemical groups. Every chemical modification or conjugation process involves the reaction of one functional group with another, resulting in the formation of a covalent bond. The creation of bioconjugate reagents with spontaneously reactive or selectively reactive functional groups forms the basis for simple and reproducible cross-linking of target molecules. Of the hundreds of reagent systems described in the literature or offered commercially, the most common chemical bond formations can be reduced to a couple dozen or so primary reactions. Amino groups are probably the most versatile functional moiety for post-synthetic derivatization. They are reactive toward isothiocyanate, isocyanate, acylazide, activated ester, sulfonyl chloride, aldehyde, epoxide, carbonate, arylating agent, imidoester, and anhydride [7]. Most usefully, reactions between an amine and a carboxylic acid or carbonate to form the corresponding amide or carbamate linkage are the most preferred derivatization strategies to conjugate biomolecules.

Amide bonds have a half-life of ca. 600 years in neutral solution at 25 °C [8]. This extraordinary stability makes amide linkages highly attractive for bioconjugation. The carboxyl group is the most abundant functional group in biomolecules. Acylation of amine with activated ester is rapid and occurs in high yield to give an amide bond. There are different ways of coupling reactive carboxyl derivatives with an amine to form amide, two of which are commonly used: (a) Carboxyl moieties can be preactivated as acyl esters, acyl azides, acylimidazoles, anhydride, etc. These reactive acylating agents are formed from the acid in a separate step followed by immediate treatment with the amine. (b) The active acylating agent is generated in situ from the acid in the presence of the amine by an addition of an activating or coupling agent [9]. Carbodiimide-based activation of carboxyl groups is the most common method for active ester preparation. Activation of carboxylic acids with carbodiimides is complicated by hydrolysis and irreversible intramolecular rearrangement of the desired reactive O-acylurea derivative to the unreactive N-acylurea derivative [10]. The latter complication is often overcome using one pot formation of active ester by the addition of an auxiliary nucleophile, e.g., 3-sulfo-1-hydroxysuccinimide (S-NHS) or 1-hydroxy-1,2,3-benzotriazole (HOBt), simultaneously with carbodiimide [11]. To conjugate polar biomolecules containing carboxylic acid, N-substituted carbodiimides are widely used. For example, 1-(3-dimethylaminopropyl)-3-ethylcarbodiimide (EDCI), a water-soluble reagent for carboxylate activation, has found general usage. This water-soluble property enables direct application of EDCI in aqueous solutions without addition of any organic solvent. During the reaction of EDCI with carboxylic group, the active intermediate O-acylisourea ester is formed. The latter reacts with primary amine, forming an amide bond.

Since the methods of amide and carbamate formation have been used widely in the bioconjugation process, this article is particularly focused on the introduction of amino groups on Morpholino oligos so that the corresponding amino groups can conjugate with biomolecules containing activated carboxylic acid or activated carbonate.

2 Amino Group at the 5′-End

There are particular sites that can be modified on the bases of Morpholino oligos to produce derivatives able to couple with a second molecule. However, modifications on those sites may impact the pairing of a Morpholino oligo with its complementary target sterically and/or electronically. Therefore, the ends of the Morpholino oligo have been the sites of choice for the introduction of functionalities.

In the solid phase synthesis of Morpholino oligos, functionalization at the 5′-end of the oligo takes advantage of the trifunctional triazine moiety whereby installation of both the amino functional group and the site for oligo elongation can be accomplished. To introduce an amino group at the 5′-end, a primary amine protected with trifluoroacetyl group was incorporated into a triazine core together with a secondary amine protected with trityl

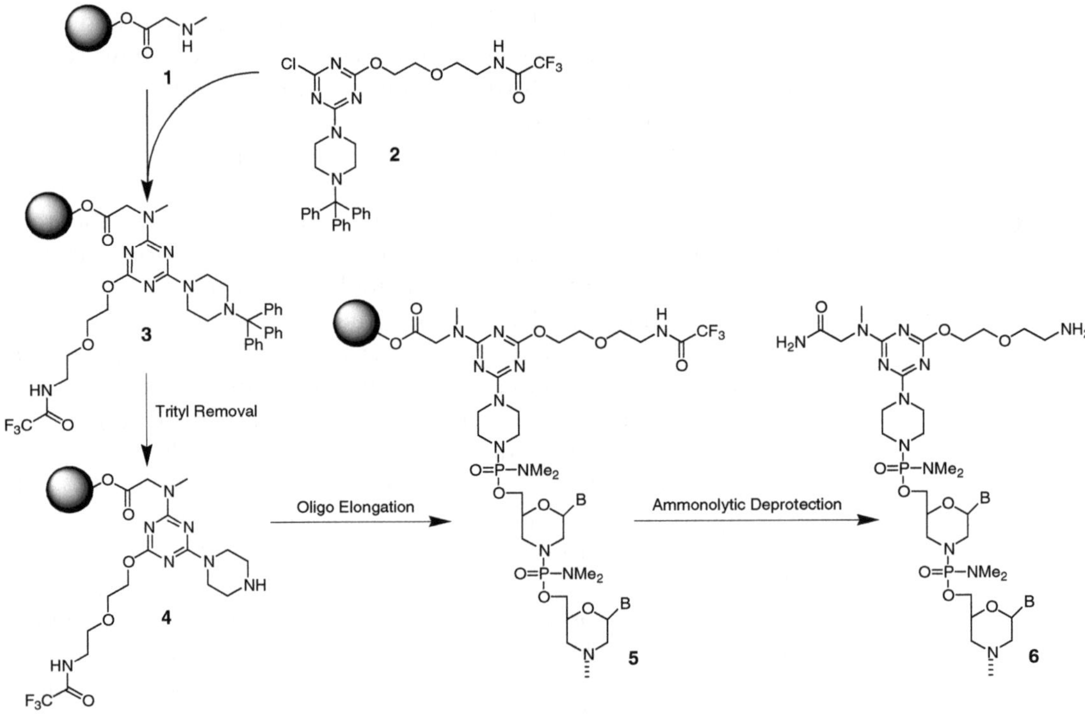

Fig. 1 Assembly of a Morpholino oligo containing the 5′-amine

group. The chloro derivative **2** (Fig. 1) provides the reactive site to add to the amino group of sarcosine **1** on the solid phase synthesis resin. The trityl group of **3** which is acid labile is then removed to generate a secondary amine **4** for oligo elongation. The trifluoroacetyl group of **5** which is base-labile is deprotected under ammonolytic conditions in the final step to complete the oligo synthesis where all the protecting groups on the subunit bases are removed and the linkage between the synthesis resin and the oligo is cleaved, releasing a Morpholino oligo with the 5′-end primary amine **6**.

3 Amino Group at the 3′-End

The secondary amine 7 (Fig. 2) of the last Morpholino subunit can, in principle, be used as an amino group for modification. As a matter of fact, most 3′-end modifications use this secondary amine as a reactive site for further installation. Its nucleophilicity is weaker than that of a normal primary amine; it was found that this secondary amine was about five times less reactive than a primary amine [12]. To achieve efficient conjugation with a second molecule, a primary amine **10** was built upon the secondary amine **7**. Trifluoroacetyl protected β-alanine in its active ester **8** was coupled onto the last Morpholino subunit to give compound **9**. Under the

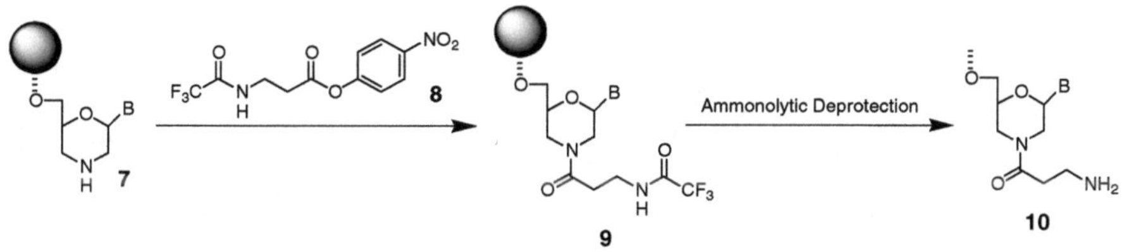

Fig. 2 Modification of a Morpholino oligo with the 3′-amine

ammonolytic deprotection conditions, the 3′-end primary amine **10** is generated with the universal deprotection of the protecting groups of the subunit bases and cleavage of the oligo from the synthesis resin.

4 Amino Group In-Between the Ends

Besides the ends of a Morpholino oligo, functionalization at the internucleoside linkage could also provide a reactive group for highly site-specific bioconjugation and maintain the minimal perturbation to the active form of the oligo. A trifluoroacetyl protected primary amine was incorporated into a special activated subunit **12** (Fig. 3). Following the standard oligo assembly process forming a short oligo **11**, the special subunit is thus attached to form the oligo **13**. After the removal of the trityl group, **14** continues the oligo elongation. The final ammonolytic deprotection not only cleaves the oligo **15** from the synthesis resin, but also removes all the protecting groups to furnish a Morpholino oligo containing the primary amine **16** in-between the ends.

5 Amino Groups at the 3′-End Together with Additional Functionalities

Double modification at the 3′-end of a Morpholino oligo 7 (Fig. 4) while still on the solid phase synthesis support employs the controllable reactivity of cyanuric chloride. This trifunctional entity can accommodate two different functional groups with the third chloride (**17s**, **18s**, **19s**, **20s**, and **21s**) to form a covalent bond with the last Morpholino subunit. After modification, acidic treatment can remove the trityl group to generate a secondary amine. At this stage, conjugation of this secondary amine with a second molecule is feasible on the synthesis resin. The final ammonolytic deprotection cleaves the oligo from the synthesis resin and also removes the trifluoroacetyl group to provide primary amines (**17** and **21**). Figure 4 shows its utility for additional installation of other functionalities:

3AB (**17**) contains a biotin on one side and a primary amine on the other side at the 3′-end. Biotin (or Vitamin H) is a small

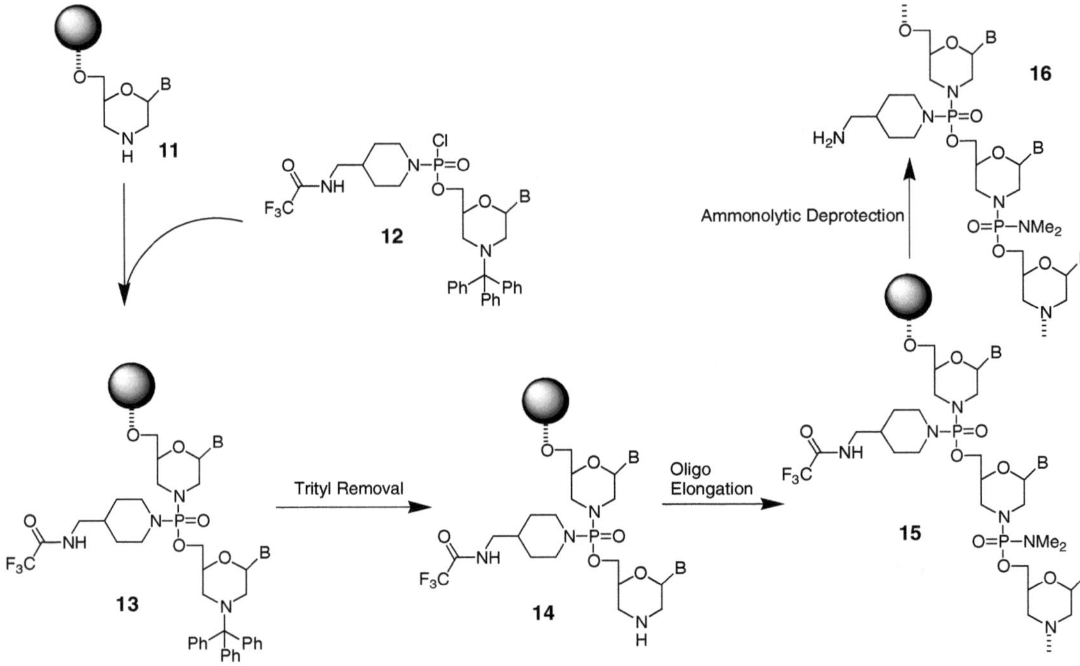

Fig. 3 Installation of a Morpholino oligo with an amino group in-between the ends

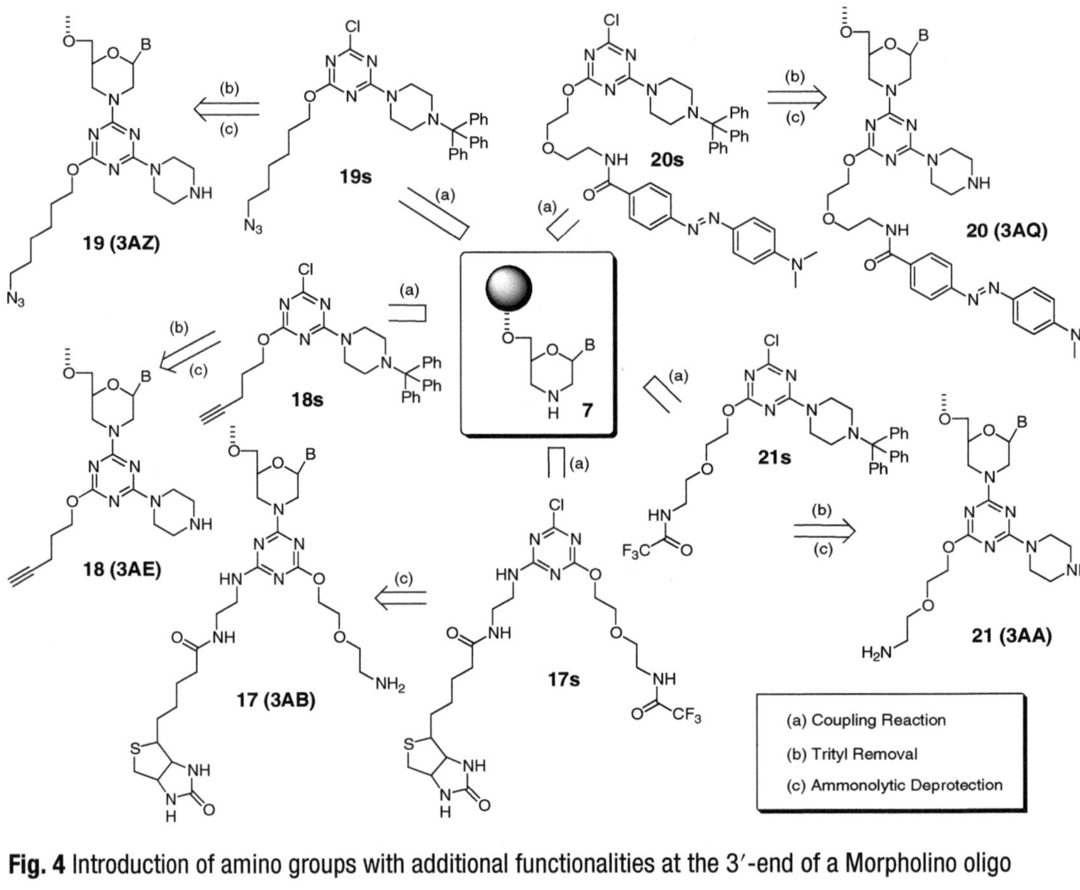

Fig. 4 Introduction of amino groups with additional functionalities at the 3′-end of a Morpholino oligo

biologically active molecule, acting as a coenzyme in living cells. With its highly specific affinity toward streptavidin, it is used in various biotechnology assays for quality and quantity testing. Biotinylated oligonucleotides can be used to attach specifically to streptavidin-enzyme conjugates, to streptavidin-protein conjugates, to streptavidin coated surfaces, or to streptavidin-dye conjugates. **3AB** provides an accompanying amino group, useful for further needed conjugation with an amine-reactive moiety [13].

3AE (**18**) and **3AZ** (**19**) both have a secondary amine on one side at the 3′-end for subsequent coupling with amine-reactive molecules. The former contains an alkyne group and the latter the azide moiety. Both can be used for further Click reactions [14]. Click coupling is a class of powerful biocompatible reactions intended to join substrates of choice with specific biomolecules. **3AE** and **3AZ** contain alkyne and azide respectively for cycloaddition to conjugate a second Clickable molecule. In addition to the Click reaction for the azide, this apparently "inert" group can undergo Staudinger ligation [15] to generate an amide bond from the azide and a specifically functionalized phosphine. This method for the selective formation of an amide bond, which does not require the orthogonal protection of the functional group, should find general utility in bioconjugate chemistry.

Dabcyl is a nonfluorescent dye predominantly used as a quencher for molecular beacons or probes [16]. If Dabcyl is coupled to an oligonucleotide in close proximity to a fluorophore, it absorbs much of the emitted light of the fluorophore. Enlarging this distance (i.e., by melting of a beacon's stem) results in an increased emission signal when the fluorophore is excited. **3AQ** (**20**) contains the Dabcyl modification and an accompanying amino group to provide a reactive site for further conjugation.

3AA (**21**) can generate two different amines in a stepwise manner. The secondary amine generated from trityl group removal can be used for conjugation with a second molecule on the synthesis resin, and the primary amine generated from the ammonolytic deprotection can be used for off-column conjugation with a third molecule. Once these two amines are generated and left for conjugation, they are chemically different but hard-to-distinguish in reactivity and can be employed for attaching two identical pieces to the 3′-end of a Morpholino oligo.

6 Amino Groups on Solid Phase Synthesis Column

Solid phase synthesis is a method in which molecules are bound on a polymer resin and synthesized step by step in a reactant solution. It provides the convenience to remove excess reactant or byproduct from the product by simply washing with fresh solvent, thereby simplifying the purification process.

To introduce the 3′-end amino group on the synthesis resin, Fmoc protected aminohexanoic acid **22** in its activated ester form can be attached on the secondary amine of the last Morpholino subunit 7 to form a stable amide bond (Fig. 5). Normally, removal of Fmoc derivative **23** is executed to generate the amine **24** just prior to the conjugation with an amine-reactive biomolecule, as the free amine **24** on storage can be capped by migration of the protecting groups from the subunit bases, compromising the subsequent conjugation. The on-column conjugation simplifies the removal of the excess reagents, eliminating the laborious purification process as compared with an off-column coupling. The prerequisite for this application is that the biomolecule **25** being coupled should survive ammonia treatment as ammonia is the reagent of choice to complete the synthesis of oligo **26**.

An orthogonally protected diamine for advanced 3′-end double modification on the solid phase synthesis resin requires a trifunctional cross-linker. A design strategy using triazine chemistry takes advantage of cyanuric chloride to build three different reactive groups in one molecule. Their reactivities can be manipulated to install two amino groups protected orthogonally and leave the third chloride to further react with N-methylaminoethanol to generate an alcohol that can be activated to form a carbonate derivative **27**. This trifunctional moiety is herein referred to as a "Y-adapter" (Fig. 6). The Y-adapter can couple with the secondary amine 7 of the last subunit of a Morpholino oligo while still on the synthesis resin to give intermediate **28**.

The chemistry on the solid support allows the convenient assembly of a vast variety of combinations where Fmoc removal under basic conditions can generate the primary amine to conjugate with amine-reactive biomolecule-A to form **29a**, followed by trityl removal under acidic conditions to generate the secondary amine to conjugate with amine-reactive biomolecule-B, or vice

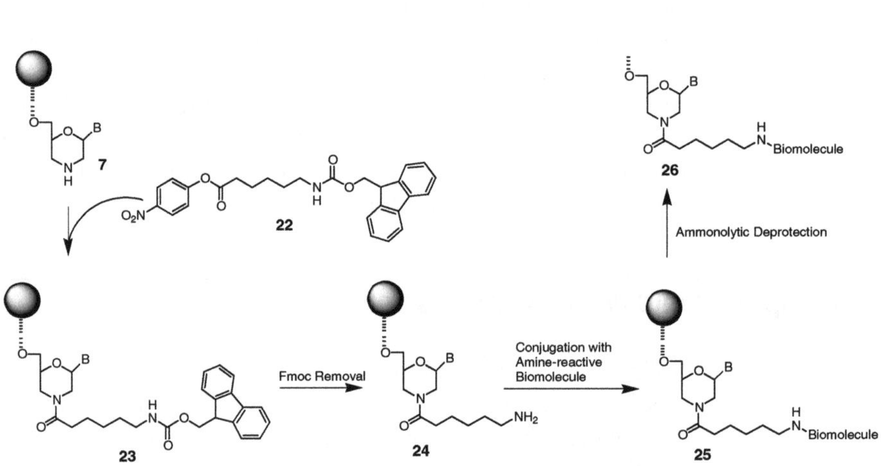

Fig. 5 On-column attachment of amino group at the 3′-end of a Morpholino oligo

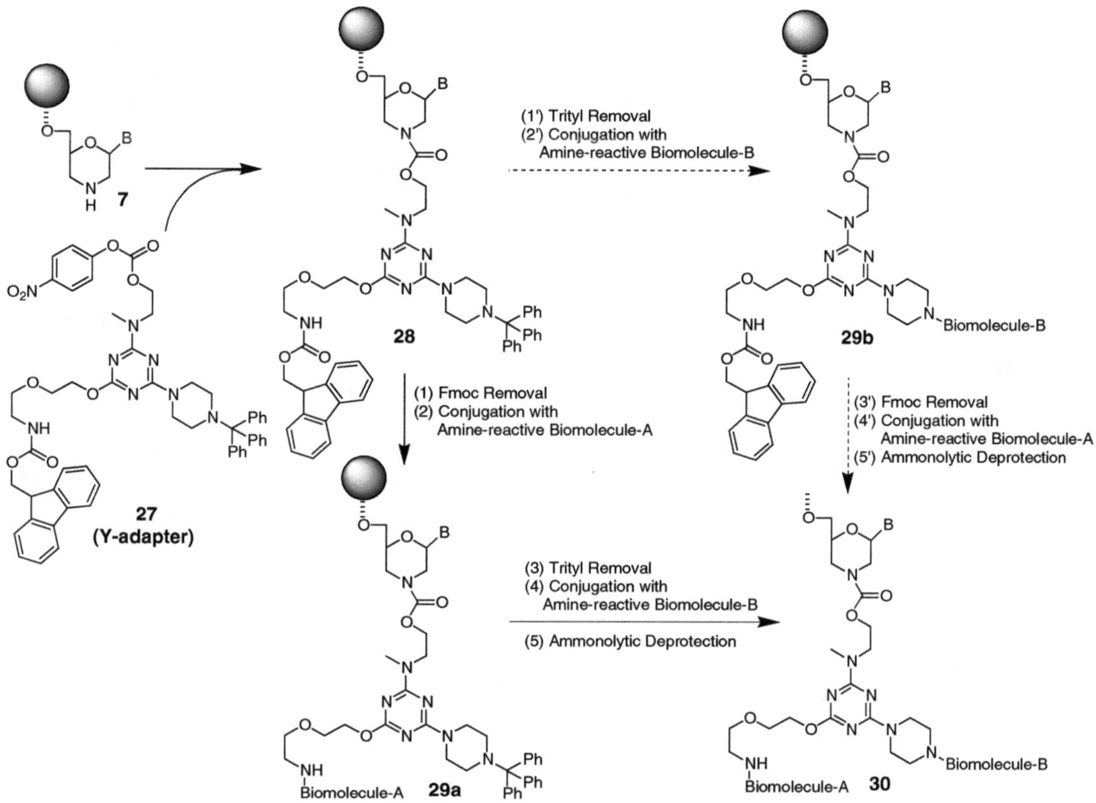

Fig. 6 On-column attachment of 3′-diamino groups orthogonally protected for further conjugation

versa via **29b**. Ammonolytic deprotection universally cleaves ammonia-sensitive groups to give rise to a Morpholino oligo **30** installed at the 3′-end with two biomolecules. Again, both biomolecule-A and B should remain intact in the presence of concentrated ammonia for a successful assembly.

7 Application of Vis-Vivo-Morpholino Assembly

A Morpholino oligo itself is difficult to physically detect in a biological assay system, and is also poorly cell-permeable in cultures and living animals. These problems have been solved individually where fluorophore-labeled Morpholinos have been constructed for fluorescent detection [6], and Vivo-Morpholino [17], a Morpholino conjugate with Vivo-Porter, a dendrimeric octa-guanidine moiety, can transport Morpholino oligo into the cytosol/nuclear compartment of cells of living animals [18]. However, the combination of both modifications in a single oligo has never been attempted before the emergence of the Y-adapter.

To enable double modifications at the 3′-end of a Morpholino oligo with functional entities (Figs. 6 and 7), the Y-adapter can

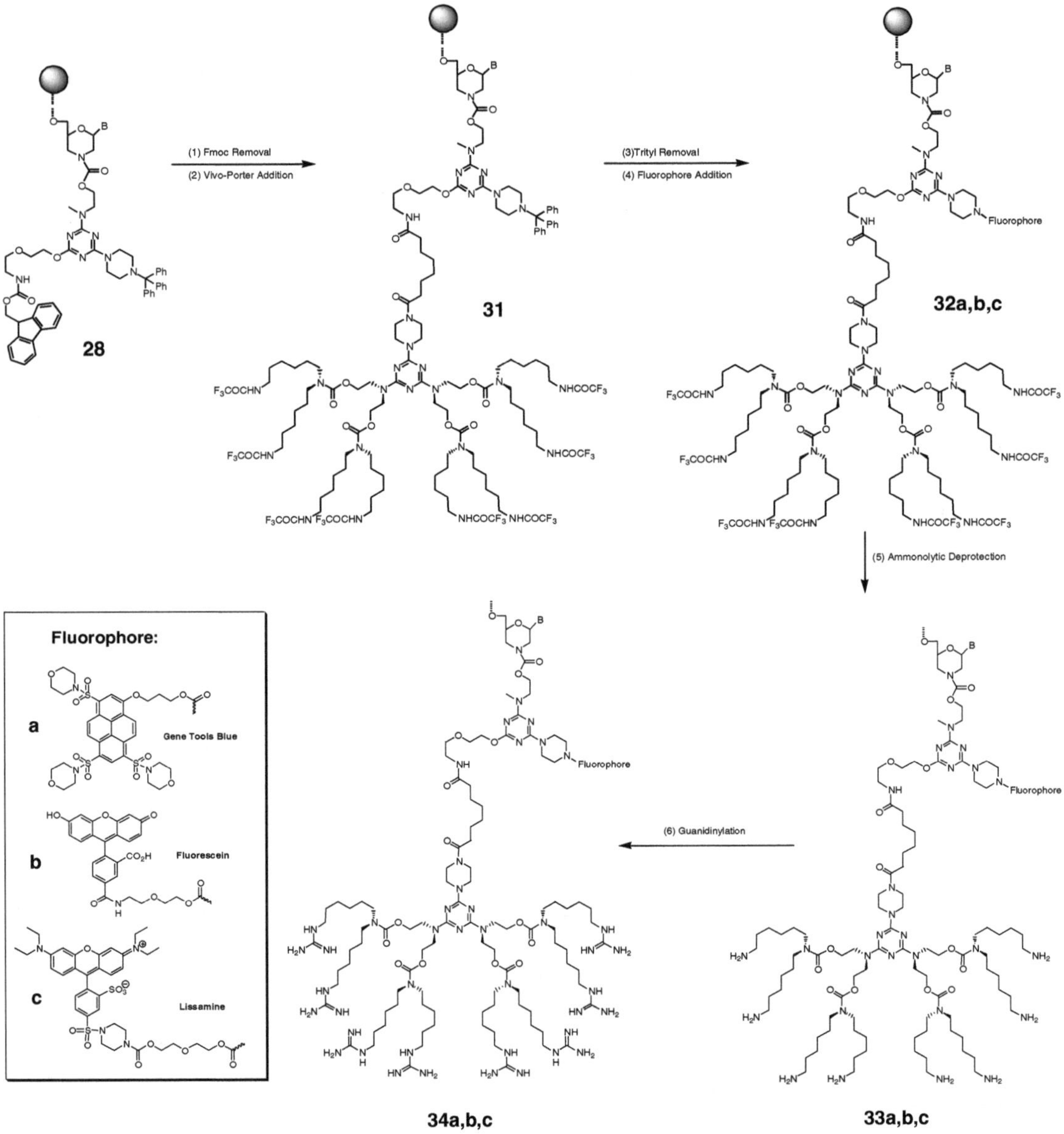

Fig. 7 Assembly of Vis-Vivo-Morpholino

serve the exact purpose. Figure 6 shows its use for further conjugation with custom-designed molecules, whereas Fig. 7 presents its application for installing fluorophores for optical detection and delivery-enabling moieties for in vivo antisense activity studies.

After the Y-adapter was attached to the last subunit of a Morpholino oligo while still on the synthesis resin (Fig. 7), removal of the Fmoc group of **28** under basic conditions generates the primary amine for further conjugation with Vivo-Porter precursor to give **31**. Acidic detritylation follows to expose the secondary amine that is reactive for the introduction of a fluorophore to give **32a,b,c**, respectively. Ammonolytic deprotection cleaves all the ammonia-sensitive bonds. The octa-amine intermediate **33a,b,c** generated undergoes guanidinylation to furnish a class of Vis-Vivo-Morpholino **34a,b,c**, a Morpholino oligo containing the Y-adapter with a fluorophore at one side and a dendrimeric octa-guanidine moiety at the other side. The oligo is now visible optically and active intracellularly on Vivo-Porter-mediated uptake from the extracellular space.

Fluorophores typically used with Morpholinos are a family of three fluorescent dyes covering the range between 400 and 600 nm emission wavelength from blue through green to red. **Gene Tools Blue** (a, Fig. 7) has excitation maximum at 421 nm and emission maximum at 465 nm. The molecular structure of the dye is nonionic, facilitating its entry into cells for intracellular detection. The spectral properties of the fluorescent dye are sufficiently different in wavelengths and intensity from fluorescein that Gene Tools Blue is remarkably useful for avoiding interference from green fluorescent protein expression in biological systems or for permitting simultaneous use of several dyes for multi-labeling applications with minimum interference. **Fluorescein** (b, Fig. 7) has an effective excitation wavelength of about 501 nm and its emission wavelength of about 524 nm. The fluorophore usually provides excellent detectability in assay systems, making it an important fluorophore for confocal laser-scanning microscopy and flow cytometry applications. The **Lissamine** form of rhodamine B (c, Fig. 7) has a maximal absorptivity at 575 nm, and its emission maximum occurs at 593 nm, emitting red luminescence. It has been used in numerous applications, including multiple-labeling techniques in microscopy, though its poor aqueous solubility is sometimes a problem.

The power and advantages of assessing intracellular processes at their most fundamental level have propelled the science of Morpholinos into bioconjugate chemistry where particular chemical groups are required to be created to effect coupling, to be modified to realize sensitive detection, or to be functionalized to achieve delivery in cell culture and animal studies. End modifications on one or both ends, and the combination thereof, yield myriad opportunities for more diverse applications. In particular, double modifications at the 3′-end using the orthogonally protected diamine (the Y-adapter) open the door for convenient assembly of advanced Morpholino conjugates. Morpholino oligos double-modified through the Y-adapter with fluorophores and an in vivo delivery moiety are now available, empowering the Morpholino oligos to be visible in optical assay systems and active in living animals. Vis-Vivo-Morpholino will be immensely helpful in understanding mechanistic aspects of numerous biological interactions and processes. With the continuous effort to innovate functionalizing techniques, Morpholino oligos will fulfill their promise for antisense applications in diagnostics and therapeutics.

References

1. Li Y-F (2014) To discover, develop and deliver a right drug: a showcase for antisense technology. J Drug Discov Dev Deliv 1:1–3
2. Sharma VK, Sharma RK, Singh SK (2014) Antisense oligonucleotides: modifications and clinical trials. Med Chem Commun 5:1454–1471
3. Uhlmann E, Peyman A (1990) Antisense oligonucleotides: a new therapeutic principle. Chem Rev 90:543–584
4. Summerton JE (2007) Morpholino, siRNA, and S-DNA compared: impact of structure and mechanism of action on off-target effects and sequence specificity. Curr Top Med Chem 7:651–660
5. Hudziak RM, Barofsky E, Barofsky DF, Weller DL, Huang SB, Weller DD (1996) Resistance of Morpholino phosphorodiamidate oligomers to enzymatic degradation. Antisense Nucleic Acid Drug Dev 6:267–272
6. Li Y-F (2016) Functionalizing Morpholino oligos for antisense drug research and development. J Drug Discov Dev Deliv 3:1021
7. Hermanson GH (2008) Bioconjugate techniques, 2 edn. Academic, San Diego, CA
8. Radzicka A, Wolfenden R (1996) Rates of uncatalyzed peptide bond hydrolysis in neutral solution and the transition state affinities of proteases. J Am Chem Soc 118:6105–6109
9. Montalbetti CAGN, Falque V (2005) Amide bond formation and peptide coupling. Tetrahedron 61:10827–10852
10. Bauminger S, Wilcheck M (1980) The use of carbodiimides in the preparation of immunizing conjugates. Methods Enzymol 70:151–159
11. Bulpitt P, Aeschlimann D (1999) New strategy for chemical modification of hyaluronic acid: preparation of functionalized derivatives and their use in the formation of novel biocompatible hydrogels. J Biomed Mater Res A 47:152–169
12. Dou S, Virostko J, Greiner DL, Powers AC, Liu G (2015) A feasible approach to evaluate the relative reactivity of NHS-ester activated group with primary amine-derivatized DNA analogue and non-derivatized impurity. Nucleosides Nucleotides Nucleic Acids 34:69–78
13. Dou SP, Virostko J, Greiner DL, Powers AC, Liu GZ (2015) Quantitative correlation of in vivo properties with in vitro assay results: the in vitro binding of a biotin-DNA analogue modifier with streptavidin predicts the in vivo avidin-induced clearability of the analogue-modified antibody. Mol Pharm 12:3097–3103
14. Shabanpoor F, Gait MJ (2013) Development of a general methodology for labelling peptide-morpholino oligonucleotide conjugates using alkyne-azide click chemistry. Chem Commun 49:10260–10262
15. Nilsson BL, Kiessling LL, Raines RT (2000) Staudinger ligation: a peptide from a thioester and azide. Org Lett 2:1939–1941
16. Chen J, Wu J, Hong Y (2016) The morpholino molecular beacon for specific RNA visualization in vivo. Chem Commun 52:3191–3194
17. Li Y-F, Morcos PA (2008) Design and synthesis of dendritic molecular transporter that achieves efficient in vivo delivery of Morpholino antisense oligo. Bioconjug Chem 19:1464–1470
18. Morcos PA, Li Y-F, Jiang S (2008) Vivo-Morpholinos: a non-peptide transporter delivers Morpholinos into a wide array of mouse tissues. Biotechniques 45:613–623

Chapter 5

Inducible Inhibition of Gene Function with Photomorpholinos

Saulius Sumanas

Abstract

Photoactivatable morpholinos (MO) allow specific temporal and spatial inhibition of gene function, which is not possible with conventional morpholino or genetic global gene knock-out approaches. Here, we describe an application of commercially available photoactivatable MO technology for specific gene inhibition in a zebrafish embryonic model and discuss the required controls related to the specificity and efficacy of this method. A similar approach should be also applicable to other model organisms.

Key words Morpholinos Caged, Photoactivatable, Zebrafish

1 Introduction

While conventional morpholinos have been widely used to inhibit protein translation or RNA splicing, they are typically delivered ubiquitously to model organisms such as zebrafish or Xenopus by microinjection or electroporation, and in most cases are not suitable to achieve temporal or spatial control of gene inhibition. Photoactivatable (caged) MOs have been designed to overcome this limitation and allow spatial or temporal control of MO function. Several different designs of caged MOs have been reported (Fig. 1). In the original design, a photocleavable linker has been used to tether a complementary MO-derived inhibitor to the 25-base targeting sequence [1]. Using this approach, the *no tail (ntl)* gene in zebrafish embryos was targeted in a spatiotemporally specific manner [2]. In a different design, a photosensitive subunit has been incorporated in the middle of MO which allows cleavage of photo-MO into two inactive fragments [3]. Such an approach allows specifically turning on genes by degrading gene inhibiting MO upon exposure to UV irradiation or turning off gene function when a caged MO is designed to inhibit an existing MO. More recently, several groups have reported circular caged MOs, where 5′ and 3′ ends are linked with a photocleavable moiety [4–6]. The covalent

Hong M. Moulton and Jon D. Moulton (eds.), *Morpholino Oligomers: Methods and Protocols*, Methods in Molecular Biology, vol. 1565, DOI 10.1007/978-1-4939-6817-6_5, © Springer Science+Business Media LLC 2017

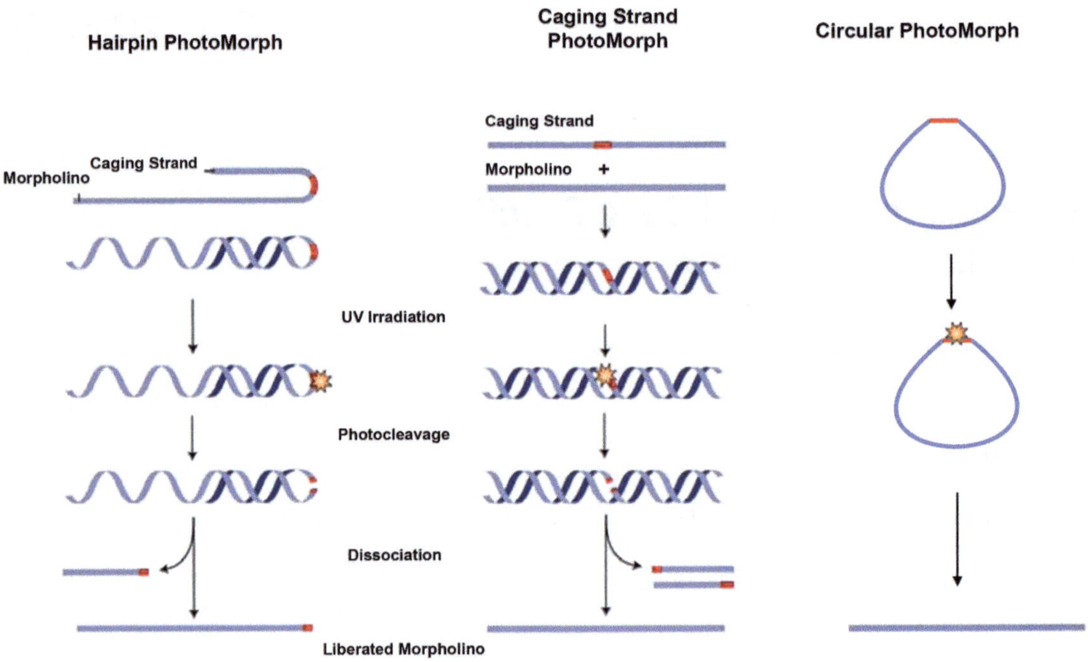

Fig. 1 Different designs of photoactivatable morpholinos. Hairpin PhotoMorph (*left*) has a complementary MO inhibitor attached to the MO with a photocleavable linker. Upon UV exposure, the linker dissociates and MO can inhibit expression of a target gene. Caging strand Photomorph (*middle*) consists of a caging strand with a photocleavable bond that inhibits MO function until it is UV irradiated. The caging strand can be RNA based (Supernova Life Sciences) or morpholino based (Gene Tools LLC.). Circular Photomorphs (*right*) incorporate photocleavable linker that is covalently attached to both ends of the MO

linkage enforces the closed circular conformation, which prevents efficient MO hybridization to target mRNA until photocleavage restores the linear, biologically active MO. Incorporation of different chemical groups allows wavelength-selective illumination to sequentially inactivate gene function [7]. In yet a different design, an excess of complementary caged RNA oligomer that contains a photosensitive moiety is used to inhibit MO activity until it is broken by UV irradiation [8]. This protocol describes a method for MO inhibition using a modified RNA oligomer (= PhotoMorph 2.0) and inducible activation of the MO using UV irradiation. We have successfully used such an approach for inducible inhibition of vascular regulator Etv2/Etsrp function in zebrafish embryos [9, 10]. A similar approach should be also applicable to other model organisms.

2 Materials

1. Caging strand solution. Design a caging strand of 25 nucleotides (SuperNova Life Science, Auburn, WA) complementary against a conventional standard MO that has been already

tested to work specifically and effectively for your gene of interest. Prepare a stock solution of 2 mM in nuclease-free water. Keep on ice when in use and store at −80 °C for long-term storage. Avoid excessive light exposure. Foil-wrapped or dark colored tubes are recommended.

2. Morpholino (Gene Tools LLC, Philomath, OR) solution. Prepare 0.3 mM solution in the nuclease-free water. Store at −20 °C (*see* **Note 1**).
3. 1× Danieau buffer: 58 mM NaCl, 0.7 mM KCl, 0.4 mM $MgSO_4$, 0.6 mM Ca $(NO_3)_2$, 5 mM HEPES, pH 7.6.
4. Yellow film, 312 Roscolux Canary (Cat. No. 2811–312), Theatre House Inc., Covington, KY.
5. 365 nm UV Bench Lamp XX-15L, Cat. No. 95-0042-07, UVP, Upland, CA.

3 Methods

1. In a foil-wrapped microcentrifuge tube, mix the caging strand with MO to the final concentrations of 500 μM for the caging strand and 50 μM of MO in 1× Danieau buffer (*see* **Note 2**).
2. Incubate at 70 °C for 30 min. This denatures any secondary structure of the caging strand and the MO (*see* **Note 3**).
3. Place the tube at 4 °C refrigerator for at least 1 h (*see* **Note 4**).
4. Prepare a standard microinjection setup. For zebrafish injections, we use air pressure injectors (PLI-90, Harvard Apparatus Inc., Cambridge, MA). The needle is backfilled with the caged MO solution and calibrated using microcapillaries. As a precaution, we use yellow film filters for the microscope and room lights.
5. Deliver the caged MO mix by microinjection. Zebrafish microinjections are performed at a 1- to 2-cell stage into the yolk or the animal blastomere similar to standard MO microinjections. We typically inject 2.5 nl of caged MO mix per embryo, but volumes 1–5 nl are possible.
6. After injection, incubate embryos at a standard temperature (28.5 °C for zebrafish). We wrap Petri dishes in foil to minimize any light exposure.
7. Perform UV photoactivation at a selected stage. We use a 365 nm "Black-Ray" UV Bench lamp (UVP). Embryos are positioned in a Petri dish in a standard fish water under the UV lamp at approximately 10 cm from the light source (Fig. 2). Remove the lids from Petri dishes as they will block UV light. Expose for 30 min, gently swirling and mixing the embryos

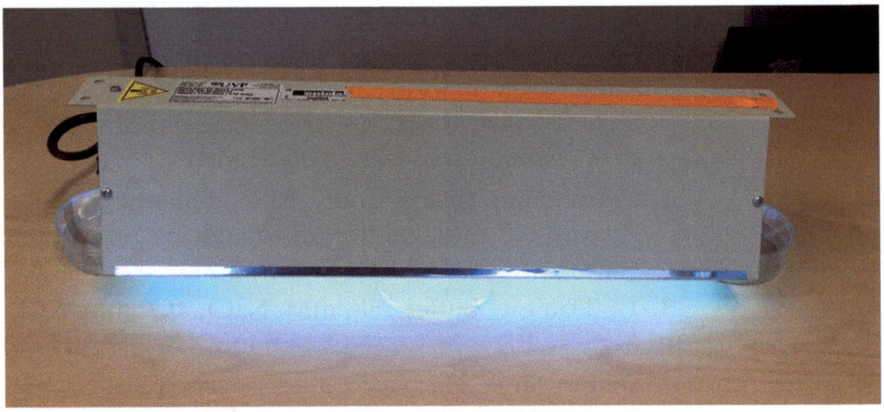

Fig. 2 Morpholino photoactivation setup. 365 nm UV Bench lamp is used for uniform photoactivation of caged MO in zebrafish embryos placed in a Petri dish

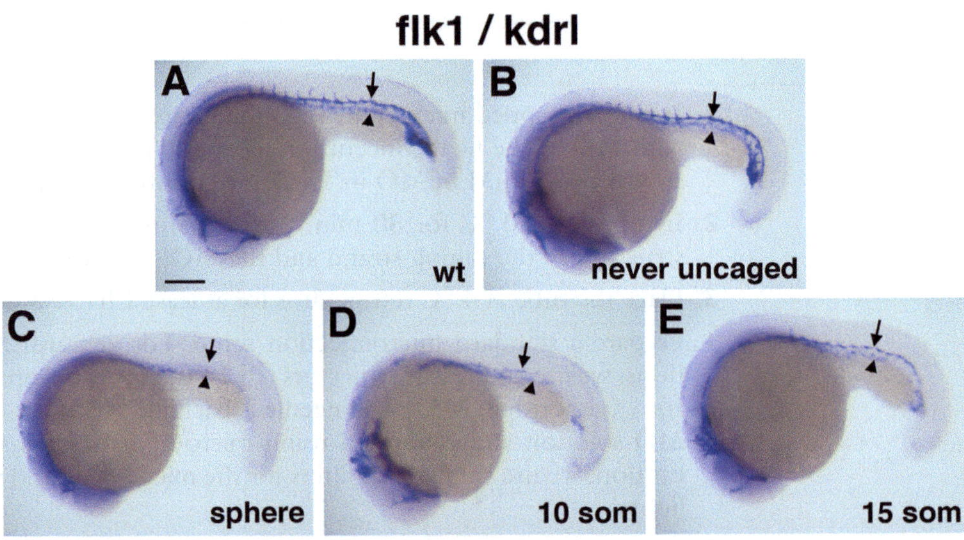

Fig. 3 An example of validation of caged Etv2 MO activity by in situ hybridization analysis of vascular endothelial marker *kdrl* expression at 22 hpf stage. Control uninjected (**a**), MO injected but never uncaged (**b**), uncaged at the sphere (4 hpf), (**c**), 10-somite (**d**), and 15-somite (**e**) stages embryos. Note that *kdrl* expression is absent in early uncaged embryos, greatly reduced in (**d**), and has only minor defects in (**e**). Never uncaged embryos (**b**) are phenotypically normal. Figure reprinted from [9]

every 10 min for uniform photoactivation throughout the embryo (*see* **Notes 5** and **6**).

8. Incubate the embryos under standard conditions and analyze for the phenotype of interest. An example of a successful experiment for Etv2 MO is shown in Fig. 3. Early uncaged embryos display absence of vascular endothelial specific *kdrl* expression, similar to regular Etv2 MO (or *etv2* genetic mutants), uncaging at the 10- to 15-somite stages results in progressively

smaller vascular defects, while never uncaged embryos are phenotypically normal (*see* **Note 7**).

4 Notes

1. Concentration of stock solutions may be adjusted depending on the effective MO concentrations and amounts. MO should also be stable at room temperature, according to the recommendations by Gene Tools, LLC.
2. Recommended effective molar ratio for the caging strand to the MO is 10:1–15:1. Effective concentrations may vary and would need to be determined experimentally. Because the caging strand is supplied in a large excess, this method can work well for high efficacy MOs (effective at 1–2 ng or lower dose per zebrafish embryos) but it may not be practical for lower efficacy MOs because it will not be possible to achieve required concentration of the caging strand. As a precaution, we perform all pipetting in a dimmed room and try to minimize the exposure to light of all solutions that contain the caging strand. We commonly prepare 5–10 μl of the caging mix fresh for each experiment (1.5 μl of 2 mM caging strand solution, 1 μl of 0.3 mM MO solution, and 3.5 μl of 1× Danieau buffer for the total volume of 6 μl). It may be possible to reuse it for a limited number of times (store at −20 °C and repeat denaturing and reannealing).
3. We perform this step in a hybridization oven which results in even heating while incubation in a heat-block will result in condensation at the cap area.
4. Do not use an ice bucket because the tube should cool down gradually. We routinely leave the solution overnight at 4 °C before injecting the next day. Alternatively, a thermocycler can be programmed for denaturation and gradual temperature decrease for annealing.
5. We have not observed any toxic effects from UV irradiation under these conditions. It is also possible to perform spatially restricted photoactivation of caged MO using localized exposure under an epifluorescent microscope as previously described [2]. It is not necessary to limit exposure to light after photoactivation.
6. It is important to include appropriate controls for each experiment. Our typical experiment includes four groups of embryos: (a) wild-type uninjected embryos, (b) experimental caged MO injected embryos that were photoactivated at a selected stage, (c) "early uncaged" embryos, which were photoactivated at the embryonic stage before the gene expression is first initiated

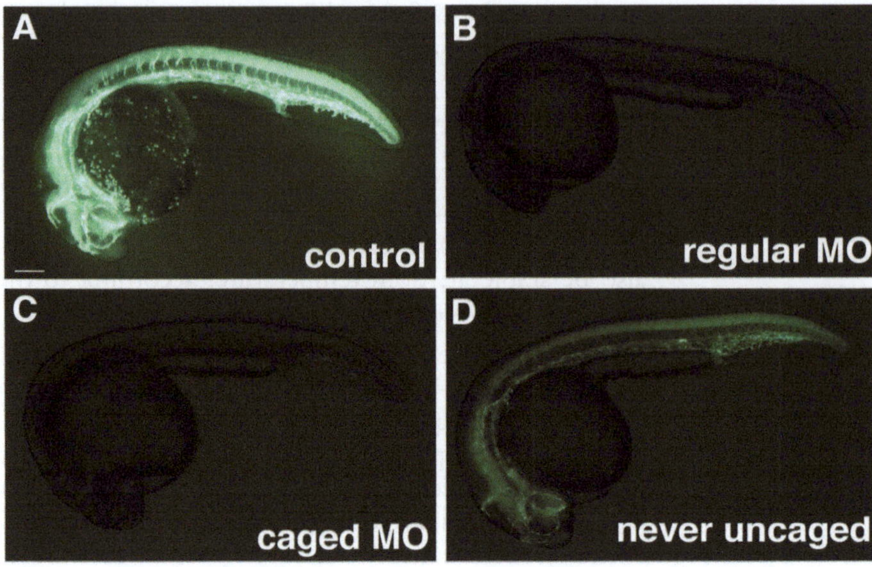

Fig. 4 Illustration of caged Etv2 MO activity and leakiness using *etv2:GFP* reporter line. (**a**) Control uninjected, (**b**) conventional *etv2* MO2 injected embryo, (**c**) caged *etv2* MO injected embryo uncaged at the sphere stage (4 hpf), (**d**) caged *etv2* MO injected embryo never uncaged. All embryos were analyzed at 24 hpf with the same imaging and acquisition settings. (**c**) Embryos that were uncaged at the sphere stage resulted in nearly complete inhibition of GFP expression, showing a similar phenotype to standard *etv2* MO injected embryos. (**d**) Caged MO injected embryos that were never uncaged showed higher GFP fluorescence than (**c**) uncaged embryos. While caged MO injected embryos that were never uncaged (**d**) showed a decrease in GFP expression in comparison to (**a**) uninjected embryos, these embryos were phenotypically normal with no apparent vascular defects. Figure reprinted from [9]

(we uncage at the shield stage for Etv2 MO), which should produce a phenotype similar to the conventional gene-specific MO, (d) "never" uncaged embryos, which are kept in the dark and not subjected to uncaging; they are expected to be phenotypically normal.

7. We commonly observe "leakiness" even in never uncaged embryos, where caged MO results in a significant inhibition of a reporter GFP expression even without any UV irradiation (Fig. 4). This is likely due to the incomplete inhibition of MO even when the caging strand is provided in tenfold molar excess. Nevertheless, in our experiments the majority of never uncaged embryos are phenotypically normal, which argues that even the reduced amount of Etv2 is sufficient for normal function in the "never uncaged" embryos.

Acknowledgment

This work was supported by NIH grant HL107369 and Cincinnati Children's Research Foundation.

References

1. Shestopalov IA, Sinha S, Chen JK (2007) Light-controlled gene silencing in zebrafish embryos. Nat Chem Biol 3:650–651
2. Shestopalov IA, Pitt CL, Chen JK (2012) Spatiotemporal resolution of the Ntla transcriptome in axial mesoderm development. Nat Chem Biol 8:270–276
3. Tallafuss A, Gibson D, Morcos P, Li Y, Seredick S, Eisen J, Washbourne P (2012) Turning gene function ON and OFF using sense and antisense photo-morpholinos in zebrafish. Development 139:1691–1699
4. Yamazoe S, Shestopalov IA, Provost E, Leach SD, Chen JK (2012) Cyclic caged morpholinos: conformationally gated probes of embryonic gene function. Angew Chem Int Ed Engl 51:6908–6911
5. Griepenburg JC, Rapp TL, Carroll PJ, Eberwine J, Dmochowski IJ (2015) Ruthenium-caged antisense morpholinos for regulating gene expression in Zebrafish embryos. Chem Sci 6:2342–2346
6. Wang Y, Wu L, Wang P, Lv C, Yang Z, Tang X (2012) Manipulation of gene expression in zebrafish using caged circular morpholino oligomers. Nucleic Acids Res 40:11155–11162
7. Yamazoe S, Liu Q, McQuade LE, Deiters A, Chen JK (2014) Sequential gene silencing using wavelength-selective caged morpholino oligonucleotides. Angew Chem Int Ed Engl 53:10114–10118
8. Tomasini AJ, Schuler AD, Zebala JA, Mayer AN (2009) PhotoMorphs: a novel light-activated reagent for controlling gene expression in zebrafish. Genesis 47(11):736–743
9. Kohli V, Schumacher JA, Desai SP, Rehn K, Sumanas S (2013) Arterial and venous progenitors of the major axial vessels originate at distinct locations. Dev Cell 25:196–206
10. Craig MP, Grajevskaja V, Liao HK, Balciuniene J, Ekker SC, Park JS, Essner JJ, Balciunas D, Sumanas S (2015) Etv2 and fli1b function together as key regulators of vasculogenesis and angiogenesis. Arterioscler Thromb Vasc Biol 35:865–876

Chapter 6

Blocking Zebrafish MicroRNAs with Morpholinos

Alex Sutton Flynt, Mahesh Rao, and James G. Patton

Abstract

Antisense morpholino oligonucleotides have been commonly used in zebrafish to inhibit mRNA function, either by inhibiting pre-mRNA splicing or by blocking translation initiation. Even with the advent of genome editing by CRISP/Cas9 technology, morpholinos provide a useful and rapid tool to knockdown gene expression. This is especially true when dealing with multiple alleles and large gene families where genetic redundancy can complicate knockout of all family members. miRNAs are small noncoding RNAs that are often encoded in gene families and can display extensive genetic redundancy. This redundancy, plus their small size which can limit targeting by CRISPR/Cas9, makes morpholino-based strategies particularly attractive for inhibition of miRNA function. We provide the rationale, background, and methods to inhibit miRNA function with antisense morpholinos during early development and in the adult retina in zebrafish.

Key words miRNA, Morpholino, Zebrafish, Development, Retina, Embryo

1 Introduction

MicroRNAs (miRNAs) are common noncoding small RNA regulators of gene expression in multicellular organisms [1]. They are short, ~22 nucleotide RNAs that regulate mRNA expression through base pairing [2, 3]. Vertebrate genomes encode hundreds of miRNA genes, many of which are highly expressed and tissue specific [4]. Investigation of miRNA function has shown critical roles in development and physiology [5, 6]. Indeed, a majority of protein coding mRNAs appear to be subject to miRNA regulation [7]. Thus, there has been intense interest in studying these riboregulators, leading to a variety of experimental strategies for manipulating miRNA function [8–10]. One strategy that has been successfully used in multiple settings is antisense technology. Synthetic oligonucleotides can be designed to interfere with miRNA function at multiple points in their life cycle [11]. In zebrafish, antisense inhibition of RNA is most commonly achieved with antisense morpholino oligonucleotides [5, 6, 12–17]. In this

Hong M. Moulton and Jon D. Moulton (eds.), *Morpholino Oligomers: Methods and Protocols*, Methods in Molecular Biology, vol. 1565, DOI 10.1007/978-1-4939-6817-6_6, © Springer Science+Business Media LLC 2017

chapter, we will discuss the design parameters used to target miRNAs with morpholinos. While the focus of the approach will be on application in zebrafish, the principles discussed will translate to other systems.

Morpholino-mediated interference of miRNA function can be designed to affect multiple steps in the life cycle of miRNAs. They can be made to inhibit the small RNA itself, its processing, or mRNA target recognition [6, 12, 13]. This offers multiple options for producing loss-of-function phenotypes, which is critical to judicious use of antisense technologies. Consequently, many of the same strategies to control for toxicity and off-target effects apply when using morpholinos to target miRNA [18]. Rules for designing morpholinos to target miRNAs take advantage of aspects of miRNA biogenesis and activity in regulatory complexes.

1.1 miRNA Biogenesis and Function

miRNAs biogenesis is characterized by sequential cleavage of hairpin RNAs to generate the mature ~22 nt form found in regulatory complexes (Fig. 1) [19]. The imperfect hairpin RNAs that are processed into miRNAs can be found in a variety of genomic contexts. The majority are products of RNA Polymerase II, either as independent transcripts or residing within protein-coding genes [20]. Initial cleavage of hairpin RNAs from nascent, primary transcripts is frequently carried out by the microprocessor complex, though there are many examples of alternative processing mechanisms [21]. "Cropping" by the microprocessor is typical for highly expressed, conserved miRNA species, thus we will focus on this group. However, many of the design rules for targeting canonical miRNAs with morpholinos would be suitable for interfering with the function of noncanonical species. The microprocessor contains two essential components: Pasha and Drosha. Pasha is a double-stranded RNA-binding protein that recognizes the base of the hairpin(s) within primary transcripts (pri-miRNA) and recruits the

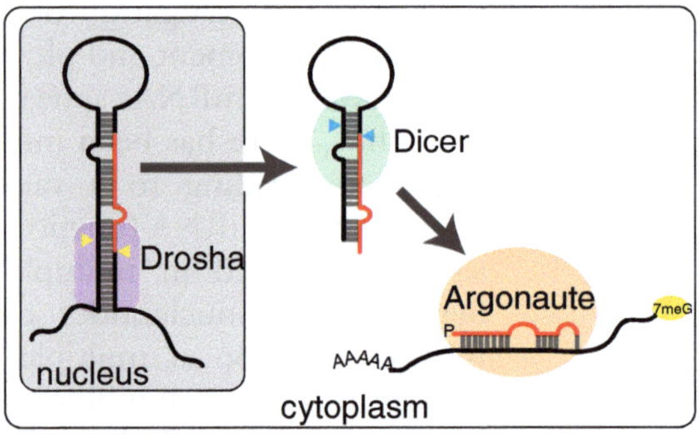

Fig. 1 miRNA biogenesis pathway. Imperfect hairpin RNAs are cleaved by the Drosha-containing microprocessor complex. After cytoplasmic export hairpins are further processed by Dicer and loaded into Argonaute proteins where they bind cognate mRNAs

RNase III enzyme Drosha [22]. Cleavage occurs approximately 10 nt from the hairpin base, yielding a ~70 nt hairpin precursor RNA (pre-miRNA).

After initial processing pre-miRNAs are "Diced" by a cytoplasmic RNase III, Dicer, again by recognition and cleavage directed by the position of hairpin loops [23]. One of the strands then associates with an Argonaute (Ago) protein as part of the RNA-Induced Silencing Complex (RISC) [24]. miRNAs exert their activity through complementary base pairing with mRNAs, typically in the 3′ UTR regions of transcripts [25]. Functional studies have demonstrated the most functional miRNA target sites contain short regions complementary to nucleotides 2–9, or the "seed" sequence at the 5′ ends of miRNAs [26]. The result of targeting is the triggering of mRNA decay via deadenylation and decapping which is accompanied by inhibition of translation [27, 28].

1.2 Inhibition of miRNA Function with Morpholinos

Noncoding RNAs, like miRNAs, are excellent candidates for targeting with morpholinos. The functional products of these genes are RNAs and therefore can be targeted directly, unlike coding genes where long-lived proteins may continue to provide some activity (Fig. 2a). Design of such morpholinos is straightforward. Currently, there are entries for 350 zebrafish miRNAs in miRbase [4]. Using the sequence information in the database, morpholinos can be created that are complementary to the mature strand of miRNAs [6]. There is very little flexibility when designing morpholinos to mature miRNAs. The number of nucleotides (21–25 nt) required to achieve specific binding with morpholinos is essentially the length of mature miRNAs (~22 nt). At this point it is unlikely that new miRNAs will be discovered in zebrafish as sequencing of small RNA clones has reached saturation in many organisms. The only small RNA species that are left to be discovered are likely those that are very lowly expressed and/or have highly atypical precursor structures falling outside the parameters used by annotation algorithms. Thus, all miRNA blocking morpholinos can be designed from database entries.

In zebrafish, the standard in the field has evolved to require the use of multiple morpholinos generating identical phenotypes [29]. The primary difficulty of using morpholinos in zebrafish is unintended toxicity, frequently manifesting as neurodegeneration or curvature of the spine [18]. These effects appear to be predominately sequence specific and not a general effect produced by the chemistry of morpholinos. Through phenocopy studies, morpholinos composed of different sequences can be examined for overlapping phenotypes. When inhibiting mRNA expression this is typically achieved through the design of morpholinos targeting splice sites, in addition to translation start sites. A similar experimental design can be applied to miRNAs. Morpholinos can be generated that can bind to segments of miRNA primary or

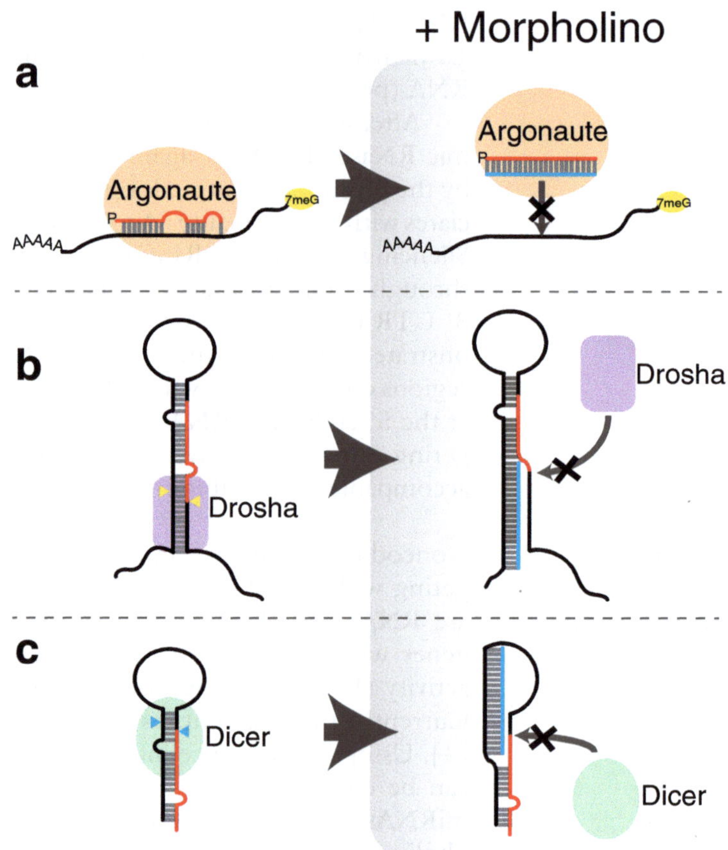

Fig. 2 Inhibition of miRNA by morpholinos. Antisense morpholinos can be designed to interfere with miRNAs directly or at different biogenesis steps. Morpholinos can be generated to bind (**a**) mature miRNAs, (**b**) Drosha cropping sites, or (**c**) Dicer cleavage sites

precursor transcripts (Fig. 2b, c). Using these strategies four additional options become available to inhibit miRNA expression [13]. Two morpholinos can be designed that are complementary to the Drosha cropping sites and two to the Dicer cleavage sites. Thus, combined with a morpholino complementary to the mature strand, five distinct morpholinos can be developed to inhibit miRNA function.

If the effect of the morpholino is specific, phenotypes induced by mature strand blocking and Drosha/Dicer blocking morpholinos should be identical. Following introduction of morpholinos designed to take advantage of either targeting strategy, a decrease in the accumulation of mature miRNA can be observed [5, 13]. Indeed, this is an essential control for demonstrating a morpholino has been effective at altering miRNA function. Another strategy that can be used to control for morpholino cytotoxicity is performing the experiment in a p53 loss-of-function genetic background

[30]. This can alleviate the effects of cycle arrest on animal phenotype that can be triggered by morpholinos but the use of p53 nulls should be used with caution, especially when studying proliferating cells.

miRNAs affect gene expression through mRNA-binding sites that reside mostly in the 3′ UTR [31]. As the sequences of nearly all zebrafish miRNAs are known, there have been multiple efforts to use this information to discern the genome-wide miRNA–mRNA interactome [31–33]. Thus, information regarding which mRNAs are bound by miRNAs is readily available. These predictions are based on functional studies showing that seed pairing is common in miRNA target recognition although there are plenty of examples where the "seed rule" is violated [34, 35]. Additionally, due to the importance of miRNA regulation to animal development, it is common to observe conservation of some target sites, implying greater significance to gene regulatory networks. Morpholinos can be used to block miRNA binding to target sites in a fashion similar to splice site morpholinos interfering with spliceosome association with exon–intron boundaries (Fig. 3). These "target protector" morpholinos inhibit specific miRNA–mRNA interactions which can be useful when studying specific phenotypes [12]. While individual miRNAs potentially regulate hundreds of mRNAs, it is common for single mRNA–miRNA interactions to be central to the role of a miRNA, depending on the specific time and place [1].

Morpholinos provide a variety of strategies for elucidating miRNA function (Fig. 4). These approaches used in combination provide a well-controlled experimental system. Initial investigation

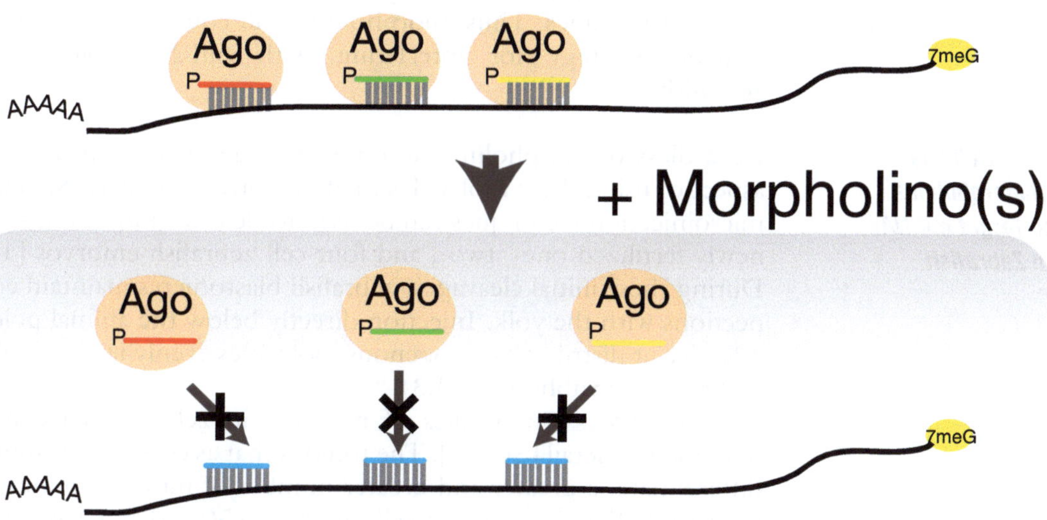

Fig. 3 Interference with miRNA targeting by morpholinos. Target protector morpholinos can be designed to bind mRNAs, and interfere with miRNA function by blocking residues involved in target recognition

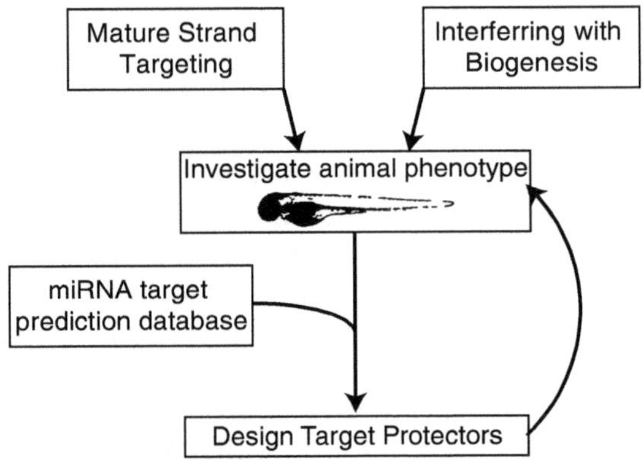

Fig. 4 Combination strategy for understanding miRNA function with morpholinos. Initial production of zebrafish phenotype with either mature strand or biogenesis inhibiting morpholinos can be used to implicate specific genes. Correlation of animal phenotype and miRNA target prediction databases informs selection of candidate genes. Target protector morpholinos can then be used to verify the importance of a specific interaction

of miRNA function typically focuses on the small RNA itself. This can be achieved using morpholinos that target either the mature miRNA or specific steps during miRNA biogenesis to generate loss-of-function phenocopies. Interpretation of phenotypes can then be used to narrow likely mRNA targets annotated in prediction databases. Target protectors can then be designed and the phenotype induced by their delivery compared to changes caused by mature miRNA or biogenesis disrupting morpholinos. Together, this approach provides a facile, time efficient method of investigating miRNAs. It avoids long-term mating strategies and maintenance of fish lines. Thus, morpholinos will continue to provide a valuable strategy for interfering with miRNAs biology in zebrafish.

1.3 Delivery of Morpholinos to Target miRNAs in Zebrafish

Regardless of morpholino-targeting strategy, delivery is a critical aspect of using these molecules to affect miRNA activity. Standard microinjection techniques can be used to deliver morpholinos into newly fertilized one-, two-, and four-cell zebrafish embryos [18]. During these initial cleavages, zebrafish blastomeres maintain connections with the yolk. Injection directly below the animal pole is effective at distributing exogenous molecules evenly in developing embryos (*see* Subheading 3.3).

The expression profiles of miRNAs in zebrafish have been described in detail [36, 37]. The trend is that as cells of the embryo differentiate, a greater and greater number of miRNA species are expressed (Fig. 5a). In one-cell embryos ~50 miRNA have been found through high-throughput sequencing. The concentration of these maternally deposited miRNAs decreases in the initial hours

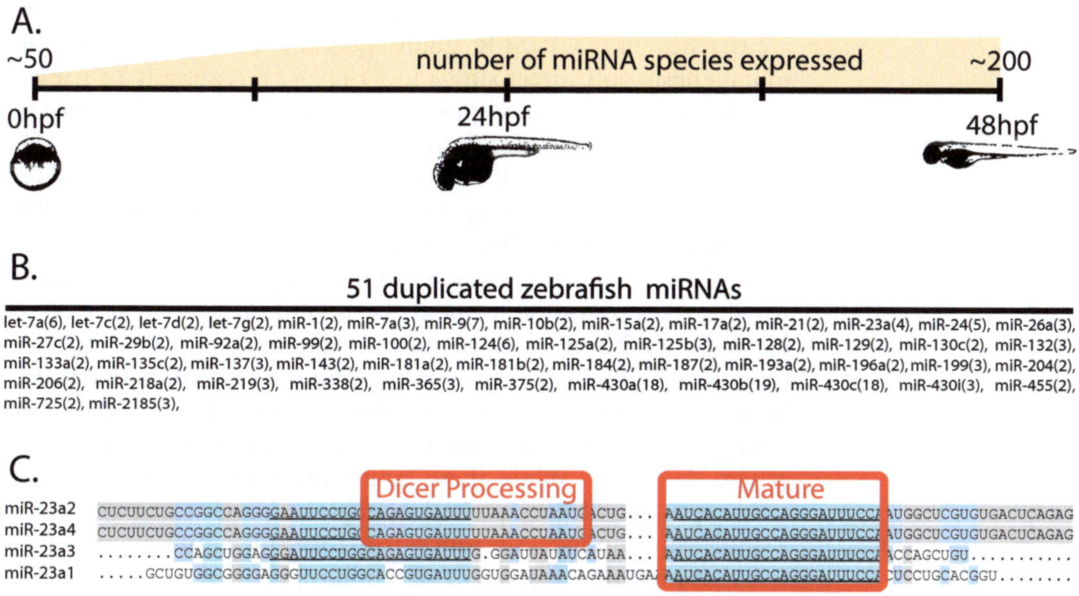

Fig. 5 Targeting miRNAs in zebrafish. (**a**) Temporal expression of miRNAs in zebrafish embryogenesis reveals a trend of increasing expression as development proceeds. (**b**) Duplicated miRNAs encoded in the zebrafish genome. The number in parentheses indicates the number of paralogs. (**c**) Duplicated miRNAs can be targeted by single morpholino oligonucleotides. The *red boxes* show targeting of the mature strand would interfere with all copies of *miR-23a*, while a morpholino preventing Dicer processing would only affect *miR-23a2* and *miR-23a4*

after fertilization. The maternal–zygotic transition is marked by expression of several large polycistronic miRNA clusters containing tandem duplications of *miR-430*, and by the end of 48 h of development, a commonly accepted window of morpholino efficacy, a few hundred miRNAs can be detected [38]. Expression data provide valuable insight into which miRNAs are available to study in embryos. Furthermore, many conserved, highly expressed miRNAs have been localized to tissues using in situ hybridization, further facilitating phenotype interpretation [39].

miRNAs can also be studied via morpholino inhibition in adult fish tissues (*see* Subheading 3, injection and electroporation into fins and eyes). This can be achieved through localized injection of morpholinos followed by electroporation [40]. Planning of experiments in mature tissues likewise can benefit from miRNA expression data. Zebrafish provide excellent models for tissue repair and homeostasis, which can be investigated by specific delivery of morpholinos. Introducing the miRNA targeting morpholinos and control morpholinos in different regions of tissues provides excellent internal controls for the effects of miRNA modulation.

Morpholinos can also be used to address a major issue facing investigation of miRNA biology: redundancy. The zebrafish genome contains multiple copies of 51 miRNAs across 179 loci

(Fig. 5b). Recovery of miRNA mutations either through new approaches like CRISPR guide genome editing or traditional mutagenesis only affect single loci [41, 42]. Assembly of triple or quadruple mutant genotypes can be burdensome and require significant time. In contrast, a single morpholino can be created that is complementary to all redundant miRNAs. This most likely is achievable when targeting the typically invariant mature RNA sequence (Fig. 5c). However, in many instances duplication of miRNAs may have left segments of the hairpin base and loop intact, which makes morpholino-mediated inhibition of biogenesis possible across orthologs with a single morpholino (Fig. 5c). However, there may still be issues caused by redundant miRNAs that cannot be resolved by single morpholinos, for example, miRNAs that belong to families that share the same seed sequence but are otherwise divergent. While these family members regulate the same transcripts they do not possess enough similarity to be affected by a single morpholino. A solution is to introduce multiple morpholinos that provide coverage of all families [5].

Morpholinos can also be used to address issues related to early lethality of miRNA mutants. As with many genetic regulators, individual miRNAs act in multiple tissues throughout zebrafish development [43]. Mutations can cause phenotypes in early development that make the study of later developmental stages problematic. Morpholinos can circumvent this issue as they can be delivered with temporal and spatial precision. Furthermore, it is becoming increasingly appreciated that like morpholinos, mutant alleles can also introduce "off-target" effects [44]. When investigating differences in phenotypes caused by CRISPR editing of the *Egfl7* gene and morpholino targeting, it was found that in the mutant, compensatory gene expression masked the true loss-of-function phenotype. Morpholino targeting, in fact, produced a cleaner phenotype. Considering the tendencies of miRNAs to participate in genetic networks featuring bistable feedback loops and fine-tuning of gene expression, mechanisms that compensate for loss of miRNA activity are likely to be in place. Inhibition of miRNAs via morpholinos could be an excellent method to control for these changes and provide a useful strategy for evaluating mutant phenotypes.

Morpholinos will continue to offer excellent strategies for altering miRNA function in zebrafish. They have well-appreciated toxicity profiles and methods to control for these effects. Targeting of miRNAs can be accomplished through multiple approaches, providing ample controls for experiments. Use of morpholinos can also circumvent issues that plague evaluation of gene function using mutant alleles, such as developmental pleiotropy and compensatory gene expression. Morpholinos will continue to be an essential tool used by zebrafish researchers to investigate miRNA biology.

2 Materials

1. One hundred percentage Tricaine (Ethyl 3-aminobenzoate methanesulfonate): 400 mg Tricaine powder, 97.9 mL ddH_2O, 2.1 mL 1 M Tris–HCl (pH 9), adjust pH to 7, dilute to 4% working concentration.
2. Plastic Spoon.
3. Paper Towels.
4. Forceps (Fine Science Tools; Dumont #55 Forceps; Item No. 11,255-20).
5. Sapphire Blade Double Edge Lancet, 0.75 mm wide, 60° (World Precision Instruments; 504,072).
6. Titanium Handle for Sapphire Blades, 13 cm, Retractable (World Precision Instruments, 501,811) (*see* **Note 1**).
7. 5 μL Syringe(5 μL Model 75 RN SYR, Small Removable NDL, 26 s ga, 2 in, point style 2; Part Number:87,930).
8. 33 G blunt end needle (33 G, Small Hub RN NDL, 1.5 in, point style 3, 6/PK; Part Number:7762-06) (*see* **Note 2**).
9. Lissamine-tagged morpholino (Gene Tools) (*see* **Note 3**).
10. Untagged morpholino (Gene Tools) (*see* **Note 3**).
11. Egg water: 60 μg Instant Ocean sea salt/1 mL ddH_2O.
12. 100 mm/15 mm Polystyrene Petri dish.
13. Electroporation paddles (Sonidel Limited; Item No: CUY650P3).
14. Electroporator (BioRad GenePulser xCell).
15. Dissecting microscope.
16. Razor blade or scalpel.
17. Glass capillaries (Thin Wall Glass Capillaries, 4″, 1/0.75 OD/ID, Filament; Item#: TW100F-4).
18. Needle puller (Sutter Instruments; Model P-97).
19. Microloader pipette tips (Eppendorf Microloader; Cat No: 930,001,007).
20. Micromanipulator (World Precision Instruments; Model number M3301R or L).
21. Compressed air.
22. Injection apparatus (Applied Scientific Instrumentation, MPPI-3; BP-15 Back Pressure Unit).
23. Injection foot pedal (Linemaster Switch Corp., Treadlite II, Cat No: T-91-PS).
24. Micrometer (Meiji Techno Japan, Microscope Micrometer, MA285).

25. Mineral oil.
26. Mating cages (*see* **Note 4**).
27. Plastic mold (Adaptive Science Tools; Item#: i34).
28. Agarose.
29. 0.5% (10×) Phenol Red: Dissolve Phenol Red sodium salt in ddH2O to 0.5%, filter sterilize, aliquot, freeze, thaw when needed and keep at room temperature.
30. Mesh strainer.
31. Transfer pipet.
32. Metal probe.

3 Methods

3.1 Retinal Injections and Electroporations

1. Anesthetize fish in 4% tricaine. Watch until gill movement slows but do not allow to gills to stop moving.
2. Using a plastic spoon, transfer anesthetized fish onto a damp paper towel. Fold towel to cover gills but do not obstruct the eye.
3. The outer cornea is a transparent membrane that surrounds the eye. It adds extra resistance that impedes electroporation. It can be seen at the periphery of the eye cup as a small translucent ridge. Locate the optic fissure, a thin black line on the ventral side of the eye. Remove the outer cornea by using forceps to pinch the tissue directly adjacent to the optic fissure. Gently pull the outer cornea clockwise around the eye cup, pulling at a low angle. If the outer cornea is dropped it can be picked up again at the same spot. One rotation around the eye cup is sufficient (*see* **Notes 5** and **6**).
4. Use a sapphire knife to make a small hole at the caudal part of the pupil.
5. Load 0.5 μL of morpholino into the syringe (*see* **Note 7**).
6. Insert a 33 G needle into the eye (*see* **Note 8**).
7. Slowly inject 0.5 μL of morpholino. Most of the liquid will remain in the eye cup but some will leak out the insertion hole.
8. Remove the needle and return the fish to a recovery tank before electroporation (*see* **Note 9**).
9. Fill the lid of a Petri dish with egg water.
10. Anesthetize injected fish briefly in 4% tricaine.
11. Use a plastic spoon to place fish in a small piece of damp paper towel. Fold paper towel over to cover the gills of fish. Keep the injected eye exposed.

12. Place fish into the egg water in Petri dish lid with the injected eye up. Use fingers to hold paper towel down around fish.
13. Lissamine is positively charged. Use the positive electrode paddle to gently press down the ventral part of the eye cup at the optic fissure to rotate and expose the dorsal part of the eye (*see* **Note 10**). Do not let the metal part of the paddle touch the eye (*see* **Note 11**).
14. Place the negative electrode paddle behind the exposed dorsal part of the eye. Keep the paddle approximately 1 mm away from the eye (*see* **Note 12**).
15. Electroporate the retina. Two consecutive 50-ms square wave pulses at 75 V with a 1-s pause between pulses. The fish will tense during each of the pulses (*see* **Note 13**).
16. Return the fish to a recovery tank and allow to recover for desired amount of time.

3.2 Fin Amputation, Injection, and Electroporation

1. Anesthetize fish in 4% Tricaine.
2. Place anesthetized fish in Petri dish filled with egg water. Lay fish on its side so that the caudal fin lays flat.
3. Use a sterile razor blade or scalpel to amputate the caudal fin (*see* **Note 14**).
4. Return fish with amputated caudal fin to a recovery tank and allow to recover (*see* **Note 15**).
5. To prepare the injection needle, insert a glass capillary into the needle puller. Have approximately equal amounts on either side of the heating filament. Run the needle pulling program (Heat = 573, Pull = 70, Vel = 55, Time = 95). Remove each of the pulled glass needles and place them in a holder where the tips will not be broken.
6. Load one injection needle with morpholino with microloader pipette tips.
7. Mount the loaded injection needle onto the micromanipulator and turn on the air supply.
8. Fill a Petri dish with egg water and insert the needle into the water without the tip touching the bottom.
9. Using forceps, pinch off the bottom third of the tapered part of the injection needle. Morpholino will begin to flow out of the needle.
10. Adjust back flow so that a small amount of morpholino flows from the tip of the needle. If egg water flows back into the injection needle, decrease the back flow. If morpholino is rapidly flowing from the tip of the needle, increase back flow. Test air pressure by pressing the pulse button or foot pedal.

11. Calibrate the injection needle by placing one drop of mineral oil on a micrometer. Insert the injection needle into the oil but do not let the tip touch the bottom. Pulse the needle once. If the diameter of the resulting drop is 0.1 mm then each pulse will inject 1 nL. If a larger volume is desired use the formula $V = 4/3\pi r^3$.
12. Anesthetize amputated fish in 4% tricaine.
13. Place fish into an empty Petri dish. Lay fish on its side so that caudal fin lays flat (*see* **Note 16**).
14. Insert the needle into the blastema, just distal to each bony ray. Do not push the needle through the fin.
15. Inject approximately 75 nL injection mix per bony ray. Inject only half the fin with a given morpholino, inject the other half with a control or do not inject to use an electroporation control (*see* **Note 17**).
16. Fill Petri dish with 4% tricaine.
17. With gloves, hold the fish dorsal or ventral side up so that the caudal fin is perpendicular to the bottom of the Petri dish. Ensure the fish is completely submerged.
18. Place 3 mm electroporation paddles on either side of the half of the fin to be electroporated. Keep paddles approximately 1 mm from the fin and do not let the paddles touch the fin.
19. Electroporate the fin. Use ten 50 ms square wave pulses, at 15 V with a 1 s pause between pulses (*see* **Note 13**). Repeat electroporation on the other half of the caudal fin.
20. Place fish in a glass slide and image caudal fin to use as a baseline to measure amount of regeneration.
21. Place fish to a recovery tank and allow to recover until desired time(s) of analysis.

3.3 Embryo Microinjection

1. In the afternoon the day before injection fill a mating cage with system water and place one female in the middle tier and a male in the top tier (*see* **Note 18**).
2. For injection molds prepare 1% agarose dissolved in egg water. Pour agarose into Petri dish base approximately half to two-thirds full. Allow to cool slightly but not to solidify. Float cleaned plastic injection mold onto molten agarose (*see* **Note 19**). Allow agarose to solidify. Carefully loosen the mold from the agarose by releasing its sides then carefully pulling the mold up at an angle (*see* **Note 20**). Cover mold with egg water completely and store at 4 °C.
3. To prepare the injection needle insert a glass capillary into the needle puller. Have approximately equal amounts on either side of the heating filament. Run the needle pulling program

(Heat = 573, Pull = 70, Vel = 55, Time = 95). Remove each of the pulled glass needles and place them in a holder where the tips will not be broken.

4. The morning of injection prepare the morpholino. Dilute the morpholino to the working concentration with ddH_2O and 1/10th the final volume with phenol red.
5. Use the microloader pipette tips to load the needle with approximately 5 μL of morpholino.
6. Mount the loaded injection needle onto the micromanipulator and turn on the air supply.
7. Fill a Petri dish with egg water and insert the needle into the water without the tip touching the bottom.
8. Using forceps, pinch off the bottom third of the tapered part of the injection needle. Morpholino will begin to flow out of the needle.
9. Adjust the back flow so that a small amount of morpholino flows from the tip of the needle. If egg water flows back into the injection needle, decrease the back flow. If morpholino is rapidly flowing from the tip of the needle, increase back flow. Test the air pressure by pressing the pulse button or foot pedal.
10. Calibrate the injection needle by placing one drop of mineral oil on a micrometer. Insert the injection needle into the oil but do not let the tip touch the bottom. Pulse the needle once. If the diameter of the resulting drop is 0.1 mm then each pulse will inject 1 nL.
11. Remove the injection needle from the mineral oil and keep it out of the way so as not to accidentally break the needle further.
12. Remove the injection mold from 4 °C and fill it with fresh, room temperature egg water.
13. Combine the male and female fish (*see* **Note 21**).
14. Wait 15–30 min for eggs to be laid (*see* **Note 22**).
15. Fill another mating cage bottom (with a solid bottom) with system water and transfer adult fish to the new container, leaving behind the embryos. Pour the embryos into a strainer with wire mesh fine enough to trap the embryos. Rinse the collected embryos with egg water to remove residual feces, scales, or other debris.
16. Flip strainer over an empty Petri dish and rinse embryos into dish using egg water.
17. Use a transfer pipette to move embryos to injection mold (*see* **Note 23**).

18. Using a probe, gently press each embryo into a groove. This will hold each embryo in place and allow for rapid injection (*see* **Notes 24** and **25**).
19. Orient the micromanipulator to insert the injection needle into the yolk of the embryos. Use the micromanipulator to bring the loaded needle close to the embryos. Orient the mold so that the needle will be inserted into the yolk of the embryo (*see* **Note 26**). Carefully use the micromanipulator to insert the needle into the yolk (*see* **Note 27**). Press the foot pedal once to inject 1 nL. If an increased amount of morpholino is desired then press the foot pedal the appropriate number of times (*see* **Notes 28** and **29**). Remove the needle from the yolk and move to the next embryo (*see* **Note 30**).
20. After the injections are complete, use a probe to gently remove embryos from the mold.
21. Use a transfer pipette to transfer the embryos to new Petri dish. Place the Petri dish containing the embryos in a 28 °C incubator and allow the embryos to grow the desired amount (*see* **Note 31**).

4 Notes

1. Sapphire knife assembly: Carefully remove sapphire knife tip from packaging, extend the retractable handle arm, screw knife tip onto handle arm, and carefully retract handle arm. The sapphire blade is very fragile. If it breaks then it will not be able to cut into the cornea. Take care not to damage the knife and if the tip breaks, replace it.
2. A 5 μL syringe comes with a beveled syringe needle that is not optimal for injections. Replacement 33 G blunt end needles are recommended. To insert the new tip, unscrew the metal cap at the end of the syringe. Remove the beveled needle and ensure that the small white plastic cylinder at the base of the needle is also removed, this sometimes becomes separated and sticks in the syringe needle port. Remove the metal cleaning wire from the 33 G blunt end needle. Insert the needle into the syringe. Replace the metal screw cap.
3. Morpholino Preparation. When choosing what morpholinos to use, first look to see if a morpholino targeting your gene of interest has already been used in published work. If not, contact Gene Tools to design a new morpholino. Their experts will work with you to design a morpholino that fulfills all the qualities of a good morpholino. Dissolve morpholinos in ddH_2O to the desired stock concentration. For electroporation into tissues such as the retina or fin, resuspend to a concentration of

3 mM. For microinjections into embryos, resuspend to a concentration of 1 mM. For electroporation use stock morpholino as a starting concentration. Titrate down to the lowest concentration that produces a phenotype without gross morphological defects. If the concentration needs to be increased, morpholinos can be lyophilized again and resuspended in a smaller volume of water. For embryo microinjections, start at a low concentration and gradually increase until a phenotype is observed. Take caution not to induce nonspecific effects or toxicity. Standard practice is to use two morpholinos against a specific target, find the concentration of each individually that produces a phenotype, then inject both together at half concentration. This lowers the possibility of toxicity or nonspecific effects, while still maintaining true phenotypes.

4. The mating cage is three tiered with the top two tiers having a mesh bottom and the bottom tier having a solid bottom. The mesh must be large enough for eggs to pass through but not adult fish.
5. The outer cornea is a transparent membrane that surrounds the eye. It adds extra resistance that impedes electroporation. It can be seen at the periphery of the eye cup as a small translucent ridge. The optic fissure can be identified as a thin black line and is located in the ventral side of the eye.
6. It is easy for the eye to become dislodged. To prevent this you can stabilize the eye with another pair of forceps. It is also easy to remove more tissue than is necessary. Though it may seem like not much tissue was removed, only complete one circuit around the eye.
7. Best practice is to withdraw more than 0.5 μL and depress the plunger to 0.5 μL. An air bubble will be between the end of the plunger and the liquid and this is the air that is in the needle. Volume in the needle should be determined by the end of the plunger, not the liquid.
8. It is easier to inject if the fish is facing away from needle so that the needle can be easily inserted into the vitreous of the eye. Use the tip of the needle to push the lens to the side as the injected liquid can sometimes sit on top of the lens instead of going behind, adjacent to the retina. Be sure not to insert the needle too far so as to avoid touching the retina.
9. Multiple fish can be lined up to inject. This method should only be used when sufficiently fast at the injection procedure.
10. Electroporation will pull the injected material into the dorsal and central parts of the retina. Lissamine is positively charged so the positive electrode will be in front of the eye and the negative electrode will be behind the eye. If other materials are injected into the eye, such as plasmid DNA, that have a net

negative charge, then the electrode paddles can be switched to accommodate this change.

11. It is difficult not to allow the metal part of the paddle to touch the eye while still keeping the other paddle far enough behind the eye to pull the morpholino into the retina. If the metal part touches the eye it is ok, as long as it is minimal. The paddle should not be lying on the eye, as this will likely cause the retina to explode from the voltage. Additionally, be sure not to press down too hard on the fish as damage to the cranium can occur.
12. The distance between the paddles can be fixed by adjusting the screw in between the two arms. Adjust the distance between the paddles to the diameter of the eye, and this will be approximately the correct distance from the eye when positioning the paddles for electroporation. Be sure the electrode paddle is not touching the eye as this may cause the retina to explode.
13. Bubbles will form on paddle electrodes, which can affect electroporation. Clean electrodes after each electroporation. Additionally, only electroporate one eye. Electroporating both eyes can cause high levels of nonspecific damage.
14. Amputations should be at the same relative proximal/distal point for each fish. For example, amputations can be done at the most proximal bony ray branching point or a similar number of bony segments from the distal or proximal end of the fin.
15. Fin injections are done when the caudal fin beings to regenerate and the blastema has formed.
16. Place fish so that needle will move across the body toward the amputated fin. Otherwise the fin may move during injections.
17. If the fish begins to move during injections, place it back into 4% tricaine until it slows down again. Pay attention to the gill movement so that it slows but does not stop. If gill movement stops that indicates the fish is dead.
18. If a large number of embryos are desired for injections, two females can be placed in the mating cage.
19. The mold must be completely free of particles and water. Any particles on the mold will compromise the quality of the agarose mold.
20. A metal spatula can be used here to facilitate separation of the agarose from the sides of the mold.
21. The fish should be combined early in the morning, approximately 30 min after the beginning of the light cycle for best results. Fish will mate later in the morning as well but the likelihood of a successful mating is best soon after the facility lights come on.

22. It is unlikely that only unfertilized eggs will be laid. If the clutch is small then viability of eggs is likely low. The number of eggs usually ranges from 100 to 200 eggs per mating but can be as many as 400 and as few as 50.
23. Keep some embryos aside, uninjected, to control for clutches with poor viability.
24. Best results occur if injections are done in the yolk directly under the cell. Some injections require injection directly into the cell. These considerations should be kept in mind when orienting each embryo in a groove.
25. Titrations or multiple groups can be injected on a single injection mold. Attention should be paid to the number of embryos in each groove.
26. An efficient way to do injections is to operate the microinjector with the dominant hand and move the mold with the opposite hand. This will allow the injector to continuously look through the microscope and have injections continue smoothly. Rapid movements should be avoided using this method as this can break the needle, requiring recalibration or using a new needle.
27. The needle must pierce the chorion covering the embryo before it can be inserted into the yolk. Sometimes the needle will bend instead of piercing the chorion. In this case try moving the mold slightly to change the angle of entry for the needle. Trying to insert the needle faster or slower may also help. If need be the needle can be rebroken to make it more rigid, however, this will also require recalibration of the needle. Be sure the needle does not pass through the organism into the agarose or even to the petri dish as this is likely to break the needle, requiring further recalibration or a new needle.
28. Embryos are sensitive to the volume injected, and injection volume should be kept to the minimum possible amount. Injecting more than 5 nL may result in many dead embryos.
29. Best results occur when injections are performed at the late 1-cell stage–early 2-cell stage. Injections can be performed as late as the 4-cell stage, when many cytoplasmic bridges exist between cells. Injection later than the 4-cell stage is not advised as cells may receive different amounts of morpholino.
30. When removing the needle after injection it is possible that a little yolk will leak out of the embryo, which is ok. If large amounts of yolk leak out or if the yolk continues to leak, it is likely the embryo will not survive. Additionally, if the bore of the broken needle is too large, removing the needle will likely cause much yolk to leak out as well as other damage to the embryo. Therefore, it is best to keep the tip of the needle small. With small needle tips it is more likely that they will get

clogged with yolk or other debris. If this occurs try pulsing the needle a few times or switch to continuous to use pressure to clear the needle. If the needle cannot be cleared this way then the tip may need to be rebroken. Recalibrate the needle after rebreaking the tip.

31. Check embryos a few hours after injection and remove dead embryos. Dead embryos will have cloudy or opaque yolk. Additionally, remove unfertilized embryos that have failed to develop. Furthermore, check embryos periodically to remove dead embryos and change egg water regularly to maintain best health of embryos.

References

1. Flynt AS, Lai EC (2008) Biological principles of microRNA-mediated regulation: shared themes amid diversity. Nat Rev Genet 9(11):831–842
2. Lagos-Quintana M, Rauhut R, Lendeckel W, Tuschl T (2001) Identification of novel genes coding for small expressed RNAs. Science 294(5543):853–858
3. Lee RC, Feinbaum RL, Ambros V (1993) The *C. elegans* heterochronic gene lin-4 encodes small RNAs with antisense complementarity to lin-14. Cell 75(5):843–854
4. Griffiths-Jones S, Grocock RJ, van Dongen S, Bateman A, Enright AJ (2006) miRBase: microRNA sequences, targets and gene nomenclature. Nucleic Acids Res 34(Suppl 1):D140
5. Flynt AS, Thatcher EJ, Burkewitz K, Li N, Liu Y, Patton JG (2009) miR-8 microRNAs regulate the response to osmotic stress in zebrafish embryos. J Cell Biol 185(1):115–127. doi:10.1083/jcb.200807026
6. Flynt AS, Li N, Thatcher EJ, Solnica-Krezel L, Patton JG (2007) Zebrafish miR-214 modulates Hedgehog signaling to specify muscle cell fate. Nat Genet 39(2):259
7. Farh KK-H, Grimson A, Jan C, Lewis BP, Johnston WK, Lim LP, Burge CB, Bartel DP (2005) The widespread impact of mammalian microRNAs on mRNA repression and evolution. Science 310(5755):1817–1821. doi:10.1126/science.1121158
8. Krutzfeldt J, Rajewsky N, Braich R, Rajeev KG, Tuschl T, Manoharan M, Stoffel M (2005) Silencing of microRNAs in vivo with 'antagomirs'. Nature 438(7068):685
9. Kloosterman WP, Wienholds E, Ketting RF, Plasterk RHA (2004) Substrate requirements for let-7 function in the developing zebrafish embryo. Nucleic Acids Res 32(21):6284–6291
10. Chen YW, Song S, Weng R, Verma P, Kugler JM, Buescher M, Rouam S, Cohen SM (2014) Systematic study of Drosophila microRNA functions using a collection of targeted knockout mutations. 31(6):784-800
11. Horwich MD, Zamore PD (2008) Design and delivery of antisense oligonucleotides to block microRNA function in cultured Drosophila and human cells. Nat Protoc 3(10):1537–1549
12. Choi W-Y, Giraldez AJ, Schier AF (2007) Target protectors reveal dampening and balancing of nodal agonist and antagonist by miR-430. Science 318(5848):271–274. doi:10.1126/science.1147535
13. Kloosterman WP, Lagendijk AK, Ketting RF, JD M, RHA P (2007) Targeted inhibition of miRNA maturation with morpholinos reveals a role for miR-375 in pancreatic islet development. PLoS Biol 5(8):e203
14. Grabher C, Payne EM, Johnston AB, Bolli N, Lechman E, Dick JE, Kanki JP, Look AT (2011) Zebrafish microRNA-126 determines hematopoietic cell fate through c-Myb. Leukemia 25(3):506–514
15. Mishima Y, Abreu-Goodger C, Staton AA, Stahlhut C, Shou C, Cheng C, Gerstein M, Enright AJ, Giraldez AJ (2009) Zebrafish miR-1 and miR-133 shape muscle gene expression and regulate sarcomeric actin organization. Genes Dev 23(5):619–632
16. Choi PS, Zakhary L, Choi W-Y, Caron S, Alvarez-Saavedra E, Miska EA, McManus M, Harfe B, Giraldez AJ, Horvitz RH, Schier AF, Dulac C (2008) Members of the miRNA-200 family regulate olfactory neurogenesis. Neuron 57(1):41–55
17. Nicoli S, Standley C, Walker P, Hurlstone A, Fogarty KE, Lawson ND (2010) MicroRNA-mediated integration of haemodynamics and

Vegf signalling during angiogenesis. Nature 464(7292):1196–1200

18. Bill BR, Petzold AM, Clark KJ, Schimmenti LA, Ekker SC (2009) A primer for Morpholino use in zebrafish. Zebrafish 6 (1):69-77. doi:10.1089/zeb.2008.0555
19. Lee Y, Jeon K, Lee JT, Kim S, Kim VN (2002) MicroRNA maturation: stepwise processing and subcellular localization. EMBO J 21(17):4663–4670
20. Rodriguez A, Griffiths-Jones S, Ashurst JL, Bradley A (2004) Identification of mammalian microRNA host genes and transcription units. Genome Res 14(10A):1902–1910
21. Ruby JG, Stark A, Johnston WK, Kellis M, Bartel DP, Lai EC (2007) Evolution, biogenesis, expression, and target predictions of a substantially expanded set of Drosophila microRNAs. Genome Res 17(12):1850–1864
22. Denli AM, Tops BB, Plasterk RH, Ketting RF, Hannon GJ (2004) Processing of primary microRNAs by the Microprocessor complex. Nature 432(7014):231–235
23. Ketting RF, Fischer S, Bernstein E, Sijen T, Hannon GJ, Plasterk R (2001) Dicer functions in RNA interfence and in synthesis of small RNA involved in developmental timing in *C. elegans*. Genes Dev 15:2654–2659
24. Liu J, Carmell MA, Rivas FV, Marsden CG, Thomson JM, Song JJ, Hammond SM, Joshua-Tor L, Hannon GJ (2004) Argonaute2 is the catalytic engine of mammalian RNAi. Science 305(5689):1437–1441
25. Lai EC (2002) Micro RNAs are complementary to 3' UTR sequence motifs that mediate negative post-transcriptional regulation. Nat Genet 30(4):363–364
26. Doench JG, Petersen CP, Sharp PA (2003) siRNAs can function as miRNAs. Genes Dev 17(4):438–442
27. Rehwinkel JAN, Behm-Ansmant I, Gatfield D, Izaurralde E (2005) A crucial role for GW182 and the DCP1:DCP2 decapping complex in miRNA-mediated gene silencing. RNA 11(11):1640–1647
28. Giraldez AJ, Mishima Y, Rihel J, Grocock RJ, Van Dongen S, Inoue K, Enright AJ, Schier AF (2006) Zebrafish MiR-430 promotes deadenylation and clearance of maternal mRNAs. Science 312(5770):75–79
29. Eisen JS, Smith JC (2008) Controlling morpholino experiments: don't stop making antisense. Development 135(10):1735–1743
30. Bedell VM, Westcot SE, Ekker SC (2011) Lessons from morpholino-based screening in zebrafish. Brief Funct Genomics 10(4):181–188. doi:10.1093/bfgp/elr021
31. Lewis BP, Burge CB, Bartel DP (2005) Conserved seed pairing, often flanked by adenosines, indicates that thousands of human genes are microRNA targets. Cell 120(1):15–20. doi:10.1016/j.cell.2004.12.035
32. Chen PY, Manninga H, Slanchev K, Chien M, Russo JJ, Ju J, Sheridan R, John B, Marks DS, Gaidatzis D, Sander C, Zavolan M, Tuschl T (2005) The developmental miRNA profiles of zebrafish as determined by small RNA cloning. Genes Dev 19(11):1288–1293
33. Kumar A, ML WAF-T, Ml T, RJ F-M, Rj M, Fau-Lefevre C (2012) Lefevre C miRNA_targets: a database for miRNA target predictions in coding and non-coding regions of mRNAs. Genomics 100(6):352–356
34. Grimson A, Farh KK-H, Johnston WK, Garrett-Engele P, Lim LP, Bartel DP (2007) MicroRNA targeting specificity in mammals: determinants beyond seed pairing. Mol Cell 27(1):91
35. Li N, Flynt AS, Kim HR, Solnica-Krezel L, Patton JG (2008) Dispatched Homolog 2 is targeted by miR-214 through a combination of three weak microRNA recognition sites. Nucleic Acids Res 36(13):4277–4285
36. Wei C, Salichos L, Wittgrove CM, Rokas A, Patton JG (2012) Transcriptome-wide analysis of small RNA expression in early zebrafish development. RNA 18(5):915–929
37. Yao Y, Ma L, Jia Q, Deng W, Liu Z, Zhang Y, Ren J, Xue Y, Jia H, Yang Q (2014) Systematic characterization of small RNAome during zebrafish early developmental stages. BMC Genomics 15(1):1–17
38. Giraldez AJ, Cinalli RM, Glasner ME, Enright AJ, Thomson JM, Baskerville S, Hammond SM, Bartel DP, Schier AF (2005) MicroRNAs regulate brain morphogenesis in zebrafish. Science 308(5723):833–838
39. Wienholds E, Kloosterman WP, Miska E, Alvarez-Saavedra E, Berezikov E, de Bruijn E, Horvitz HR, Kauppinen S, Plasterk RHA (2005) MicroRNA expression in zebrafish embryonic development. Science 309(5732):310–311
40. Thatcher EJ, Paydar I, Anderson KK, Patton JG (2008) Regulation of zebrafish fin regeneration by microRNAs. Proc Natl Acad Sci 105(47):18384–18389
41. Hwang WY, Fu Y, Reyon D, Maeder ML, Tsai SQ, Sander JD, Peterson RT, Yeh JRJ, Joung JK (2013) Efficient genome editing in zebrafish using a CRISPR-Cas system. Nat Biotechnol 31(3):227–229. doi:10.1038/nbt.2501
42. Sertori R, Trengove M, Basheer F, Ward AC, Liongue C (2015) Genome editing in zebrafish: a practical overview. Brief Funct Genomics 15(4):322–330

43. Weinholds E, Koudijs MJ, van Eeden F, Cuppen E, Plasterk R (2003) The microRNA-producing enzyme Dicerl is essential for zebrafish development. Nat Genet 35(3):217–218

44. Kok Fatma O, Shin M, Ni C-W, Gupta A, Grosse Ann S, van Impel A, Kirchmaier Bettina C, Peterson-Maduro J, Kourkoulis G, Male I, DeSantis DF, Sheppard-Tindell S, Ebarasi L, Betsholtz C, Schulte-Merker S, Wolfe Scot A, Lawson Nathan D (2015) Reverse genetic screening reveals poor correlation between morpholino-induced and mutant phenotypes in zebrafish. Dev Cell 32(1):97–108. doi:10.1016/j.devcel.2014.11.018

Chapter 7

Using Morpholinos to Examine Gene Function During Fin Regeneration

Ryan Thummel and M. Kathryn Iovine

Abstract

In the zebrafish regenerating fin, specific gene-targeting morpholinos have been widely utilized to assess gene function. Unlike in embryos, injection of standard morpholinos in the adult regenerating fin is not sufficient for cellular uptake. Rather, morpholinos are first injected extracellularly into the blastemal compartment, followed by electroporation for cellular uptake. Knockdown phenotypes are evaluated 1–4 days post electroporation. This chapter provides a description of the reagents, equipment, and procedure for successful injection and electroporation of morpholinos into the regenerating fin.

Key words Zebrafish, Fin regeneration, Electroporation, Gene knockdown

1 Introduction

Defining the function of an essential gene in adult tissues requires conditional inhibition of the gene of interest, such as through use of a chemical inhibitor, a temperature-sensitive allele, or by morpholino-mediated gene knockdown. The advantages of the latter strategy are that morpholinos may be synthesized against any target, and evaluation of phenotypes can proceed rapidly. Disadvantages include the possibility that the morpholino exhibits off-target effects and that gene function cannot be completely eliminated through knockdown methods. Alleviating concerns of off-target effects can be minimized by testing a second nonoverlapping targeting morpholino for similar phenotypes, and by demonstrating that the intended target is indeed reduced. Correlating mutant and morphant phenotypes can also demonstrate morpholino specificity. This is possible when hypomorphic alleles for the gene of interest are available or when the gene of interest is not essential [1, 2].

The zebrafish fin has a relatively simple structure which is fully restored following amputation. Regeneration proceeds in stages beginning with wound healing, formation of a proliferative blastema, and fin outgrowth [3, 4]. A multitude of phenotypes

Hong M. Moulton and Jon D. Moulton (eds.), *Morpholino Oligomers: Methods and Protocols*, Methods in Molecular Biology, vol. 1565, DOI 10.1007/978-1-4939-6817-6_7, © Springer Science+Business Media LLC 2017

may be evaluated following knockdown such as regenerate length [5], cell proliferation [5], changes in gene expression [6], changes in the length of bony fin ray segments [1], and changes in extracellular matrix components or protein distribution [7, 8].

This chapter provides the detailed methods for completing morpholino-mediated gene knockdown during fin regeneration by using injection followed by electroporation for cellular uptake. Vivo-morpholinos, which are capable of crossing the plasma membrane without electroporation, have also been utilized [9]. However, electroporation of standard morpholinos has the advantage of targeting a region within the tissue without the concerns of "leakage" of gene knockdown into adjacent regions (i.e. blood stream or adjacent tissue). For this procedure, morpholinos are injected into the regenerating blastema at 2–3 days post amputation (dpa), followed by immediate electroporation to facilitate cellular uptake (*see* also [10]). Modification with a fluorophore (fluorescein or lissamine) permits confirmation of success at 24 h post electroporation (dpe) (*see* Fig. 1). Confirmation of target

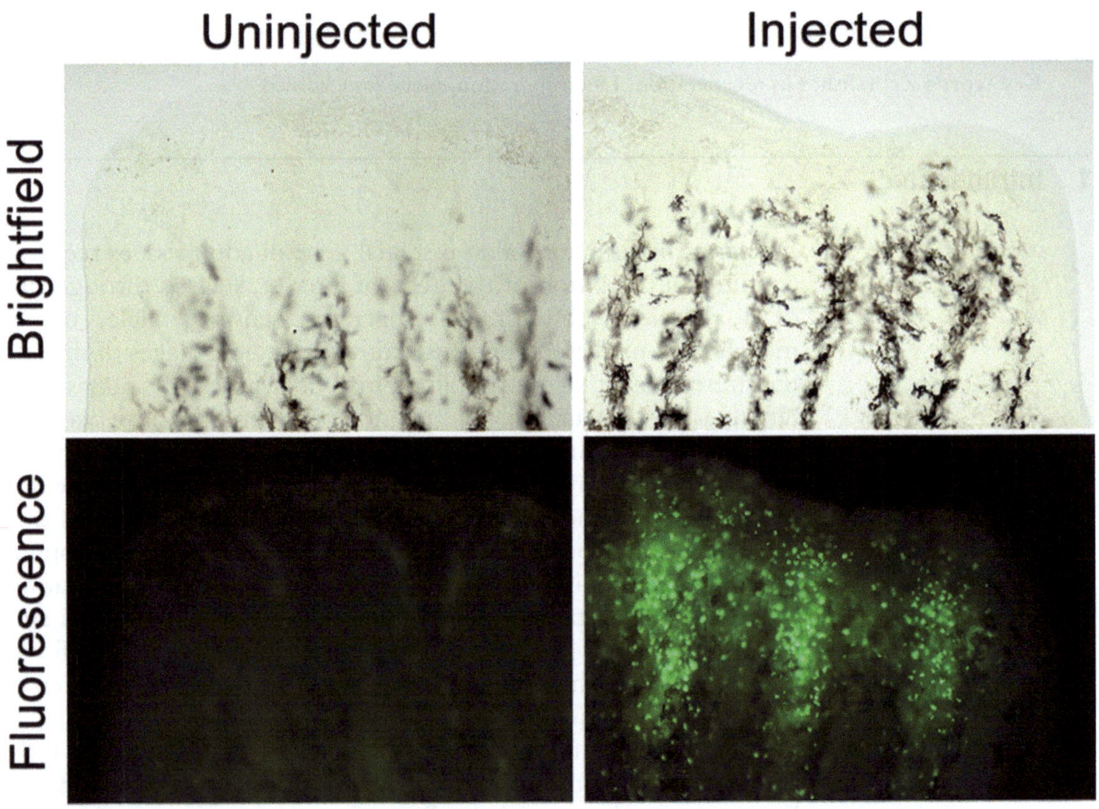

Fig. 1 Uptake of morpholino following successful injection and electroporation. The images on the *left* show a fin that was not injected with morpholino but was electroporated. Only background autofluorescence is apparent. The images on the *right* show a fin that was injected with a fluorescein-modified morpholino, and immediately electroporated. The green signal identifies cells that took up the morpholino. Images were collected 24 hpe. Three fin rays are shown

knockdown through immunological detection (for ATG-blocking morpholinos) or through RT-PCR (for splice-blocking morpholinos) is performed at 1 dpe. Phenotypes of interest are typically monitored 1–4 dpe.

2 Materials

Anesthesia—either 0.2 mg/ml tricaine methanesulfonate or 0.8 mg/l 2-phenoxyethanol.

Sterile razor blades or scalpels.

Agarose.

250 ml flask.

Microwave.

60 mm Petri dishes.

120 mm Petri dishes, 20 mm thick.

Morpholinos of interest (targeting and nontargeting).

Nuclease-free water.

Dissecting microscope (for fin amputations and morpholino injections).

Microloader tips for loading needles (Eppendorf).

Femtotips II (Eppendorf).

Microinjector with air regulator (such as Narishige IM 300).

CUY21 square wave electroporator with 3-mm diameter tweezer electrode (Protech International).

Fluorescent microscope for visualizing successful procedure.

Kimwipes.

Plastic spoon or other means of gently transferring anesthetized fish.

3 Methods

3.1 Amputate Caudal Tail Fin

1. Anesthetize the fish by transferring to a small bowl of either 0.2 mg/ml tricaine methanesulfonate or 0.8 mg/l 2-phenoxyethanol until gill movement decreases significantly. Then, transfer anesthetized fish to a sterile, dry, and clean surface, such as the inside of a Petri dish.
2. If necessary, gently fan out the caudal fin so that it is completely flat.
3. Using a sterile razor blade make a cut perfectly perpendicular to the body axis below the first lepidotrichial branching point. This is best achieved by pressing the razor blade down firmly and evenly through the tissue, and then while it is still firmly against the hard surface of the Petri dish, make two back and forth movements of the blade perpendicular to the body axis. Perform this using a dissection microscope if necessary (*see* **Note 1**).
4. Return fish to fresh system water, typically for 2–3 dpa.

3.2 Create Injection Plate

1. Heat 2% agarose in sterile water until melted, as one would for an electrophoresis gel.
2. Pour melted agarose solution in a Petri dish (120 mm diameter, 20 mm thick) filling ~1/3 of the height of the dish. Allow to cool and harden.
3. Add a second layer of fresh, melted 2% agarose solution on top of the hardened first layer. Fill to nearly the top of the Petri dish (~4/5 of the height of the dish). Allow to cool and harden.
4. Using a sterile razor blade carefully cut out a rectangular-shaped well that removes the second layer of agarose, but leaves the bottom layer intact. The well should be in the center of the dish and have a size that corresponds to the length of the razor blade by ~1/2 the length of the razor blade.
5. Finally, on one of the short sides of the rectangle, cut out a small triangular portion of the second layer that would approximate the size and shape of the head of the adult fish. The final shape of the well should appear as shown below.

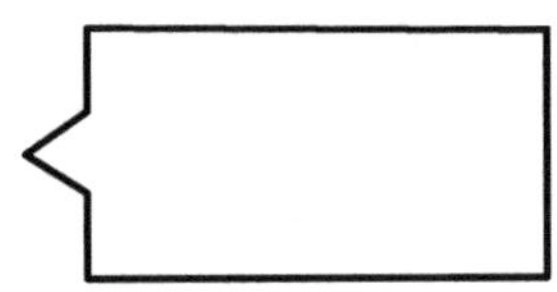

6. Keep injection plate wet with system water prior to use (*see* **Note 2**).

3.3 Prepare Morpholino Solution

1. To create a 3 mM stock solution of the morpholino, dilute 300 nM of morpholino into 100 μl of nuclease-free water in a nuclease-free microcentrifuge tube. Seal the tube with paraffin and store at room temperature (*see* **Note 3**).
2. Dilute stock solution with nuclease-free water immediately prior to usage to obtain working solution (usually 0.6–1.2 mM).

3.4 Prepare Microinjection Rig

1. A standard injection rig for the microinjection of zebrafish embryos can typically be used or modified for the microinjection of fin tissue. To test an existing rig, anesthetize an adult zebrafish and transfer to the injection plate. Place the head of the fish in the triangular portion of the well and center the body axis of the fish on the length of the well. Place the injection plate on an existing microinjection apparatus and ensure that the Petri dish is flat and stable. Ensure that there is adequate depth of field to visualize the microinjection and that the angle and placement of the microinjection needle can reach the location of the fin tissue in the well at an angle of ~45°. Modify the rig as needed and return the fish to system water.

2. Load the working solution of the morpholino into standard microinjection needles and load the needle onto the injection rig (*see* **Note 4**). To determine the volume per injection, inject once into a small petri dish of oil. Measure the diameter of the droplet (i.e. 1 pulse/click) to calculate the volume (use the equation, $V = 4/3\pi r^3$). Adjust the output volume to ~ 5 nl per injection.

3.5 Morpholino Injection in Regenerating Fins

1. Anesthetize a single fish from Subheading 3.1 and transfer to the injection plate as described in Subheading 3.4. Remove any excess liquid and make sure that the fin is flat.
2. Place the injection plate with the fish on the injection rig and orient the microinjection needle appropriately. The body axis of the fish needs to be oriented parallel the needle with the head of the fish facing toward the needle. Using the micromanipulator controls on the injection rig, adjust the needle so that it is just distal to the cut bony fin ray and hovering above the presumptive blastema of the regenerated tissue. Then, under the microscope, find the location of the needle and fine tune the location as needed.
3. Insert the microinjection needle in the proximal-most blastema and slowly insert the needle into the tissue until it is centered in the presumptive blastema (*see* **Note 5**). Remove your hand from the micromanipulator controls so as not to accidently move the needle during injection.
4. Inject ~75 nl (15 clicks) of the morpholino into the blastema pausing ~1 s in between injection clicks. The morpholino can be visualized as a "puff" in the tissue immediately following each click.
5. Remove the needle and readjust the rig as necessary to inject the adjacent blastema. To determine the effect of the gene knockdown on fin outgrowth, inject the morpholino into only the blastemas from one half of the fin (i.e. the dorsal side). This allows you to compare the effect of the morpholino-injected side to the uninjected side.

3.6 Electroporation of Fin Tissue

1. Following injection of the morpholino, transfer the injection plate with the fish to the bench and fill the Petri Dish with the anesthesia solution.
2. Orient the fish onto its ventral side (i.e. right side up) and gently sustain this orientation by holding the fish between the index finger and thumb. Make sure that the caudal fin is exposed and completely submerged in the solution.
3. Place tweezer electrodes (CUY650-P3, Protech International) on either side of the dorsal half of the fin, centered around the area of injection. Place electrodes as close as possible to the tissue without touching it (*see* **Note 6**).

4. Electroporate the tissue using the following parameters: ten consecutive 50-ms pulses at 15 V, with a 1 s pause in between pulses (CUY21 Square Wave Electroporator, Protech International). Gently wipe off any bubbles that stick to the electrodes with a Kimwipe. Repeat the electroporation on the ventral half of the fin as a control.
5. If performing analysis of regrowth 24 h post electroporation (hpe), a 0 hpe "before" image needs to be obtained. Quickly transfer the fish to a glass slide and image the animal such that the entire caudal fin is captured in the image. Make sure to note the dorsal/ventral halves of the fin in the image.
6. Return the fish to system water to revive.
7. Repeat procedure with additional animals to achieve a desired N (typically 6–10).

3.7 Analysis of Fin Regrowth

1. At 24 hpe (3 dpa), anesthetize the fish and reimage the whole caudal fin in the identical dorsal/ventral orientation used the day prior (*see* **Notes** 7 and **8**).
2. Match the 3 dpa image with the corresponding 2 dpa image and trace the areas of regrowth in both the dorsal and ventral halves using NIH Image.
3. Calculate the difference in regrowth:

   $(\text{Dorsal}_{3\text{dpa}} - \text{Dorsal}_{2\text{dpa}})/(\text{Ventral}_{3\text{dpa}} - \text{Ventral}_{2\text{dpa}}) \times 100 = \%$ area.

4 Notes

1. In order to compare different individuals, it is important that amputations are completed at the same level. In addition, with some animals the caudal fin is not perfectly perpendicular to the body axis; in this case the fin should be cut perpendicular to its own axis, not that of the body.
2. Molds can be reused; cover with water; lid and Parafilm; store at 4 °C.
3. If using fluorescently tagged morpholinos, store stock solution away from light.
4. Ensure that the needle bore is not too thick so as to damage tissue, but not so narrow as to easily become clogged.
5. Take care to avoid pushing the needle through the fin tissue.
6. Take care not to touch the fin tissue with the electrodes during electroporation. This will cause necrosis (this is detectable as dark discoloration of the tissue at 24 hpe).
7. Negative controls include the generic "standard control" morpholino, which has a sequence that is predicted not to target zebrafish genes, or a custom-paired "five mismatch" control.

8. It is useful to compare both targeting and control morpholinos to uninjected/electroporated fins in order to rule out if the control morpholinos themselves have any effect.

Acknowledgements

The authors would like to thank Raj Banerji for helpful comments during the preparation of the chapter and for providing the images used to illustrate a successful procedure.

References

1. Hoptak-Solga AD, Nielsen S, Jain I, Thummel R, Hyde DR, Iovine MK (2008) Connexin43 (GJA1) is required in the population of dividing cells during fin regeneration. Dev Biol 317:541–548
2. Ton QV, Iovine MK (2013) Identification of an evx1-dependent joint-formation pathway during FIN regeneration. PLoS One 8:e81240
3. Poss KD, Keating MT, Nechiporuk A (2003) Tales of regeneration in zebrafish. Dev Dyn 226:202–210
4. Akimenko MA, Mari-Beffa M, Becerra J, Geraudie J (2003) Old questions, new tools, and some answers to the mystery of fin regeneration. Dev Dyn 226:190–201
5. Thummel R, Bai S, Sarras MP Jr, Song P, McDermott J, Brewer J et al (2006) Inhibition of zebrafish fin regeneration using in vivo electroporation of morpholinos against fgfr1 and msxb. Dev Dyn 235:336–346
6. Yin VP, Poss KD (2008) New regulators of vertebrate appendage regeneration. Curr Opin Genet Dev 18:381–386
7. Bhadra J, Iovine MK (2015) Hsp47 mediates Cx43-dependent skeletal growth and patterning in the regenerating fin. Mech Dev 138(Pt 3):364–374
8. Govindan J, Iovine MK (2014) Hapln1a is required for connexin43-dependent growth and patterning in the regenerating fin skeleton. PLoS One 9:e88574
9. Chablais F, Jazwinska A (2010) IGF signaling between blastema and wound epidermis is required for fin regeneration. Development 137:871–879
10. Hyde DR, Godwin AR, Thummel R (2012) In vivo electroporation of morpholinos into the regenerating adult zebrafish tail fin. J Vis Exp 61:3632

Chapter 8

Using Morpholinos to Probe Gene Networks in Sea Urchin

Stefan C. Materna

Abstract

The control processes that underlie the progression of development can be summarized in maps of gene regulatory networks (GRNs). A critical step in their assembly is the systematic perturbation of network candidates. In sea urchins the most important method for interfering with expression in a gene-specific way is application of morpholino antisense oligonucleotides (MOs). MOs act by binding to their sequence complement in transcripts resulting in a block in translation or a change in splicing and thus result in a loss of function. Despite the tremendous success of this technology, recent comparisons to mutants generated by genome editing have led to renewed criticism and challenged its reliability. As with all methods based on sequence recognition, MOs are prone to off-target binding that may result in phenotypes that are erroneously ascribed to the loss of the intended target. However, the slow progression of development in sea urchins has enabled extremely detailed studies of gene activity in the embryo. This wealth of knowledge paired with the simplicity of the sea urchin embryo enables careful analysis of MO phenotypes through a variety of methods that do not rely on terminal phenotypes. This article summarizes the use of MOs in probing GRNs and the steps that should be taken to assure their specificity.

Key words Morpholino, Specificity, Sea urchin, Perturbation, Gene regulatory network, Knockdown, Mutant, Morphant, CRISPR

1 Introduction

Sea urchins are a long time favorite model organism for developmental biologists and over the last 15 years have become a prime subject for the study of gene regulatory networks (GRNs). The goal of GRN research is to uncover the genomic control functions that direct embryonic development [1]. Many technical advances, such as the sequencing of the sea urchin genome and the ability to query transcription in a high-throughput manner, have contributed to the success of the GRN approach in sea urchins [2, 3]. But quite possibly no advance has been as important as the arrival of morpholino antisense oligonucleotides (MOs) as knockdown agents, which enabled experimental perturbation of embryonic development in a gene-specific manner [4, 5].

Hong M. Moulton and Jon D. Moulton (eds.), *Morpholino Oligomers: Methods and Protocols*, Methods in Molecular Biology, vol. 1565, DOI 10.1007/978-1-4939-6817-6_8,

© Springer Science+Business Media LLC 2017

No single gene by itself directs development and the main motivation for the assembly of GRNs is to determine how regulatory genes work together. Their interactions set cells on paths to particular cell fates and set up spatial expression patterns that precede the morphogenetic events that eventually unfold. It is essential to identify and characterize the expression of these regulatory factors, but their role can only be revealed through perturbation of their function [3]. However, phenotypic readouts are often not sufficient as the appearance of gross deficiencies is delayed relative to when regulatory genes are first activated. Instead, their function is more directly revealed by quantifying changes in expression levels and by examining changes in spatial expression patterns of potential downstream targets [6]. Together, this allows the assembly of detailed models that integrate the contribution of different regulatory genes to a common outcome [7, 8].

Gain-of-function methods or pharmacological inhibition have been used with great success, but a more direct way to probe the function of a gene is usually through loss-of-function approaches. The long generation time of sea urchins prevents the establishment of mutant lines. It thus was not until the availability of MOs that gene-specific loss-of-function methods could be used routinely. This success is reflected by the number of papers (~160) that report their use in sea urchin and other echinoderms (Gene Tools, MO publication database: pubs.gene-tools.com).

While mutants carry an aberration in the genomic sequence that can be identified, MO function relies on recognition of their sequence complement in transcripts, causing a block of translation or splicing. However, all knockdown methods that rely on sequence recognition are prone to off-target binding, which may have undesired side effects that may be mistaken for true knockdown phenotypes if not properly controlled. Despite an overall good agreement between morphant and mutant phenotypes, use of MOs is still often viewed critically [9]. This applies particularly to defects that emerge late in development, e.g., during organ formation in zebrafish [10]. Owing in part to the high complexity of the zebrafish embryo, it can be difficult in these cases to conclusively demonstrate that such defects are specific and not due to off-target effects, or due to specific effects at an earlier time in development.

In this regard, sea urchins offer some advantages for MO analysis relative to zebrafish. The comparatively simple morphology of sea urchin embryos throughout early larval development paired with the slow kinetics of gene expression allows for better control compared to other systems. The wealth of knowledge about the progression of gene expression in space and time can be leveraged to assess both consequences *and* specificity of MO-mediated knockdown. To this end, several orthogonal approaches must be taken to ascertain the validity of a newly designed MO. This includes a test of efficiency, independent confirmation of the perturbation effects, and a comprehensive assessment of the effects on expression of other genes.

If carefully controlled, MOs are an easy to use, precise, and efficient tool for the study of gene networks. But more generally, the strategy to establish specificity of MOs also applies to other perturbation methods, such as genome editing or CRISPR interference, that are currently under evaluation.

2 Advantages and Limitations

MOs are oligomers with a usual length of 25 bases, modified from ribonucleotides. They bind to their sequence complement in transcripts and inhibit translation or block access of the splicing machinery [11]. They are nontoxic to the embryo at concentrations sufficient to cause specific defects. In sea urchins and other echinoderms, MOs are usually delivered by microinjection into the zygote and are active for several days postfertilization. Their stability is a considerable advantage and extends their reach over other approaches, e.g., mRNA overexpression, where much of the RNA is degraded during the first day. Microinjection into the fertilized egg allows translation inhibition of both maternal and zygotic transcripts, but splice-blocking MOs can only be used to target zygotic transcripts because maternally deposited transcripts are already processed.

As a quick review of the literature reveals, translation blocking MOs are used most commonly in sea urchins although splice-blocking MOs have also been successfully applied [12–14]. This bias may be due to the sequence constraints in designing an effective splice-blocking MO, which is usually aimed at the splice donor site. In addition, the exact consequences of applying a splice-blocking MO can be difficult to predict. Splice-blocking MOs may cause exon-skipping events that create altered but functional gene products, in particular, for genes with a more complex exon structure. In contrast, isoforms using different transcription and translation start sites can be knocked down specifically and independently [15, 16].

In contrast to other knockdown methods such as RNAi, MOs are independent of host components. Processing of exogenous double-stranded RNAs into siRNAs is dependent on Dicer and other endogenous proteins of the small interfering RNA pathway [17]. Although Dicer has been reported to be required for embryonic development in sea urchins [18], use of siRNA for gene knockdown has not been systematically explored.

A significant reason for the success of MOs is that beyond the sequence of the transcript to be targeted nothing needs to be known about its biological function. The sequencing of the sea urchin genome provided high-quality gene prediction that significantly accelerated the study of gene function in development. In contrast, although dominant negative proteins are among the most potent perturbation agents, not least because they operate at the protein level, they are difficult to design without prior knowledge of biological function.

One of the main limitations of MO-mediated knockdown is its delivery by microinjection. While it makes the MO available throughout the embryo, it does not allow for control in space and time. This is an issue for genes with more complex or composite expression patterns. For example, Nodal signaling is first employed in establishing the primary axis of the sea urchin embryo [19], but later it is required for spreading left/right asymmetry [20]. MO application affects the earliest expression phase causing embryos to radialize and thus leaves Nodal function in the second phase inaccessible. For genes with more complex, disjointed spatial expression patterns it can be difficult to disentangle gene function in one territory from that in another.

In some instances, this problem can be overcome by use of pharmacological inhibitors. However, many inhibitors interfere with intercellular signals by binding to their associated receptors. Even if their binding to a particular type of receptor is well established, their effects are often pleiotropic and may equally affect all members of a receptor family. They are thus limited in their ability to disentangle the contributions of individual factors as has been accomplished by MO-mediated knockdown for the Wnt family of signaling factors [7].

An alternative to traditional MOs is the use of vivoMOs, which has been applied successfully in sea urchins [21, 22]. These MOs are connected to an octa-guaidinium dendrimer that enables transfer across the membrane [23]. In principle, vivoMOs can be applied like pharmacological inhibitors. But their limited solubility in seawater may be insufficient to effectively knockdown targets with the same success as regular MOs delivered by microinjection into the zygote. Injection into the larvae or adult rudiment is possible, if cumbersome [22]. In other systems, light-activated MOs have been used successfully to control knockdown [24]. However, the high light intensities required to uncage these MOs cause photo damage, preventing their application in sea urchins.

3 Efficacy

The success of a MO-mediated knockdown experiment depends on the ability of the MO to efficiently block the production of the target protein while minimizing nonspecific side effects.

The only direct way to test whether a MO effectively blocks protein synthesis of the target is to determine its absence using immunohistochemistry or detection via Western blot. This can be challenging in sea urchins, as in many other species, where large collections of antibodies are not available and custom antibodies will have to be generated. If no specific antibody is available, the efficacy of the MO can be gauged indirectly. But only direct measurement of the target protein will reveal if a complete knockdown has been achieved, which is an important insight into the limits of the experiment. Although partial knockdowns may result in informative phenotypes, they may give an incomplete picture of the role of the targeted protein. Furthermore, absence of the targeted protein is an essential indicator that a MO causes specific defects. When direct evidence for efficient knockdown is not available, it is essential that the perturbation effects are confirmed independently by multiple methods.

A test of efficacy that is sometimes reported is coinjection of a translation-blocking MO with an in vitro-transcribed mRNA encoding a fluorescent protein. If the mRNA contains the MO target site upstream of the start codon, presence of the MO should abolish fluorescence. However, this experiment does not test whether translation and transcription start sites have been correctly identified and is thus inferior to detection of the endogenous protein. Although available genomic resources have made it easier to map the coordinates of new transcripts, it is still a good idea to confirm these independently via methods such as RACE-PCR or PCR amplification of alternative splice forms.

The efficacy of splice-blocking MOs can be tested by PCR amplification of a region spanning the exon that is targeted and its flanking introns. Quantitative PCR can be used to determine the reduction of correctly spliced transcript compared to controls. But depending on the complexity of the exon structure, the altered splicing pattern can be difficult to predict and may need to be determined empirically.

One additional complication for the design of MOs when using sea urchins, compared to inbred model organisms, is the high degree of polymorphism. Sea urchin experiments are usually conducted on animals collected in the wild and single-copy DNA may vary by as much as 5% [2]. To minimize the risk of directing a MO against a site with known variability, genomic resources (available through Echinobase: echinobase.org) should be inspected for sequence variants.

4 Minimizing Off-Target Hybridization

If a knockdown phenotype is to be informative, the MO must be effective in preventing protein production (or function in case of a splice-blocking MO) and it must be specific.

In contrast to mutants that carry permanent disabling changes in their genomic sequence, MOs interact with the *transcript* in a sequence-specific manner. However, as in all nucleotide hybridization reactions, MOs will bind to mismatched targets, albeit more fleetingly. This may have unintended, yet reproducible, consequences that could erroneously be ascribed to the loss of the intended target. It is thus vital to carefully conduct and control any MO knockdown experiment to establish its specificity.

When designing a new translation-blocking MO, the first step is to map the translation start site. Then, it must be confirmed that the suggested oligo sequence uniquely matches the gene of interest and has no subpar hits elsewhere in the transcriptome. Since the sequencing of the sea urchin genome, this is a routine operation and can be accomplished via BLAST search on the Echinobase website or on the sea urchin genome site at NCBI (*S. purpuratus* species ID:86) [25]. Care must be taken to set the length for the initial exact match, from which BLAST builds alignments, to a word size of 7 bp so that hits can be identified. If BLAST does not automatically adjust for search of short sequences adjust parameters to: word size—7; filter setting—off, expect value—1000 (BLAST by default starts with longer initial matches and will not report short alignments).

It is a matter of debate how many mismatches in a MO are required to prevent it from binding potential secondary targets. MOs are usually designed to a length of about 25 bp, and it is assumed that four mismatches will render it nonfunctional [11]. However, sea urchins live at relatively low ambient temperatures (15 °C for *Strongylocentrotus purpuratus*) and this may lead to an increased tolerance of mismatches and thus a larger risk of off-target binding. A warning example is provided by a study aimed at knocking down the transcription factor Runt1, a member of the well-studied Runx family [26, 27]. Two different MOs were injected that caused seemingly specific phenotypes, but were later found to be due to a loss of expression of two histone isoforms. One MO could be aligned to the histone H4 transcript in a consecutive stretch of 18 bp, although this also included four mismatches. This was sufficient to affect histone expression and led to arrested development at late blastula stage [26]. Thus, MO candidates, where the majority of bases can be aligned with *any* other transcripts (without gaps, mismatches notwithstanding) should be avoided.

5 Establishing Specificity

Once a MO has been found to be effective, several orthogonal approaches must be taken to establish that the loss of the intended target is indeed the underlying cause of the phenotype. The most important among these are an independent confirmation of the perturbation effects via alternative means and a comprehensive assessment of the effects at the transcript level.

5.1 Phenotypic Effects

A first indication that a MO causes specific defects is an otherwise normal progression of development. If a gene that is expressed only zygotically is targeted, no adverse effects should be noticeable before the onset of transcription in wild-type controls. Thereafter, the phenotypic consequences will depend on the expression and function of the target. Knockdown of cell autonomous transcription factors will primarily affect expression of downstream targets within their own expression domain. Knockdown of signaling factors may affect expression in distant cells where the signal is received. The effects of a MO injection should be in line with the expected function of the target. Catastrophic phenotypes such as embryo death should be viewed critically, in particular, if the target is spatially restricted in its expression, which is usually the case for network candidates.

Sea urchin embryos are able to tolerate moderate levels of MOs without suffering from general toxicity and injection of control MOs, which are discussed in more detail later, does not adversely affect embryonic development. In *Strongylocentrotus purpuratus* the starting MO concentration in the injection solution is 300 μM (given an injection volume of ~ 8 pl). This is usually sufficient to effectively knockdown protein expression and will rarely have to be exceeded. If injection at this concentration results in severe developmental delay or death, the MO likely has off-target effects that cause defects, as in the Runt1 example discussed earlier. Although the concentration may be lowered and more specific phenotypes may be revealed, it is likely that the off-target effects remain significant, even if they no longer overtly affect development. Such MOs should be discarded. In zebrafish, a generic stress response leading to p53-mediated cell death has been attributed to nonspecific off-target effects and can be overcome by coinjection of p53 MO [28]. However, in sea urchins no equivalent responses are known.

5.2 Injection of Control Morpholino

Injection of control MOs is an important mock treatment, in particular, in GRN analyses, where the MO effect on transcript levels of potential downstream targets is usually evaluated quantitatively. MO injection can cause stress and may lead to mild developmental delays, depending on the sensitivity of a particular batch of embryos. This can be accounted for by comparison to control injected embryos.

A standard control MO, whose sequence complement is not found in the sea urchin genome, is available from Gene Tools. It can be injected into sea urchins without any obvious effects on morphology or progression of development. However, in quantitative gene expression analysis it was found to consistently alter gene expression of some skeletogenic differentiation genes (Stefan Materna and Paola Oliveri, unpublished) and may thus have unwanted, if mild, effects. An alternative is to inject a random control MO (IUPAC sequence: N_{25}). This MO is in essence a mix of many MOs, where each particular species is present at such low concentration that it is unlikely to be of any consequence. Random control MOs have been used as mock controls in many GRN studies and no adverse effects have been observed [6].

While these generic control MOs are essential as mock treatments, they are not useful to distinguish between true effects due to the loss of the target protein and nonspecific side effects. A different control that is occasionally used is to inject a second MO that is changed at several positions to prevent it from binding its target transcript. Four nucleotide substitutions should in theory suffice to abolish the translation blocking ability of the MO. However, as the Runt1 example demonstrates, in sea urchin this may not fully prevent the MO from functioning, in particular, when targeting highly abundant transcripts. Furthermore, changing the MO sequence raises the possibility that it may now bind other unintended targets with unpredictable consequences. Thus, while this control may establish that the effects of the MO are due to its exact sequence, it does not reveal whether the effects are attributable to the loss of the intended target.

5.3 Independent Confirmation

Because of the possibility for secondary targets it is necessary to apply a second, independent method to disrupt expression or function of the targeted protein. Independent confirmation is particularly important where "weak" phenotypes are generated, such as mild morphogenetic defects [29], which may occur as unspecific side effects of microinjection. An independent loss-of-function method may simply be a second MO, which must not overlap with the first MO to prevent hybridization with the same potential off-targets. Alternatively, embryos could be treated with a pharmacological inhibitor, or a dominant-negative protein could be expressed. To demonstrate that knockdown of a signaling molecule is specific, the receptor that it is thought to bind could be knocked down. Similar outcomes are a good indication of a specific effect. However, if both MOs cause relatively generic defects, such as developmental delay or death, the results should be viewed skeptically since both may elicit similar off-target effects.

Figure 1 shows the results of different perturbations disrupting the Delta/Notch signaling pathway, one of the best-studied signaling pathways in sea urchin embryonic development [30, 31]. The Delta ligand is membrane bound and presented by skeletogenic cells starting at blastula stage. It is received by the Notch receptor in the neighboring cells, where it induces mesodermal fate [32, 33]. Treatment with Delta or Notch MO prevents formation of all mesodermal cell types, which is easily recognized phenotypically by the absence of pigment cells. The MO results are virtually identical to those obtained following pharmacological inhibition of Notch function by treatment with the γ-secretase inhibitor DAPT, or expression

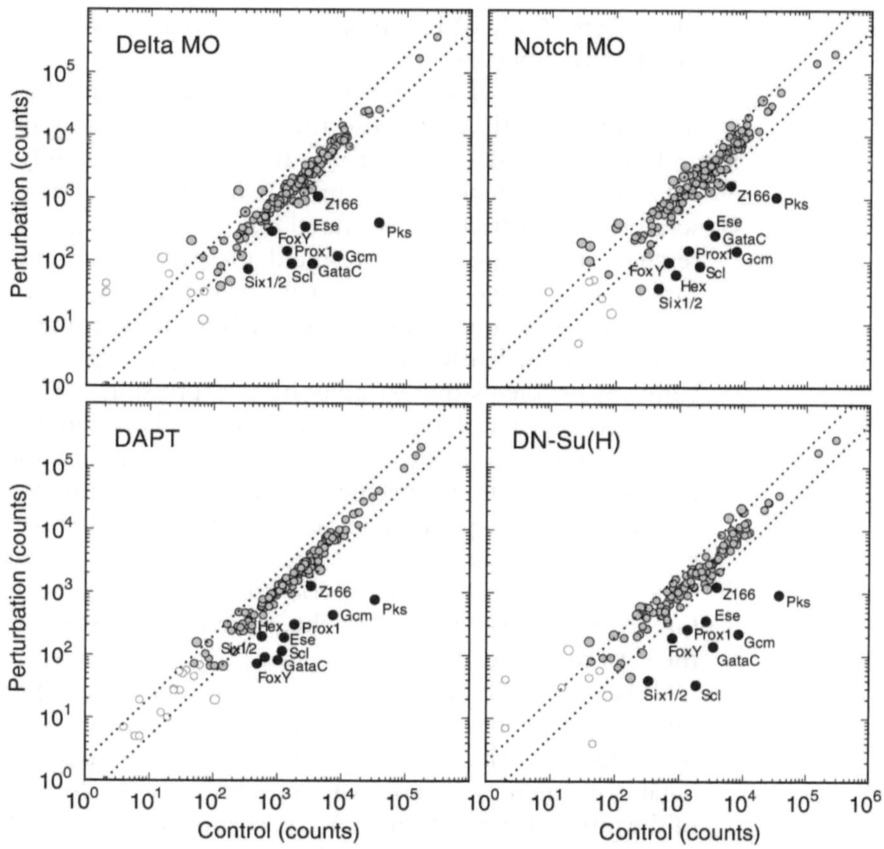

Fig. 1 Quantitative analysis of experiments perturbing the Notch pathway. *Upper panels*: Sea urchin embryos were injected with a MO targeting either the Delta ligand or Notch receptor. Embryos were lysed at 24 hpf and transcript prevalence for 182 genes was measured with the Nanostring nCounter. The counts obtained were plotted against those of control MO-injected embryos. Both MOs affect the same set of mesodermal genes (*labeled dots*) and cause a comparable decrease of their transcript prevalence. *Lower panels*: Disruption of Notch function by treatment with either DAPT, a known Notch inhibitor, or overexpression of a dominant negative version of the downstream effector Su(H) had almost identical effects as MO mediated knockdown. *Dashed lines* denote a twofold significance threshold between experiment and control. Genes considered to be expressed insignificantly in both experiment and control are depicted with open gray circles. Data as reported in [31]

of a dominant-negative form of Suppressor of Hairless (Su(H)), the downstream transcriptional effector. The striking agreement of these four different and complementary perturbation methods firmly establishes the specificity of the Delta and Notch MOs.

5.4 Rescue and Reversal of Phenotype

Specificity of a MO can be further demonstrated by rescue experiments, where the MO is delivered together with an mRNA that encodes the target protein but lacks the MO target sequence. In some cases, the MO-induced phenotype can be fully reversed with more than 95% efficiency [34]. However, such striking outcomes are rare and most frequently coinjection of mRNA will lead to either a partial rescue or produce an overexpression/gain-of-function phenotype.

The effects of mRNA injection will thus have to be studied in parallel in wild-type embryos.

The chance of success of a rescue experiment declines with developmental time, because mRNA is delivered at fertilization and is subject to degradation. It is difficult to fully rescue a MO-induced phenotype by mRNA injection if a targeted transcription factor becomes activated only at larval stages. Thus, such rescue experiments are best suited for genes activated relatively early in development, or maternally, or broadly expressed genes.

If the gene under study is spatially restricted, injection of in vitro-transcribed mRNA will lead to ectopic expression. To circumvent problems due to ubiquitous expression, rescues may be attempted in a cell type-specific manner. This can be achieved through injection of constructs in which the protein under study is expressed in a lineage-specific manner. The rescue transcript must not contain the sequence complement of the MO to be immune to its effects. Because DNA incorporation results in mosaicism, it is necessary to coinject a second construct that drives expression of a fluorescent marker instead of the protein under study to identify the cells that may have been rescued [35].

Overexpression experiments may produce striking phenotypes, but they are prone to artifacts, particularly at high concentration of transcript [6, 36]. Nevertheless, a clear-cut phenotypic response in a gain-of-function experiment can inform the biological function of the target protein. If mRNA expression elicits the exact opposite effect of the MO, it supports the idea that the MO response is, at least partially, due to the loss of the target protein. But gain-of-function experiments are limited in revealing whether certain aspects of a MO phenotype are artifactual.

5.5 Quantitative Evaluation

An excellent way to evaluate the specificity of a MO is quantitative analysis of its effects on gene expression in a comprehensive manner.

One central assumption, which has been validated in many GRN studies, is that only a small number of genes are affected by loss of an upstream regulator [37]. Of course, the effects will ripple through the GRN and may eventually (and indirectly) affect many dozens or more genes, resulting in aberrant phenotypes. However, the initial effects will be limited to the cell type in which a cell autonomous transcription factor is expressed or to those cells that are first exposed to a signal. Thus, when the expression levels of a comprehensive collection of sea urchin genes are analyzed, the effect of a MO is expected to be limited in scope and can be used to distinguish specific MO effects from nonspecific side effects.

A case in point is the Delta/Notch perturbations shown in Fig. 1. Here, transcript abundance was quantified for a large number of locally restricted regulatory genes, covering all distinct expression domains of the embryo [31]. These data were plotted as a function of counts obtained for control embryos. Out of 182 genes only expression of mesodermal genes is lost using different perturbation methods. But perhaps more importantly, this experiment provides more than 150 controls that clearly indicate that the embryo, with the exception of the mesoderm, develops more or less normally.

An oft-observed side effect of MO injection is a delay in development. Slight delays are normal side effects, which are controlled for by comparison to control MO-injected embryos. If severe delays are observed in repeat experiments, this is a red flag that may indicate off-target effects. For analyses of GRNs, it is necessary to minimize delays between experiment and control as much as possible, because even a few hours may result in significant changes in transcript abundance. This may increase experimental noise, or, at worst, lead to the erroneous interpretation of an observed effect as specific. This is of particular concern if changes in transcript levels of only a few genes are tested.

Inclusion of a larger gene set can help assure proper developmental timing, because a significant delay will increase the scatter of data points along the diagonal, corresponding to no change in transcript abundance. This is illustrated in Fig. 2 using wild-type expression data for mismatched time points (this gene set is identical to the one used in Fig. 1) [38]. The effects are larger at the

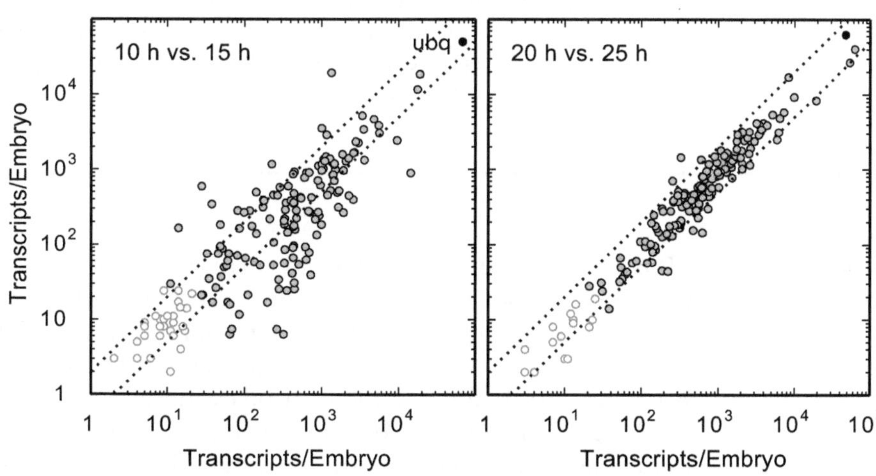

Fig. 2 Effect of developmental delay in quantitative perturbation analysis. Transcript levels of genes expressed throughout embryogenesis were plotted for two time points 5 h apart. At blastula stage (10 h vs. 15 h) a 5-h delay in experiment compared to control leads to large differences in transcript abundance. This is a particularly dynamic period as many genes included in this data set are activated during this time. Just prior to gastrulation proper (20 h vs. 25 h), the impact of developmental delay on prevalence in this gene set is less dramatic, but many genes still change several-fold during this time. The substantial scatter along the diagonal indicates that developmental stages are mismatched. The poly-ubiquitin gene, which has been used as a reference in many studies, is marked by a *black dot*. Its abundance is not significantly affected by developmental delay. *Symbols* as in Fig. 1. Expression data taken from [38]

earlier time point, because many of the genes included in this set are activated at blastula stage. Transcript abundance of many genes still changes several-fold between 20 and 25 h post fertilization. Accordingly, the scatter along the diagonal remains substantial and is thus indicative of the severe delay between the experiment and control.

Toward the end of gastrulation mismatching time points for this particular gene set does not lead to large scatter. The genes included in this experiment are expressed more or less steadily during this period. Thus, for quantitative analysis to clearly indicate delay at this stage a different, more informative gene set needs to be chosen.

Microinjection is an invasive technique that adversely affects development. This can be seen in the Notch perturbation results in Fig. 1. While Notch MO and expression of DN-Su(H), which are both delivered by microinjection, cause some variation in transcript abundance, noninvasive disruption of Notch function with DAPT treatment results in less scatter and close alignment of data points along the diagonal.

In the past, transcript abundance has often been queried for a few informative genes by quantitative PCR. While this is a great method to test effects on select genes, this analysis is limited in scope and insufficient to demonstrate specificity of a newly designed MO. It is necessary to analyze larger gene sets, e.g., with the Nanostring nCounter, or the entire transcriptome via RNAseq [39, 40]. The ability to gain insights in a comprehensive manner into what is, and what is not, affected by MO treatment adds a new layer of confidence in the validity of the reagent.

6 Delivery

6.1 Dilution and Storage

MOs are delivered in lyophilized form and have to be reconstituted before use. This can be done by adding nuclease-free water or a buffered solution of 10 mM Tris, pH 8.0 to a final MO concentration of 5 mM [34]. Reconstituted MO solutions may be stored at −20 °C for extended periods without any loss of efficacy. For long-term storage, MOs may be lyophilized in small aliquots. Before use, MO solutions should briefly be heated to 65 °C to assure the MO is fully dissolved.

6.2 Injection into Zygotes

Several detailed protocols for microinjection into sea urchins are available [41, 42]. Microinjection is an invasive technique that may introduce variability depending on the skill of the experimenter and the sensitivity of any particular batch of embryos. For injection into *S. purpuratus*, volumes should not exceed 10 pl as injection of larger volumes has adverse effects on development; most papers report injection volumes between 2 and 4 pl. For larger bubble sizes, a starting concentration of about 300 μM MO in the injection solution is a good rule of thumb and will usually be sufficient to cause effective knockdown. At this concentration the MO will be present at several magnitudes molar excess to moderately expressed genes. For smaller injection volumes, the concentration can be adjusted accordingly. Much higher concentrations cause adverse effects such as developmental delay.

When a new MO is injected for the first time, it is a good idea to establish a dose–response relationship. Thereafter, the MO should be used at the minimal concentration at which ≥80% of embryos show the defect caused by the embryo (if overt phenotypic defects occur). As discussed earlier, severe delays or even death at intermediate concentrations are a sign of unspecific side effects.

6.3 Application of VivoMorpholinos

VivoMO are reconstituted in deionized water and added to a final concentration of ≤20 μM in seawater containing sea urchin embryos. At concentrations >20 μM vivoMOs may precipitate in seawater and are toxic to embryos [21]. As with regular MOs a dose–response relationship should be established and the minimal effective concentration should be used. Although vivoMOs have been applied successfully in sea urchins, the window where they can be used successfully is relatively small. It is limited by solubility/toxicity in one direction and efficient knockdown in the other direction. For example, a vivoMO directed against BMP2/4 recapitulated zygotic delivery of a regular MO at a concentration of 5 μM [21]. But two different vivoMOs directed against proteins involved in skeletogenesis did not affect skeletal rod formation at 5 μM, although they did at 10 and 15 μM. One of these was a splice-blocking vivoMO and altered splicing was observed at the effective concentration [22]. However, the effective concentration is close to the upper limit of what is feasible. Thus, whether vivoMOs can be employed successfully must be established on a per case basis.

7 Morphants Versus Mutants

A serious challenge to the validity of MO-induced phenotypes comes from reports of major discrepancies between the phenotypes of morphants and zebrafish mutants that were generated by genome editing [9, 43]. Despite the good agreement of morphant and mutant phenotypes in many well-studied cases [44], this has led to calls that MOs should only be considered reliable if the MO phenotype recapitulates that of an established mutant [45]. Since mutant lines are not available for sea urchins this is a burden of proof that cannot easily be met. In response, it has been pointed out that it is uncertain whether all mutants in these studies were indeed true loss-of-function mutants [9]. Regardless, this renewed discussion will not fade quickly.

The now most commonly used method for creating specific alterations in the genome uses CRISPR/Cas technology [46]. In this method, a nuclease (usually Cas9) is directed to a particular genomic location with high specificity by forming a complex with a short guide RNA, which is complementary to the target sequence. Once bound it generates a double-strand break, which is repaired by nonhomologous end joining. This results in short insertions or deletions that potentially produce a deleterious mutation, which can be confirmed by sequencing.

Due to its high efficiency CRISPR/Cas can also be used as an acute knockout tool; it has recently been used in sea urchins to target the *nodal* gene [47]. Several guide RNAs targeting the coding region were tested, resulting in radialized embryos, just like MO-mediated knockdown of *nodal* or pharmacological inhibition. One potential problem is that short deletions may leave the coding region in frame, thus creating a mosaic embryo. Only two-thirds of gene copies may thus truly be defective, even at 100% cutting efficiency. This may be sufficient for a gene like *nodal* where the sea urchin embryo is supremely sensitive to loss of its function resulting in global deficiencies (a MO targeting *nodal* transcript can be used at a tenfold lower concentration than those targeting most other genes [48]). Whether it can be applied with similar success to other targets remains to be seen.

An interesting alternative may be CRISPR interference. Here a nuclease-dead Cas9 protein is guided to the 5′ region just downstream of the transcriptional start of a target gene where it blocks transcript elongation [49]. For maximum efficiency guide RNAs can be used combinatorially [50], or Cas9 can be augmented with a repression domain [51]. In contrast to recombinant transcription factors containing activation or repression domains, CRISPRi complexes bind with high specificity due to the long recognition site of the guide RNA. Efficacy can be established by quantifying transcription of the target. But because side effects due to off-target binding cannot per se be excluded, specificity will have to be demonstrated just as when using MOs.

Another recent study reported discrepancies between morphant and mutant phenotypes [50]. However, the authors could show that this is due to genetic compensation in the mutant rather than an erroneous result caused by off-target binding of the MO. While the mechanism of compensation is unknown, this study suggests that morphants and mutants may yield different insights. Thus, mutants are not inherently superior over morphants when the goal is to probe the consequences of an acute loss of protein expression.

8 Outlook

Gene-specific knockdown using MOs has been tremendously successful in sea urchins and enabled the mapping of GRNs on an embryo-wide scale. For the foreseeable future, it will remain one of the main approaches for revealing gene function.

The main limitation for advancing the study of gene networks is its most common way of delivery, i.e., injection into the zygote. This restricts the use of MOs to the first few days of development or the first expression phase of the targeted proteins. This is a serious issue for genes with complex or disjointed expression patterns that span multiple domains, a common occurrence for regulatory factors. In fact, by the end of gastrulation the vast majority of transcription factor genes (and more than 40% of the genes in the genome) have been activated at least once and will likely be employed again at a later time in a different context [2, 52]. This represents an enormous challenge as GRN approaches in sea urchins expand into larval development and formation of the adult. More widespread use of vivoMOs will help to overcome this obstacle, at least in part. It will allow more systematic exploration of later events that have been accessible primarily through pharmacological inhibition.

While MO technology is limited in how its activity can be controlled in space and time, this is just beginning to be explored for CRISPR-based methods. The potential of this technology is obvious even if its impact on GRN research in sea urchins is difficult to predict. But in addition to its developmental potential CRISPR may have another critical advantage over MOs: While MOs are relatively affordable, the cost per oligo is nonetheless substantial. In contrast, a guide RNA may be generated for less than a tenth the cost. It may thus become feasible to use CRISPR combinatorially, or target larger numbers of genes in an unbiased way, similar to RNAi screens in other systems. If nothing else, CRISPR-based techniques may become an additional and complementary tool to probe gene networks in sea urchin.

Regardless of the exact technique, all gene-specific perturbation methods in sea urchin are based on sequence recognition in the early embryo, either through binding to DNA or RNA. All methods, whether they involve MOs or CRISPR, are prone to artifacts due to off-target binding. It is thus crucial to establish that the observed phenotypic effects are due to the loss of the target protein. The controversy surrounding the discrepancy between morphant and CRISPR mutant phenotypes in fish underlines the necessity to use best practices and appropriate controls. To this end, several orthogonal approaches should be taken that together conclusively demonstrate specificity and efficacy to allow genuine insights into the regulation of development.

Acknowledgements

I am grateful to the many members of the sea urchin community who shared their experiences. Many thanks to Drs. Jonathan Rast, Eric Erkenbrack, Lauren Shipp, and Stephanie Woo for detailed comments on the manuscript. In memory of Eric Davidson.

References

1. Davidson EH, Peter I (2015) Genomic control process. Academic, San Diego, CA
2. Sea Urchin Genome Sequencing Consortium, Sodergren E, Weinstock GM, et al (2006) The genome of the sea urchin *Strongylocentrotus purpuratus*. Science 314:941–952. doi:10.1126/science.1133609
3. Davidson EH (2015) Genomics, "discovery science," systems biology, and causal explanation: what really works? Perspect Biol Med 58:165–181. doi:10.1353/pbm.2015.0025
4. Howard EW, Newman LA, Oleksyn DW et al (2001) SpKrl: a direct target of beta-catenin regulation required for endoderm differentiation in sea urchin embryos. Development 128:365–375
5. Davidson EH, Rast JP, Oliveri P et al (2002) A genomic regulatory network for development. Science 295:1669–1678. doi:10.1126/science.1069883
6. Materna SC, Oliveri P (2008) A protocol for unraveling gene regulatory networks. Nat Protoc 3:1876–1887. doi:10.1038/nprot.2008.187
7. Cui M, Siriwon N, Li E et al (2014) Specific functions of the Wnt signaling system in gene regulatory networks throughout the early sea urchin embryo. Proc Natl Acad Sci U S A 111:E5029–E5038. doi:10.1073/pnas.1419141111
8. Peter IS, Faure E, Davidson EH (2012) Predictive computation of genomic logic processing functions in embryonic development. Proc Natl Acad Sci U S A 109:16434–16442. doi:10.1073/pnas.1207852109
9. Kok FO, Shin M, Ni C-W et al (2015) Reverse genetic screening reveals poor correlation between morpholino-induced and mutant phenotypes in zebrafish. Dev Cell 32(1):97. doi:10.1016/j.devcel.2014.11.018
10. Bedell VM, Westcot SE, Ekker SC (2011) Lessons from morpholino-based screening in zebrafish. Brief Funct Genomics 10:181–188. doi:10.1093/bfgp/elr021
11. Eisen JS, Smith JC (2008) Controlling morpholino experiments: don't stop making antisense. Development 135:1735–1743. doi:10.1242/dev.001115
12. Röttinger E, Saudemont A, Duboc V et al (2008) FGF signals guide migration of mesenchymal cells, control skeletal morphogenesis [corrected] and regulate gastrulation during sea urchin development. Development 135:353–365. doi:10.1242/dev.014282
13. Materna SC, Ransick A, Li E, Davidson EH (2013) Diversification of oral and aboral mesodermal regulatory states in pregastrular sea urchin embryos. Dev Biol 375:92–104. doi:10.1016/j.ydbio.2012.11.033
14. Haillot E, Molina MD, Lapraz F, Lepage T (2015) The maternal Maverick/GDF15-like TGF-β ligand panda directs dorsal-ventral axis formation by restricting nodal expression in the sea urchin embryo. PLoS Biol 13:e1002247. doi:10.1371/journal.pbio.1002247
15. Ransick A, Davidson EH (2012) Cis-regulatory logic driving glial cells missing: Self-sustaining circuitry in later embryogenesis. Dev Biol:1–9. doi:10.1016/j.ydbio.2012.02.003
16. Andrikou C, Pai C-Y, Su Y-H, Arnone MI (2015) Logics and properties of a genetic regulatory program that drives embryonic muscle development in an echinoderm. eLife 4. doi:10.7554/eLife.07343
17. Ketting RF (2011) The many faces of RNAi. Dev Cell 20:148–161. doi:10.1016/j.devcel.2011.01.012
18. Song JL, Stoeckius M, Maaskola J et al (2012) Select microRNAs are essential for early development in the sea urchin. Dev Biol 362:104–113. doi:10.1016/j.ydbio.2011.11.015
19. Duboc V, Röttinger E, Besnardeau L, Lepage T (2004) Nodal and BMP2/4 signaling organizes the oral-aboral axis of the sea urchin embryo. Dev Cell 6:397–410
20. Su Y-H (2014) Telling left from right: left-right asymmetric controls in sea urchins. Genesis 52:269–278. doi:10.1002/dvg.22739
21. Luo YJ, Su YH (2012) Opposing nodal and BMP signals regulate left–right asymmetry in the sea urchin larva. PLoS Biol 10(10):e1001402. doi:10.1371/journal.pbio.1001402.g006
22. Heyland A, Hodin J, Bishop C (2014) Manipulation of developing juvenile structures

in purple sea urchins (*Strongylocentrotus purpuratus*) by morpholino injection into late stage larvae. PLoS One 9:e113866. doi:10.1371/journal.pone.0113866

23. Morcos PA, Li Y, Jiang S (2008) Vivo-Morpholinos: a non-peptide transporter delivers Morpholinos into a wide array of mouse tissues. Biotechnology 45:613–618
24. Payumo AY, Walker WJ, McQuade LE et al (2015) Optochemical dissection of T-box gene-dependent medial floor plate development. ACS Chem Biol 10:1466–1475. doi:10.1021/cb5010178
25. Cameron RA, Samanta M, Yuan A et al (2009) SpBase: the sea urchin genome database and web site. Nucleic Acids Res 37:D750–D754. doi:10.1093/nar/gkn887
26. Coffman JA, Dickey-Sims C, Haug JS et al (2004) Evaluation of developmental phenotypes produced by morpholino antisense targeting of a sea urchin Runx gene. BMC Biol 2:6. doi:10.1186/1741-7007-2-6
27. Coffman JA, Kirchhamer CV, Harrington MG, Davidson EH (1996) SpRunt-1, a new member of the runt domain family of transcription factors, is a positive regulator of the aboral ectoderm-specific CyIIIA gene in sea urchin embryos. Dev Biol 174:43–54. doi:10.1006/dbio.1996.0050
28. Robu ME, Larson JD, Nasevicius A et al (2007) p53 activation by knockdown technologies. PLoS Genet 3:e78. doi:10.1371/journal.pgen.0030078
29. Shipp LE, Hill RZ, Moy GW et al (2015) ABCC5 is required for cAMP-mediated hindgut invagination in sea urchin embryos. Development 142:3537–3548. doi:10.1242/dev.126144
30. Sherwood DR, McClay DR (2001) LvNotch signaling plays a dual role in regulating the position of the ectoderm-endoderm boundary in the sea urchin embryo. Development 128:2221–2232
31. Materna SC, Davidson EH (2012) A comprehensive analysis of Delta signaling in pregastrular sea urchin embryos. Dev Biol 364:77–87. doi:10.1016/j.ydbio.2012.01.017
32. Sherwood DR, McClay DR (1999) LvNotch signaling mediates secondary mesenchyme specification in the sea urchin embryo. Development 126:1703–1713
33. Sweet HC, Gehring M, Ettensohn CA (2002) LvDelta is a mesoderm-inducing signal in the sea urchin embryo and can endow blastomeres with organizer-like properties. Development 129:1945–1955
34. Angerer LM, Angerer RC (2004) Disruption of gene function using antisense morpholinos. Methods Cell Biol 74:699–711
35. Solek CM, Oliveri P, Loza-Coll M et al (2013) An ancient role for Gata-1/2/3 and Scl transcription factor homologs in the development of immunocytes. Dev Biol 382:280–292. doi:10.1016/j.ydbio.2013.06.019
36. Prelich G (2012) Gene overexpression: uses, mechanisms, and interpretation. Genetics 190:841–854. doi:10.1534/genetics.111.136911
37. Davidson EH (2010) Emerging properties of animal gene regulatory networks. Nature 468:911–920. doi:10.1038/nature09645
38. Materna SC, Nam J, Davidson EH (2010) High accuracy, high-resolution prevalence measurement for the majority of locally expressed regulatory genes in early sea urchin development. Gene Expr Patterns 10:177–184. doi:10.1016/j.gep.2010.04.002
39. Geiss G, Bumgarner R, Birditt B et al (2008) Direct multiplexed measurement of gene expression with color-coded probe pairs. Nat Biotechnol 26:317–325. doi:10.1038/nbt1385
40. Tu Q, Cameron RA, Davidson EH (2013) Quantitative developmental transcriptomes of the sea urchin *Strongylocentrotus purpuratus*. Dev Biol:1–8. doi:10.1016/j.ydbio.2013.11.019
41. Cheers MS, Ettensohn CA (2004) Rapid microinjection of fertilized eggs. Methods Cell Biol 74:287–310
42. Stepicheva NA, Song JL (2014) High throughput microinjections of sea urchin zygotes. J Vis Exp:e50841. doi:10.3791/50841
43. van Impel A, Zhao Z, Hermkens DMA et al (2014) Divergence of zebrafish and mouse lymphatic cell fate specification pathways. Development 141:1228–1238. doi:10.1242/dev.105031
44. Nasevicius A, Ekker SC (2000) Effective targeted gene "knockdown" in zebrafish. Nat Genet 26:216–220. doi:10.1038/79951
45. Schulte-Merker S, Stainier DYR (2014) Out with the old, in with the new: reassessing morpholino knockdowns in light of genome editing technology. Development 141:3103–3104. doi:10.1242/dev.112003
46. Sander JD, Joung JK (2014) CRISPR-Cas systems for editing, regulating and targeting genomes. Nat Biotechnol 32:347–355. doi:10.1038/nbt.2842
47. Lin C-Y, Su Y-H (2016) Genome editing in sea urchin embryos by using a CRISPR/Cas9 system. Dev Biol 409:420–428. doi:10.1016/j.ydbio.2015.11.018
48. Nam J, Su Y-H, Lee PY et al (2007) Cis-regulatory control of the nodal gene, initiator of the sea urchin oral ectoderm gene network. Dev Biol 306:860–869. doi:10.1016/j.ydbio.2007.03.033

49. Qi LS, Larson MH, Gilbert LA et al (2013) Repurposing CRISPR as an RNA-guided platform for sequence-specific control of gene expression. Cell 152:1173–1183. doi:10.1016/j.cell.2013.02.022
50. Rossi A, Kontarakis Z, Gerri C et al (2015) Genetic compensation induced by deleterious mutations but not gene knockdowns. Nature 524:230–233. doi:10.1038/nature14580
51. Mandegar MA, Huebsch N, Frolov EB et al (2016) CRISPR interference efficiently induces specific and reversible gene silencing in human iPSCs. Cell Stem Cell 18:541–553. doi:10.1016/j.stem.2016.01.022
52. Howard-Ashby M, Materna SC, Brown CT et al (2006) High regulatory gene use in sea urchin embryogenesis: Implications for bilaterian development and evolution. Dev Biol 300:27–34. doi:10.1016/j.ydbio.2006.10.016

Chapter 9

Manipulating Gene Expression in the Chick Embryo

Octavian Voiculescu and Claudio D. Stern

Abstract

The chick embryo is a well-established representative of amniote embryos, which has been used to make many discoveries, including many major concepts, which have moved our knowledge of developmental biology in hugely important ways. The chick has a relatively compact genome and is easily amenable to embryological manipulations and in vivo imaging. Morpholino gene manipulations have been used in a variety of contexts, and constitute a quick and versatile molecular tool. Here we describe methods to deliver morpholinos to chick embryos, allowing targeting of specific cell populations at defined developmental stages, using two stages as examples: the epiblast of the embryo in the first day of incubation, when the primary germ layers of the embryo are specified, and in ovo electroporation of the neural tube as an example of a later stage. With slight modifications, these general methods can be used to target other embryonic tissues.

Key words Chick embryo, Electroporation, Morpholino, Epiblast, Neural plate, Ectoderm, Mesoderm, Endoderm

1 Introduction

The chick embryo has been one of the top choices as a model for classical and modern embryology, and was key for the discovery of many major concepts including the discovery of the circulation, the first oncogene, the mechanisms of limb development, the first genes underlying left-right asymmetry, the mechanisms of somite formation and their development into the musculoskeletal system, neural development, heart development and many other processes (for review *see* [1]). Chick embryos are easily available year round and not very expensive, and lend themselves particularly well to dynamic observations and surgical manipulations. It is also particularly useful because amniote embryos (including the human) are different from anamniote embryos in several aspects, including size, shape, and internal organization. The chick genome, sequenced more than a decade ago, is relatively compact and sits at a convenient distance between anamniotes and mammals, which is useful to dissect out gene regulatory circuits. Reverse genetics

Hong M. Moulton and Jon D. Moulton (eds.), *Morpholino Oligomers: Methods and Protocols*, Methods in Molecular Biology, vol. 1565, DOI 10.1007/978-1-4939-6817-6_9,

© Springer Science+Business Media LLC 2017

techniques (such as knock-outs and gene replacements using CRISPR-Cas9) are now starting to emerge for avian embryos. The chick embryo is probably the best system for molecular manipulations targeted to particular cells or tissues with precise spatial and temporal control, by using electroporation, a technique introduced by Harukazu Nakamura [2] about 20 years ago. Electroporation can be used for gain-of-function experiments by introduction of DNA plasmids. For loss-of-function, morpholinos are an excellent tool and have been used extensively and successfully in a large variety of contexts, most often without off-target effects [3]. Electroporation of both DNA and Morpholinos is done most easily in epithelial tissues (such as the epiblast, the neural tube, otic vesicle or somites) but can also be used for mesenchymal ones. Here, we describe how to target morpholinos to any region of the flat epiblast before and at primitive-streak stages in embryo culture [4] (Method A), and into the neural tube once this closes, in ovo (Method B). Method A allows targeting precursors of specific mesodermal lineages (e.g. notochord, somites, lateral plate) in the epiblast, before they gastrulate, as well as regions destined to either ectoderm, neural tube, the neuro-ecotodermal border or neural crest. Method B allows targeting one side of the neural tube, which can be modified to direct the Morpholino to different dorsoventral populations of cells. With slight modifications, these procedures can be adapted to other tissues and developmental stages.

2 Materials

Prepare all reagents for the electroporation mixture using ultrapure water (18 MΩ cm at 25 °C). Embryo saline solutions can be prepared with deionized water and stored at room temperature.

2.1 Chick Embryos and Materials for Both In Ovo or In Vitro Manipulations

Store the fertile hens' eggs at 14–18 °C for up to 2 weeks, incubate at 38 °C for the embryos to attain the desired stage. The methods described here are useful for pre-primitive streak (Eyal-Giladi and Kochav [5] stages X–XIV) and Hamburger and Hamilton [6] stages 2–15. For incubations longer than 1 day, lay the eggs on their side.

2.2 Electroporation Solution and Application Micropipette

1. Carboxyfluorescein- or lissamine-tagged morpholinos: dissolve in ultrapure water to 1 mM, as per manufacturer's instructions (*see* **Note 1**).
2. Fast Green FCF: dissolve 4 mg in 1 mL water.
3. Sucrose: dissolve 600 mg in 1 mL water.
4. Tris-HCl 1 M, pH 7.5.
5. Pure plasmid DNA, at least 2 mg/mL in ultrapure water or TE (store at −20 °C).

6. On the day of each experiment, calculate the total volume needed to electroporate all the embryos collected and intended for each particular morpholino; less than 0.5 μL of electroporation solution per embryo should suffice. In a typical experiment, make 10 μL of electroporation solution to electroporate 10–20 embryos. Mix in a 0.2 mL PCR tube: 5 μL stock morpholino, 1 μL Fast Green, 1 μL sucrose, 1 μL Tris-HCl (*see* **Note 2**), plasmid DNA to a final concentration of 0.2 mg/mL (*see* **Note 3**) and make up to 10 μL with ultrapure water.
7. Application micropipette. Using a pipette puller and borosilicate glass capillary tubing (e.g. Sigma 50 μL aspirator tubes, P1049), pull a fine micropipette with a relatively long, tapered end. Break just the tip of the pipette gently with a watchmaker's forceps. Attach the pipette to a mouth tube (e.g. Sigma A5177). It is important that the bore of this capillary be of the right size (*see* **Note 4**). Check that saline can be aspirated or expelled from the pipette with relative ease, but significant pressure needs to be exerted so that only 1–2 μL can be drawn into the pipette within a few seconds.

2.3 Electroporator

Intracel TSS20 (http://intracel.co.uk/) pulse electroporator; other makes and models can be used successfully but the particular pulse settings need to be adjusted (*see* **Note 5**).

2.4 Materials for Electroporation in Embryo Culture (Method A)

1. Make stock solutions A and B for the Pannett-Compton saline. For solution A, weigh 121 g NaCl, 15.5 g KCl, 10.42 g $CaCl_2{\cdot}2H_2O$, 12.7 g $MgCl_2{\cdot}6H_2O$, dissolve in 1 L water, and autoclave. For solution B, weigh 2.365 g $Na_2HPO_4.2H_2O$, 0.188 g $NaH_2PO_4{\cdot}2H_2O$, dissolve in 1 L water and autoclave. On the day of the experiment, make the final Pannett-Compton solution: measure 120 mL of solution A in a graduated cylinder and pour into a 5 L plastic container. Add 2700 mL water. Measure 180 mL of solution B in the graduated cylinder and add to the container. The order in which these components are added is important because mixing undiluted A and B will cause precipitation of Ca and Mg phosphates. Use the Pannett-Compton solution within 1 day.
2. Glass baking tray, approximately 20 × 30 × 4 cm or larger.
3. Watch glasses, 5–8 cm diameter for Method B.
4. Glass rings, 25–28 mm diameter, 1.5–2.0 mm wall thickness and 2 mm high.
5. Two pairs of coarse forceps.
6. Two pairs of watchmaker's forceps, no. 4 or 5.
7. One pair of small scissors.
8. Pasteur pipettes 150 mm. Gently flame the thin end of one pipette and attach a rubber teat at the other end; this will be

used to clean the vitelline membrane and embryo of adherent albumen and yolk. With a glass cutter, remove the thin end of another Pasteur pipette at the level of the shoulder, flame the cut end and insert a rubber teat at the top (uncut end); this will be used to transfer embryos from one dish to another.

9. Plastic Petri dishes, 35 and 60 mm diameter.
10. Small beaker (100–250 mL).
11. Sealable plastic box (a sandwich box is ideal).
12. Dissection microscope with transmitted light base. A microscope with fluorescence optics will also be needed.
13. Commercial (e.g. BEX, Japan) or homemade chambers and electrodes. Both the chamber and the movable electrode should be flat, with a face diameter of 1–3 mm (*see* **Note 6**). The chamber should allow placing the embryo over a well smaller in diameter than the embryo, at the bottom of which sits the positive electrode at a depth of about 2 mm.
14. Coarse manipulator to position the negative electrode.

2.5 Electroporation In Ovo (Method B)

1. Make stock Tyrode's 10× solution. Weigh 80 g NaCl, 2 g KCl, 2.71 g $CaCl_2 \cdot 2H_2O$, 0.5 g $NaH_2PO_4 \cdot 2H_2O$, 2 g $MgCl_2 \cdot 6H_2O$, 10 g sucrose, dissolve in 1 L water and autoclave. On the day of the experiment, add 10 mL of this stock solution to 90 mL water.
2. One pair of small scissors.
3. Syringes, 1 and 5 mL capacity.
4. Needles, 21- and 30-G.
5. Indian ink (preferably Pelikan Fount India; other makes are suitable as long as they do not contain phenol or other toxic organic solvents). On the day of the experiment, dilute ink 1:10 into Tyrode's solution.
6. Sellotape (Scotch tape in USA) or PVC (electrical) tape
7. Egg support. Knead plasticine (PlayDo or Claymation in USA) into a ball 3 cm diameter, then flatten it and gently impress an egg into it.
8. Dissection microscope with lateral illumination. A microscope with fluorescence optics will also be needed.
9. Commercial (e.g. Intracel, UK) or homemade electroporation electrodes. The latter can be made of Ag or Pt round wires (0.5 mm diameter). Bend the tips (2 mm long) at a 45° angle. Insulate the rest of the wires with nail varnish—this is very important, to prevent current from leaking outside the intended area for electroporation. Make a holder to keep the electrodes with the tips running parallel, 3 mm apart.

3 Methods

3.1 Electroporation of Young Embryos in Culture (Method A)

1. Pour Pannett-Compton solution in the baking tray, to a height of at least 3 cm (a deeper dish and more saline makes the operation easier). Open the eggs using the coarse forceps to remove the top of the egg shell. Tip the thin albumen into a small beaker and remove the thick albumen by pulling it out of the shell using the forceps. Collect 8–20 intact egg yolks (depending on the size of the baking dish) in the saline, making sure they are completely submerged. Align the yolks in columns of 5–6 on one side of the tray.
2. Place a series of six watch glasses and glass rings at the opposite end of the tray.
3. Position each yolk with the embryo facing uppermost and use the scissors to cut around the vitelline membrane just below the equator of the yolk. Keep the cut yolks tidy in the baking tray, in the same 6-column arrangements on one side. One by one, peel off the membrane from each yolk, and transfer the membrane (with its yolk-facing surface uppermost) to a watch glass. Place a glass ring over the membrane and gently pull the membrane so that the ring sits over it. Most of the time, the embryo will remain attached to the membrane (and will now face upwards); if the embryo remained on the yolk, it can be retrieved from there later (*see* **step 4** below). For a demonstration, *see* [7]. Place the assembled rings with stretched vitelline membranes, sitting on watch glasses, on the bench. Make sure that there is enough saline both inside and outside the ring so the embryo and membrane do not dry out.
4. Using the finer (uncut) Pasteur pipette, free each embryo from either the vitelline membrane or the yolk (the latter if the embryo remained attached to the yolk when the membrane was peeled off). Collect embryos in a 60 mm Petri dish containing Pannett-Compton solution.
5. Repeat the operation until all embryos have been collected and a corresponding number of vitelline membranes stretched on the glass rings are assembled. About 18–24 embryos and culture assemblies can be produced within 2 h.
6. Place the electroporation dish under the dissecting microscope, fill it with Pannett-Compton solution, and bring the negative electrode 2–3 mm directly above the well. Connect the electrodes to the electroporator (top electrode to the negative pole, bottom one to the positive pole) and apply a series of 3–5 10 V pulses to make sure the circuit is closed. Reverse the polarity and apply another series of pulses (*see* **Note 7**), then reverse the polarity again to electroporation configuration. Set the parameters of the Intracel TSS20 electroporator appropriately: for

pre-streak stage embryos, 2.9 V, 5 pulses, 50 ms width, 500 ms space; for streak-stage embryos, 5.7 V, 3 pulses, 50 ms width, 500 ms space. Push the negative electrode away, in the horizontal plane, just enough to expose the well (*see* **Note 5**).

7. Using the mouth tube, aspirate about 2 μL electroporation solution into the tip of the micropipette (*see* **Note 4**). Place the tube with micropipette attached on the side of the microscope, making sure the tip does not break during the subsequent manipulations.
8. Place an embryo over the well in the centre of the electroporation dish, with the epiblast facing up. Adjust the position of the embryo over the well so the area to be electroporated sits over the centre of the well. Very gently push down on this area, with the aid of the forceps and small puffs of saline, to create a small depression.
9. Using the micropipette and mouth tube, dispense about 0.3–0.5 μL electroporation solution over the area to be targeted. The tip of the capillary should be as close to the surface of the embryo as possible, without touching or scoring it.
10. Bring the negative (top) electrode directly above the well and trigger the electroporator (*see* **Note 8**). Push the negative electrode away, in the horizontal plane, just enough to expose the embryo and the well.
11. Apply very gentle puffs of saline using the small-mouthed Pasteur pipette, until all the electroporation mixture is removed from the top of the embryo. Transfer the electroporated embryo into another 60 mm Petri dish filled with Pannett-Compton. Embryos can be checked immediately for fluorescence (from the carboxyfluorescein or lissamine attached to the morpholino).
12. Repeat steps 3–5 above, re-filling the micropipette with electroporation solution as needed.
13. Assemble embryo cultures. Dispense 1.5–2.0 mL thin egg albumen to a number of 35 mm Petri dishes equal to the number of embryos to be cultures. Using the large-mouthed Pasteur pipette, transfer each embryo to the inside of a glass ring. Under the dissection microscope, place the embryo in the middle of the vitelline membrane, with the epiblast facing the membrane and the hypoblast/endoderm up. Gently remove the Pannett-Compton solution from inside the ring, making sure the embryo stays in the centre during this operation. Slide the glass ring horizontally over the watch glass and place it slowly over the albumen in one 35 mm Petri dish; as the ring sinks into the albumen and the centre of the vitelline membrane rises, ensure that the embryo remains in the centre. Using two pairs of forceps, push down the ring against the

bottom of the Petri dish and hold in place for a few seconds, until the ring sticks in place. Wet the inside of the lid of the dish evenly with albumen and discard the excess, then place the lid over the Petri dish and rotate it about one turn, until a continuous film of albumen forms between the lid and the dish. The purpose of doing this is to leave a thin hydrophilic film on the lid to prevent droplets of condensation from forming which could fall on the embryo, and also to help seal the lid.

14. Place assembled cultures in a humidified, sealed plastic box and incubate at 38 °C.

3.2 Electroporation Older Embryos In Ovo (Method B)

1. Place an egg in the egg support, keeping the same orientation as during the incubation. Orient the egg with the sharp end to the left. In most cases, the embryo will sit just under the shell, with the head pointing away from the investigator.
2. Apply a square piece of Sellotape/Scotch tape, 2.5 cm × 2.5 cm, over the highest portion of the shell.
3. Aspirate the diluted India ink into the 1 mL syringe, attach a 30-G needle. Expel all air from the ink, syringe and needle. If desired, bend the needle by 45°, with the face of the opening pointing away.
4. Insert a 21-G needle, attached 5 mL syringe, into the blunt end of the egg, downward at an angle of 45°. Slowly remove 3 mL albumen—this will lower the embryo under the shell.
5. Using the scissors, cut a 1.5 cm diameter window into the tape-protected shell. Keep the tip of the scissors as flat as possible during this operation, to avoid damaging the yolk and the embryo. The purpose of the tape is to prevent small pieces of shell from falling on the embryo.
6. Add back 2 mL albumen into the egg through the window, to raise the yolk again and help to manipulate the embryo.
7. Gently insert the bent needle of the syringe with India ink into the edge of the extraembryonic membranes around the embryo, at a very shallow angle. Once the needle is under the middle of the embryo, slowly dispense enough ink to give contrast to the embryo. Gently remove the needle.
8. Apply 1–2 drops of Tyrode's solution over the embryo to cover it by saline. Using the tip of a 21-G needle, cut a small hole into the vitelline membrane to expose the embryo around the neural tube.
9. Aspirate 0.5 μL of electroporation mixture into the capillary attached to the mouth pipette.
10. Insert the capillary needle into the neural tube, in the head region, with the tip pointing towards the trunk. Dispense enough electroporation mixture to fill the lumen of the neural tube. Retract the capillary gently but as quickly as possible.

11. Place the electrodes on each side of the neural tube (parallel to its axis) and press down on either or both of them to orient the field as desired. Trigger the train of electric pulses with the aid of the foot pedal. Good results can be obtained by applying 5 pulses of 12 V, 50 ms long at an interval of 1 s. Successful morpholino electroporation can be checked immediately by fluorescence.
12. Using the same 5 mL syringe with 21-G needle and hole in the flat end of the shell, remove 1–2 mL albumen to lower the yolk and make sure the embryo is covered by yolk. Cover the window with Sellotape/Scotch or PVC tape and place the operated egg into a tray.
13. Return all electroporated eggs to the incubator and allow them to develop for the desired length of time.

4 Notes

1. The efficacy of each morpholino should be tested for each application, using either antibodies against the corresponding protein or RT-PCR to detect altered splicing. With the protocol described above, the final concentration of morpholinos in the delivery solution is 500 μM. Efficient knock-down can be achieved at much lower concentrations (100 μM). Conversely, if higher morpholino concentrations are needed, the stock carboxyfluorescein-conjugated morpholinos can usually be prepared at 2–5 mM.
2. Buffering the electroporation solution with Tris–HCl pH 7.5 seems to be particularly important when using lissamine-conjugated morpholinos. Otherwise the area electroporated becomes grayish in appearance and the cells are soon ejected from the epiblast forming clumps above it during culture. Similar effects are seen with plasmid DNA dissolved in water. Tris-HCl pH 7.5 does not appear to interfere with electroporation of carboxyfluorescein-conjugated morpholinos or plasmid DNA dissolved in TE, and is therefore recommended in all cases.
3. The addition of plasmid DNA is only necessary to electroporate the embryonic epiblast (prospective ectoderm and neural plate) at primitive streak stages. The plasmid does not need to contain eukaryotic promoters if used only to promote morpholino delivery with high efficiency (more than 90 % of the cells in the targeted area). At the concentrations indicated here, very few cells (if any) receive enough plasmid to express genes even when using strong eukaryotic promoters and enhancers. If both morpholino and plasmid DNA are to be

introduced for experimental purposes (e.g. a rescue of the effect of the morpholino), the plasmid concentration should be elevated to 1–2 mg/mL in the final electroporation solution, but the number of cells receiving useful amounts of plasmid DNA will always be several-fold lower than that of cells electroporated with morpholino (about a third of cells in the electroporated area). This can be useful because it allows rescued cells (receiving both DNA and morpholino) to be distinguished from those receiving either alone [8]. The extra-embryonic area opaca at primitive streak stages, as well as the epiblast at pre-primitive streak stages, can be electroporated successfully without adding plasmid DNA to the electroporation solution.

4. As a rule of thumb, the bore of the capillary tip should be similar or just larger than for lipophilic dye (e.g. DiI) labeling. If microliters can be aspirated into the pipette effortlessly within 1 s, the opening of the capillary is too large, with two undesired effects. First, saline will enter the pipette and dilute the electroporation solution. Second, neither the amount of morpholino nor the region where it is applied can be controlled, and either too much or too little solution will be applied to the desired embryonic location.
5. Other electroporator models, including BTX ECM 830 (https://www.btxonline.com/ecm-830-for-in-vivo--applications/) and BEX CUY21EX (http://www.bexnet.co.jp/english/product/device/in-vivo/cuy21ex.html) can also be used successfully with chick embryos, but optimum settings vary with the model and the geometry of the electrodes. For example, the BEX CUY21EX achieves the same results described above for streak embryos using the following settings: pre-pulse 50 V, 0.05 ms duration, and a set of 5 pulses of 8 V, on 75 ms, off 50 ms.
6. Bespoke electrodes can be manufactured easily to suit each particular application. For example, 100–250 μm Pt wire can be used to make parallel electrodes and electroporate "lines" across the embryo. The same wire can be coiled in a flat, circular pattern; a small bead of epoxy glue can be applied on the back of this coil to insulate it and provide mechanical support and make a flat-end electrode. In all cases, both electrodes should always be kept at least 2 mm away from the embryo; this avoids burning the tissues and there is no gain in electroporation efficiency when approaching the electrodes closer to the embryo.
7. Salt deposits and gas bubbles can affect the conductivity of electrodes. Electrodes should be cleaned after each experiment with soapy water. If poor electroporation is seen, passing strong

pulses with reversed polarity is a good way to clean the surface of the electrodes—this loosens salt deposits. Once this is done, reset the electroporator to the settings used for embryos. Remove the gas bubbles formed on the electrodes applying puffs of saline with the small-mouthed Pasteur pipette.

8. Diffusion of the electroporation solution away from the site to which it is applied can be limited by gently depressing the target area (**step 3**) and applying the electrical pulses as quickly as possible. The latter is best achieved using a foot pedal, available for all electroporator models.

References

1. Stern CD (2005) The chick; a great model system becomes even greater. Dev Cell 8:9–17
2. Nakamura H, Katahira T, Sato T, Watanabe Y, Funahashi J (2004) Gain- and loss-of-function in chick embryos by electroporation. Mech Dev 121:1137–1143
3. Mende M, Christophorou NA, Streit A (2008) Specific and effective gene knock-down in early chick embryos using morpholinos but not pRFPRNAi vectors. Mech Dev 125:947–962
4. New DAT (1955) A new technique for the cultivation of the chick embryo invitro. J Embryol Exp Morph 3:320
5. Eyal-Giladi H, Kochav S (1976) From cleavage to primitive streak formation: a complementary normal table and a new look at the first stages of the development of the chick. I. General morphology. Dev Biol 49:321–337
6. Hamburger V, Hamilton HL (1951) A series of normal stages in the development of the chick embryo. J Morphol 88:49–92
7. Voiculescu O, Papanayotou C, Stern CD (2008) Spatially and temporally controlled electroporation of early chick embryos. Nat Protoc 3:419–426
8. Sheng G, dos Reis M, Stern CD (2003) Churchill, a zinc finger transcriptional activator, regulates the transition between gastrulation and neurulation. Cell 115:603–613

Chapter 10

Inhibition of Bacterial Growth by Peptide-Conjugated Morpholino Oligomers

Seth M. Daly, Carolyn R. Sturge, and David E. Greenberg

Abstract

Morpholino oligomers (MOs) are antisense molecules designed for sequence-specific binding of target mRNA. In bacteria, inhibition is hypothesized to occur by preventing translation initiation. Cell-penetrating peptides may be conjugated to the 5′- or 3′-termini of an MO to enhance cellular entry and therefore inhibition. Here we describe the three standard microbiological assays to assess in vitro antibacterial MO efficacy.

Key words Antisense, Morpholino oligomers (MOs), Minimum inhibitory concentration (MIC), Minimum bactericidal concentration (MBC), Checkerboard synergy

1 Introduction

Assessing the susceptibility or resistance of bacteria to a particular antimicrobial treatment has been done for more than 100 years. In modern times, this practice has been standardized by the Clinical and Laboratory Standards Institute (CLSI) to allow clinical labs across world to produce accurate information for the health care community. These CLSI guidelines have defined the minimal inhibitory concentration (MIC) as the lowest concentration of an antimicrobial agent that prevents visible growth of a microorganism in an agar or broth dilution susceptibility test [1]. Clinically, these assays are used to assess whether a specific bacterium is resistant or sensitive to therapeutic antimicrobials. The MIC method described here is based upon the CLSI broth microdilution method.

MICs assays assess growth inhibition but are not themselves indicative of bacterial death. However, in addition to the MIC method, the CLSI has developed guidelines for testing the minimal

Hong M. Moulton and Jon D. Moulton (eds.), *Morpholino Oligomers: Methods and Protocols*, Methods in Molecular Biology, vol. 1565, DOI 10.1007/978-1-4939-6817-6_10, © Springer Science+Business Media LLC 2017

bactericidal concentration (MBC), which is defined as at least a 99.9% (≥3-log) reduction in bacterial burden [2]. Again, the MBC method described here is based upon the CLSI bactericidal method.

Increasingly, bacterial inhibition using multiple antimicrobials has been used as a way to determine the potential interactions (synergistic, additive, or antagonistic) of two agents. Reasons for assessing these interactions include the increasing incidence of multidrug-resistant bacteria. Although not adequately tested in clinical trials, using multiple therapeutics has several proposed benefits and drawbacks (reviewed in ref. [3]). Proposed benefits include: (a) broader coverage during empiric therapy, (b) synergistic or additive interactions between compounds, and (c) inhibiting multiple targets at once might decrease the emergence of resistance; while proposed drawbacks include: (a) increased toxicities, (b) potential for drug–drug interactions, and (c) increased costs [3]. The synergy assay described here is a modified MIC assay in which each antibacterial agent is diluted to generate every twofold dose combination. The resulting growth inhibition of their combinations is assessed to determine the interaction.

MOs are conjugated to cell-penetrating peptides to enhance cellular entry [4, 5]. We have found that the most efficient cell-penetrating peptide depends on the organism. For example, the most efficient peptide in *Burkholderia* to date is $(RFF)_3RXB$ while in *Acinetobacter* and *Pseudomonas* to date it is $(RXR)_4XB$ [6–8]. Review of the literature may provide insight for novel organisms, otherwise empiric testing will be required. However, even with cell-penetrating peptides MOs have trouble penetrating the cell membranes of *Pseudomonas aeruginosa*. To circumvent this entry defect we have found that sub-inhibitory concentrations of the cyclic portion of polymyxin B, polymyxin B nonapeptide (PMBN), enhance entry of MOs. For example in *P. aeruginosa* PAO1 the MIC of PMBN is >16 μg/mL and 2 μg/mL of PMBN can enhance MO MICs by four- to eight-fold [8].

When testing essential genes we have found that growth media can play a significant role. Standard CLSI guidelines for MIC testing are in cation-adjusted Mueller-Hinton II medium, but careful consideration should be given to experimental conditions most similar to the physiologic environment that would be encountered by the pathogen and antisense molecule in a therapeutic setting. For example, in *P. aeruginosa* the MIC in a minimal media was up to eightfold lower than in MHII [8]. It should also be noted that different media might be required based on the auxotrophic requirements of different organisms.

Finally, the procedures described below are specifically for aerobic, nonfastidious organisms that grow optimally at 37 °C. Alterations in culture conditions will be required for organisms that do not fit these parameters.

2 Materials

Morpholino oligomers are usually supplied as lyophilisates and can be stored at −20 °C. Stocks should be dissolved in ultrapure water (≥18 MΩ cm) and stored at 4 °C. While the MO itself is quite stable, the cell-penetrating peptides are much less stable. Therefore they should be kept as concentrated stocks, and all solutions should be kept refrigerated or on ice while setting up any assays.

2.1 MIC Assay

1. 1 mM morpholino oligomer (MO) stock.
2. 96-well plate spectrophotometer capable of reading optical density at 600 nm (*see* **Note 1**).
3. 96-well polystyrene plates (*see* **Note 2**).
4. Gas permeable plate sealing film (*see* **Note 3**).
5. Mueller-Hinton II (MHII) broth (cation-adjusted).
6. MHII agar (alternate agar may be substituted).
7. Phosphate Buffered Saline (PBS).

2.2 MBC Assay

MIC Assay materials.

2.3 Synergy Assay

1. MIC Assay materials.
2. Antibiotic stock solutions prepared as per manufacturer.

3 Methods

3.1 Minimum Inhibitory Concentration (MIC) Assay

The volumes described are sufficient for one 96-well plate, capable of testing six MOs in duplicate (experimental replicates).

1. Bacterial colonies should be freshly obtained from −80 °C laboratory stocks by streaking onto an appropriate agar plate and incubating overnight at 37 °C.
2. Inoculate 3 mL of MHII and incubate 16–18 h at 37 °C on a 220 rpm shaker (*see* **Note 4**). If the culture will not be used immediately, place on ice.
3. Dilute the overnight culture in PBS until the optical density at 600 nm (OD_{600}) is between 0.07 and 0.1. This is roughly equivalent to 1×10^8 CFU/mL (*see* **Note 5**).
4. Add 60 μL of the OD_{600} stock to 12 mL of MHII to attain 5×10^5 CFU/mL. Vortex vigorously.
5. Serially dilute and plate this stock to determine CFU/mL of the inoculum. (*see* **Note 6**)
6. Aliquot 442.8 μL of the 5×10^5 CFU/mL stock into individual tubes and add 7.2 μL of each 1 mM MO stock to be tested. Vortex.

This results in a final concentration of 16 μM, and is typically the highest preliminary concentration tested (*see* **Note 7**).

7. Add 200 μL of each MO stock from **step 6** in duplicate wells to row A of a 96-well plate (*see* Fig. 1a).
8. Aliquot 100 μL of the 5 × 10^5 CFU/mL to rows B-G and in wells H7-12 (positive control). Aliquot 100 μL of MHII to wells H1-6 (negative control).
9. Transfer 100 μL from row A into row B and mix by pipetting. Continue diluting to row G. This results in a twofold serial dilution from (16–0.25 μM).
10. Seal the plate with a gas-permeable membrane and incubate 18–20 h (*see* **Note 4**) at 37 °C and 220 rpm.
11. Remove the gas permeable membrane and measure OD_{600}. The MIC is defined as the lowest concentration of MO with an average well OD_{600} within three standard deviations of the negative control wells (i.e.,—if the negative control is 0.048 ± 0.005, then the MIC would be wells with $OD_{600} \leq 0.063$, *see* Fig. 1b).

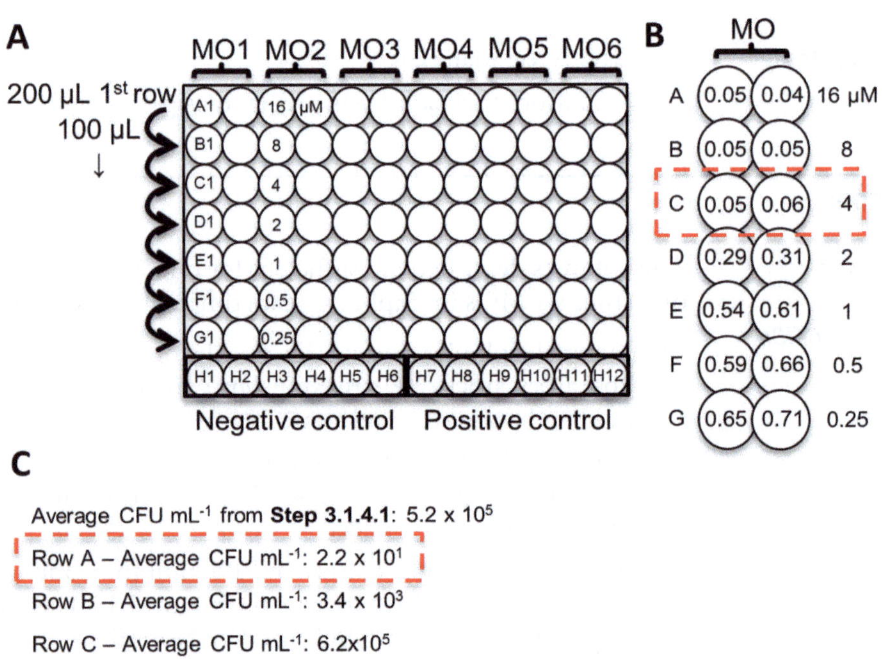

Fig. 1 Minimum Inhibitory (MIC) and Minimum Bactericidal Concentration (MBC) Assays. (**a**) MIC assay plate layout for testing six different morpholino oligomers (MOs) in duplicate from 16 μM to 0.25 μM. 200 μL of MO is added in duplicate to row A and diluted by serial twofold dilutions (100 μL transfer) to row G. Row H contains the positive and negative controls. (**b**) Mock OD_{600} sample data for a single MO. The dashed box indicates the MIC of 4 μM (lowest dilution with mean $OD_{600} \leq 0.063$). (**c**) Mock CFU/mL data for rows A-C in (b). The inoculum concentration was 5.2 × 10^5 CFU/mL. The MBC is defined as a 3-log reduction (≤5.2 × 10^2 CFU/mL), so the MBC is 16 μM

3.2 Minimum Bactericidal Concentration (MBC) Assay

1. Conduct an MIC assay as in Subheading 3.1.
2. Serially dilute and plate all wells with no growth ($OD_{600} \leq 0.063$ in this example, rows A–C in example Fig. 1b).
3. Incubate plates overnight at 37 °C.
4. Determine average CFU/mL for each MO concentration. The MBC is defined as a 99.9% (3-log) reduction in bacterial number from the starting inoculum **step 5** of Subheading 3.1 (*see* Fig. 1c).

3.3 Synergy Assay

The synergy assay is basically two MIC assays performed perpendicularly on the same plate. Unlike the MIC assay, the bacteria are added after serial dilutions of the MO and antibiotic for ease of calculations. The MIC values for the MO and antibiotic of interest should be determined beforehand. The example volumes below are for one assay and mock MICs of 8 μM for the MO and 2 μg/mL for the antibiotic (10 mg/mL stock). The antibiotic is diluted first and added at 16 times the MIC (32 μg/mL) and the MO is diluted second and added at 8 times the MIC (64 μM). This is done so that well A1 is twofold higher than the expected MIC of each agent. This assures the MIC is validated on the plate, as it is common for the MIC to be variable by twofold [1].

1. Add 50 μL of MHII to every well of a 96-well plate.
2. Prepare bacterial culture as in **steps 1–3** of Subheading 3.1.
3. Add 120 μL of OD_{600} stock to 12 mL of MHII to attain 1×10^6 CFU/mL. Vortex vigorously and place on ice.
4. Serially dilute and plate to determine CFU/mL as in **step 5** of Subheading 3.1 and **Note 6**.
5. Prepare antibiotic and MO stocks. (*see* **Notes** 7 and **8**).
6. Antibiotic will be diluted first, the desired final starting concentration is 4 μg/mL, which is twofold higher than the expected MIC of 2 μg/mL. Therefore a 16-fold concentration of the MIC is prepared (32 μg/mL). Add 2.4 μL of the 10 mg/mL antibiotic stock to 747.6 μL of MHII. Vortex.
7. MO will be diluted second, the desired final starting concentration is 16 μM, which is twofold higher than the expected MIC of 8 μM. Therefore an eightfold concentration of the MIC is prepared (64 μM). Add 28.8 μL of the 1 mM MO stock to 421.2 μL of MHII. Vortex.
8. Add 50 μL of antibiotic stock to wells A1-A11 (*see* Fig. 2a).
9. Perform twofold dilutions by adding 50 μL of A1-11 to B1-11, mix. Repeat to row G1-11, skipping row H.
10. Add 50 μL of MO stock to wells of column 1 (*see* Fig. 2a).
11. Perform twofold dilutions by adding 50 μL to column 2, mix. Repeat to column 10, skipping columns 11 and 12.

A

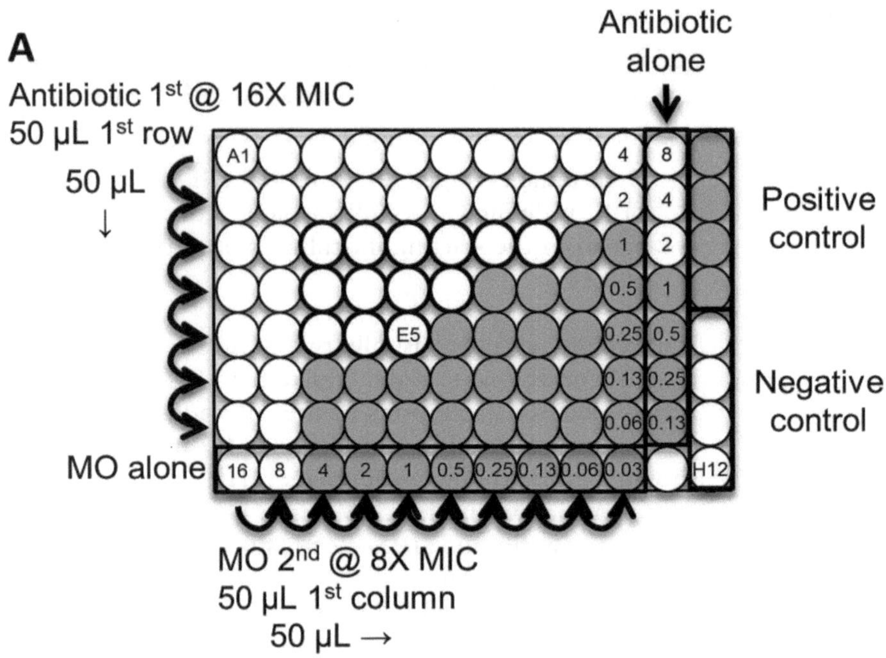

B

$$FIC\ Index = \left(\frac{MIC\ of\ MO\ in\ Combination}{MIC\ of\ MO\ alone}\right) + \left(\frac{MIC\ of\ Antibiotic\ in\ combination}{MIC\ of\ Antibiotic\ alone}\right)$$

$$FIC\ Index = \left(\frac{1}{8}\right) + \left(\frac{0.25}{2}\right)$$

$$FIC\ Index = 0.25$$

Fig. 2 Synergy Assays. (**a**) Synergy assay plate layout for testing a morpholino oligomer (MO) in combination with an antibiotic. 50 μL of antibiotic at 16 times the MIC is added to wells A1-11 and then serially diluted two-fold (50 μL transferred) to row G. 50 μL of MO at eight times the MIC is added to wells column 1 and then serially diluted twofold (50 μL transferred) to column 10. 50 μL of a 1×10^6 CFU/mL stock is added to all but the negative control wells which receive an additional 50 μL of MHII. After plating the bacteria is at 5×10^5 CFU/mL and well A1 contains MO at 16 μM and antibiotic at 4 μg/mL, while well G10 contains MO at 0.03125 μM and antibiotic at 0.0625 μg/mL. Note that the antibiotic only column is at twofold higher concentration than the other antibiotic-containing wells. Shaded wells represent mock growth wells. Dark outline wells indicate wells in which inhibition is below the MIC alone values. (**b**) Well E5 is the best interaction well. Demonstration of calculating the FIC Index for well E5

12. Add 50 μL of 1×10^6 CFU/mL bacterial stock to all wells except E12-H12. Add 50 μL of MHII to E12-H12 (negative controls). *See* Fig. 2a and **Note 9**.
13. Seal the plate with a gas-permeable membrane and incubate 18–20 h (*see* **Note 4**) at 37 °C and 220 rpm.
14. Remove the gas permeable membrane and measure OD_{600} as in **step 11** of Subheading 3.1.
15. Assess synergy by locating the lowest dilution combination well in which there is no growth ($OD_{600} \leq 0.063$ in this example, *see* Fig. 2a, b). Calculate the fractional inhibitory concentration index (FIC index) using the equation in Fig. 2b (*see* **Note 10**):

4 Notes

1. A spectrophotometer is not required and CLSI recommends visual comparisons using the unaided eye [1]. In the authors' experience a 96-well plate reader is simply more efficient for reading multiple plates quickly and accurately.
2. Sterile untreated round-bottom 96-well plates provide the most optimal growth conditions. However, in the authors' experience flat-bottom and tissue culture plates do not significantly alter MIC values. Due to the charge of tissue culture plates, novel cell-penetrating peptide-conjugated MOs should be compared in both plate types. This is due to the theoretical potential for charge-mediated interactions between MO and tissue culture treated plate.
3. The gas permeable membrane prevents evaporation in the wells. An 18 h incubation in a humid (water pan) incubator does not require the membrane, however it is recommended for longer incubations or when performed in nonhumid incubators.
4. These incubation times are for bacteria that reach late logarithmic or early stationary phase by 16–20 h. Incubation times should be increased for slower growing bacteria to the time required for a control culture to reach this growth stage.
5. As a note, the commercially available 0.5 McFarland standard may also be used as its optical density is also roughly equivalent to 1×10^8 CFU/mL.
6. The ideal inoculum is 5×10^5 CFU/mL and is based off the OD_{600} measurement in **step 3** of Subheading 3.1. Each organism's growth rate is different, so if the inoculum is not between 1×10^5 CFU/mL and 1×10^6 CFU/mL an adjustment is required. Divide the CFU/mL obtained in **step 5** of Subheading 3.1 by the OD_{600} in **step 3** of Subheading 3.1 to get CFU mL^{-1} OD^{-1}. This number can then be used to estimate CFU/mL more accurately. e.g.,—The OD_{600} was 0.08 and yielded 3×10^6 CFU/mL. **Step 4** of Subheading 3.1 is a 1:200 dilution, so the stock was 6×10^8 CFU/mL yielding 7.5×10^9 CFU mL^{-1} OD^{-1}. For this specific organism the OD_{600} can be multiplied by 7.5×10^9 CFU mL^{-1} OD^{-1} to determine the CFU/mL. This value can be used to dilute to the desired 5×10^5 CFU/mL.
7. If there is no inhibition, increase the starting concentration of MO or antibiotic. However, do not add more than 10% of drug to the media (i.e.,—don't add more than 45 μL of 1 mM to 405 μL of MHII). If required by volume constraint, increase the MO or antibiotic stock concentration utilized (e.g., 10 mM).

8. Antibiotics are added to row A in the synergy assay because they are generally more soluble and less costly than MO. However, the compounds can be loaded in alternate orientation. Remember that to reach the correct final concentration the first compound to be added should be prepared at eight times the desired starting concentration (2× the expected MIC in the example) and the second at four times the desired starting concentration. Also remember that the first compound to be diluted will be at 2× the concentration in the "alone" column (column 11 in the example).
9. Refer to the layout in Fig. 2a. After **step 10** of Subheading 3.3, all wells contain 5×10^5 CFU/mL of bacteria, well A1 contains 16 μM MO and 4 μg/mL antibiotic while G10 contains 0.03125 μM MO and 0.0625 μg/mL antibiotic with every combination in the remaining wells. Column 11 and row H are antibiotic and MO only, respectively, to confirm MIC values. The antibiotic only column is at a twofold higher concentration due to one less dilution step (The MO dilution series skips this column). Column 12 contains positive (A12-D12) and negative (E12-H12) growth controls.
10. There are many interpretations of FIC indices in the literature [9]. We suggest a synergistic interaction when the FIC index is ≤0.5, an additive interaction when FIC index is between 0.5 and 1.0, and an antagonistic interaction when the FIC index is ≥1.0.

References

1. CLSI (2015) Methods for dilution antimicrobial susceptibility tests for bacteria that grow aerobically; approved standard-tenth edition. Clinical and Laboratory Standards Institute, Wayne, PA
2. CLSI (1999) Methods for determining bactericidal activity of antimicrobial agents; approved guideline. Clinical and Laboratory Standards Institute, Wayne, PA
3. Moellering RC Jr (1983) Rationale for use of antimicrobial combinations. Am J Med 75:4–8
4. Wu RP, Youngblood DS, Hassinger JN, Lovejoy CE, Nelson MH, Iversen PL et al (2007) Cell-penetrating peptides as transporters for morpholino oligomers: effects of amino acid composition on intracellular delivery and cytotoxicity. Nucleic Acids Res 35:5182–5191
5. Good L, Awasthi SK, Dryselius R, Larsson O, Nielsen PE (2001) Bactericidal antisense effects of peptide-PNA conjugates. Nat Biotechnol 19:360–364
6. Geller BL, Marshall-Batty K, Schnell FJ, McKnight MM, Iversen PL, Greenberg DE (2013) Gene-silencing antisense oligomers inhibit acinetobacter growth in vitro and in vivo. J Infect Dis 208:1553–1560
7. Greenberg DE, Marshall-Batty KR, Brinster LR, Zarember KA, Shaw PA, Mellbye BL et al (2010) Antisense phosphorodiamidate morpholino oligomers targeted to an essential gene inhibit Burkholderia cepacia complex. J Infect Dis 201:1822–1830
8. Howard JJ, Sturge CR, Daly SM, Gill M, Labandeira-Rey M, Harbour L et al (2016) Inhibition of *Pseudomonas aeruginosa* by peptide-conjugated phosphorodiamidate morpholino oligomers. ASM Microbe, Boston, MA
9. Elion GB, Singer S, Hitchings GH (1954) Antagonists of nucleic acid derivatives. VIII. Synergism in combinations of biochemically related antimetabolites. J Biol Chem 208: 477–488

Chapter 11

Use of Translation Blocking Morpholinos for Gene Knockdown in *Giardia lamblia*

Jana Krtková and Alexander R. Paredez

Abstract

Giardia lamblia, a major parasite, is an emerging model organism due to its compact genomic arrangement and composition. The most popular reverse genetic technique, RNAi, is ineffective in *Giardia*. In contrast, protein depletion by translation blocking morpholinos is suitable for most gene targets and provides up to 80% depletion of the target protein. The method is fast, reliable, and specific. After antisense morpholino oligomer delivery into *Giardia* trophozoites by electroporation, the cells can be used for many subsequent analyses 8–48 h after treatment. In this chapter, suitable gene tags, plasmids, and techniques necessary for proper morpholino targeting are described.

Key words Morpholino, Knockdown, *Giardia*, Integration, Quantitative Western blotting

1 Introduction

Giardia lamblia (syn. *G intestinalis* and *G. duodenalis*), a parasite from the Excavates supergroup, is an emerging model organism for cell biology, and is currently the most divergent eukaryote that can be manipulated with molecular techniques [1]. *Giardia* has a minimalistic genome composed of only ~6000 genes; the extent to which this can be attributed to its early origins or parasitic reductionism is not clear [2]. In either case, studies in *Giardia* are informative for basic research because the variances in its biology offer an alternative perspective on biological processes. Ultimately, we expect that these differences can be leveraged for the development of new therapeutic treatments. Up to 20% of cases are resistant to the frontline clinical treatments metronidazole and tinidazole [3]. However, serious side effects are associated with these treatments, resulting in a critical need for new therapeutics. To understand gene function or to determine whether a specific protein would make a suitable drug target, genetic disruption is a powerful method to gain inferences about essentiality and

Hong M. Moulton and Jon D. Moulton (eds.), *Morpholino Oligomers: Methods and Protocols*, Methods in Molecular Biology, vol. 1565, DOI 10.1007/978-1-4939-6817-6_11, © Springer Science+Business Media LLC 2017

function. However, gene knockouts remain challenging, since *Giardia* is a binucleate tetraploid. RNA interference, commonly used in higher eukaryotes, has not been successfully used in *Giardia*, despite of the presence of RNA-dependent RNA polymerase, Dicer and Argonaute, and other known components of the RNA interference machinery [4]. Long dsRNA was reported to cause potent and selective interference in endogenous genes (medium subunits of adaptins, the RNA-dependent RNA polymerase) as well as exogenous GFP [5]. However, other tested genes, such as highly expressed tubulin and cyst wall protein 2 were unaffected, suggesting that dsRNA antisense inhibition has limited use [5]. Carpenter and Cande first demonstrated that antisense translation-blocking morpholinos could be used in *Giardia* [6]. Morpholinos remain the fastest and most reliable method to target genes of interest for functional studies.

Translation blocking morpholinos are modified antisense oligomers, typically of 25 bp targeted to the start codon of the gene of interest. They block translation through steric interference with the translation initiation complex and thus prevent loading of the mRNA into ribosomes for translation. Morpholinos contain A, C, G, or T nucleic acid bases, but because the ribose ring is replaced by a morpholine ring and a nonionic phosphorodiamidate linkage replaces the anionic phosphodiester linkage [7], morpholino oligomers cannot be degraded by cellular nucleases. These properties make morpholinos highly stable and effective.

Before delving into experimental design, it is necessary to confirm that a suitable morpholino can be designed to target your gene of interest (*see* Subheading 3.1 for instructions). Once confirmed, a control morpholino must be selected. The standard negative control offered by Gene Tools LLC does not match the base composition of any gene in *Giardia*, and so it is our preferred control because we only need to keep a single control on hand. Other options such as scrambled sequences and five base pair mismatches are also available and in some cases may be better controls. For an extensive discussion of considerations relevant to making this decision, *see* [8]. After confirming that a morpholino can be designed to target your gene of interest it is necessary to consider how you will verify knockdown. Since morpholinos work through translation blocking and not RNA degradation, western blotting, rather than qRT PCR, is used to confirm target depletion. In most cases, specific antibodies are not available for *Giardia* proteins; therefore, we have provided extensive advice about epitope tagging genes at their endogenous locus in order to monitor depletion (*see* Subheading 2.6, 3.2 and **Notes 1–4**).

2 Materials

2.1 Equipment

1. GenePulserXcell, BioRad, Hercules, CA for electroporation of *Giardia* trophozoites.
2. Gel electrophoresis equipment for DNA, SDS-PAGE, and transfers.
3. Plate reader (optional for measuring protein concentrations).
4. Thermal cycler.
5. Gel Doc capable of imaging multiplex Western blots (e.g., Chemidoc-MP BioRad, Hercules, CA).

2.2 Supplies

1. Immobilon-FL (Millipore) low fluorescence PVDF membrane.
2. Falcon plastic screw-cap tubes Falcon 16 × 120 mm.
3. 0.4 cm cuvette for electroporation.
4. 8-well plates.

2.3 Biological Material

1. *Giardia lamblia*, strain WB-C6.
2. *E. coli* DH5α.

2.4 Buffers and Media

1. Media for Giardia culture: TYDK 1L, pH 7.1. 20 g Casein digest (N-Z-Case Plus, Sigma-Aldrich), 10 g Yeast extract (BD), 10 g Dextrose, 2 g NaCl, 0.6 g KH_2PO_4, 1.3 g K_2HPO_4 $3H_2O$, 2 g l-cysteine (Sigma-Aldrich), 0.2 g l-Ascorbic acid, 1 mL Ferric Ammonium Citrate (12 mg/mL, 1000×), 10 mL Bovine Bile (13 g/250 mL, filter sterilized and frozen; Sigma-Aldrich B3883), 100 mL Bovine Serum, 890 mL ddH_2O, adjust pH and filter sterilize media using 0.22 μm filter (Millipore), Puromycin stock solution: 10 mg/mL, G418 stock solution: 100 mg/mL. G418 (geneticin) is used for selection of the neo resistance gene, neomycin phosphotransferase.
2. Ethanol precipitation of DNA: 3 M sodium acetate, pH 5.2, 100% ethanol, 70% ethanol.
3. Protein isolation and Western blot:
   (a) HBS (10×) 1 L: 80 g NaCl, 3.7 g KCl, 2.57 g $Na_2HPO_4 \cdot 5H_2O$, 10 g glucose, 50 g Hepes, adjust pH to 7.0 with NaOH, filter sterilize (do not autoclave)
   (b) Lysis buffer: 50 mM Tris–HCl, pH 7.5, 150 mM NaCl, 7.5% glycerol, 0.25 mM $CaCl_2$, 0.25 mM ATP, 0.5 mM DTT, 0.5 mM PMSF, 0.1% Triton X-100, Halt protease inhibitors (Pierce).
   (c) DC lysis buffer: 62.5 mM Tris–HCl, 2% SDS, 1× Halt Protease inhibitor, 0.2 mM PMSF.

(d) 5× sample buffer: 0.225 M Tris–HCl, pH 6.8, 50% glycerol, 5% SDS, 0.05% bromophenol blue, 0.25 M DTT.

(e) TBS (10×) 1 L: 84 g NaCl, 24.2 g Tris–HCl in 1 L, pH 7.6 with HCl and autoclave.

(f) TBST: 1× TBS + 0.05% Tween-20.

(g) Blocking solution: 5% dry milk, 0.05% Tween-20 in TBS.

2.5 Antibodies

To quantify protein depletion you will need both primary and secondary antibodies suitable for detecting the target protein and a loading control such as tubulin. In the absence of custom antibodies to detect the target protein, refer to Subheading 2.6 and **Notes 1–4** for instructions on epitope tagging the target gene. Table 1 indicates primary antibodies for loading controls as well as epitope tag detection. Table 2 indicates secondary antibodies for detecting the listed primary antibodies; these have been extensively validated for use in *Giardia*.

Table 1
Useful primary antibodies

Primary antibody	Source/type	Reference/manufacturer	Dilution in blocking solution	Comments
Anti-HA (HA7)	Mouse monoclonal; IgG1	Sigma-Aldrich, Cat.# A2095	1:2500	Excellent for IFA and WB
Anti-VSVG, clone P5D4	Mouse monoclonal; IgG1	Sigma-Aldrich, Cat.# V5507	1:2500	Good for IFA and WB
Anti-HA (HA.C5)	Mouse monoclonal; IgG3	Abcam, Cat. # Ab18181	1:1000	Good for WB
Anti-c-Myc, clone 9E10	Mouse monoclonal; IgG1	Sigma-Aldrich Cat. # M4439	1:2500	Fair for IFA and WB
Anti-giActin 28 PB+1	Rabbit polyclonal	[11]	1:2500	For loading Ctrl
Antiacetylated tubulin clone 6-11B-1	Mouse monoclonal; IgG2b	Sigma-Aldrich, Cat.# T 6793	1:2500	For loading Ctrl
Anti-CWP1 647-conjugated	Mouse monoclonal, IgG1	Waterborne Inc.	1:200	For encysting cells

Table 2
Useful secondary antibodies

Secondary antibody	Specificity	Manufacturer	Dilution
Alexa 488	Isotype specific anti-mouseIgG2b	Life Technologies	1:200
Alexa 555	Isotype specific anti mouseIgG1	Life Technologies	1:200
Alexa 647	Anti-rabbit	Life Technologies	1:200
Alexa 488	Anti-rabbit	Life Technologies	1:200
Alexa 555	Anti-mouse H+L	Life Technologies	1:200
Alexa 488	Isotype specific anti mouseIgG3	Life Technologies	1:200
Horseradish peroxidase-linked	anti-mouse	Bio-Rad	1:7000
Horseradish peroxidase-linked	anti-rabbit	Bio-Rad	1:7000

2.6 Plasmid Suitable for Endogenously Tagging Your Gene of Interest

In the absence of a custom antibody, we recommend using a pKS vector carrying a versatile tag (in our order of preference: 3×HA, 3×VSVG, 3×Myc) and a resistance cassette (puromycin, neomycin, blasticidin) to epitope tag the target protein [9]. For epitope tagging, *see* **Notes 1–4**. Epitope tags are listed in Table 3.

2.7 A Custom Antisense Morpholino Designed Against the Target Gene and a Control Morpholino

See Subheading 3.1. GENE TOOLS, LLC, Oregon, will evaluate the suitability of your gene of interest for morpholino knockdown and if possible design a suitable morpholino oligo. An online oligo design website is available at https://oligodesign.gene-tools.com/request/.

3 Methods

3.1 Morpholino Design

1. A major limitation of morpholino use in *Giardia* is the lack of introns in most genes, which constrains the use of morpholinos to targeting the translation initiation site. Occasionally, no suitable morpholino can be designed due to intermolecular interactions such as morpholino dimers or hairpins; these will prevent the morpholino from interacting with their target (*see* Table 4). Morpholinos can be targeted from −22 to +25 bp of the target gene (for examples of successfully used morpholinos, *see* Table 5). We recommend that you use as little of the 5′ UTR as possible, since *Giardia* tends to have short and variable 5′ UTR lengths. The UTR length can be determined using 5′ RACE; however, we tend to submit requests that only include 10 bp of UTR and send a note requesting minimal use of the UTR.
2. To retrieve your sequence of interest, go to GiardiaDB.org and under "Tools," select "SequenceRetrieval." In our example we

Table 3
Epitope tags regularly used for immunofluorescence and immunoblotting and thus ideal for monitoring morpholino knockdown efficacy

Tag	Common name	AA sequence	Nucleotide sequence	MW [Da]
3×HA (BEST)	Hemagglutinin triple tag	EYPYDVPDYAEYP YDVPDYAEYPYDV PDYAGS	gaatatccttatgacgtcccagactacgcggagtacccatacgatgtacctg attatgctgagtacccatacgacgtaccagattacgctggatcc	3802
3×VSVG (GOOD)	*Vesicular stomatitis virus* (*VSV*-G) *tag*	YTDIEMNRLGKG SYTDIEMNRLGK GSYTDIEMNRLGK	TATACTGATATTGAAATGAATCGCTTAGGTAAA gggtccTACACCGACATCGAGATGAACCGCTTG GGCAAGggctctTATACAGACATAGAAATGAACAG ACTTGGAAAA	4271
3×Myc (FAIR)	Myc tag	ASMQKLISEEDLL RSEEQKLISEEDLLRSEE QKLISEEDLL	gcatcaatgcagaagctgatctcggaggaggatctgctgcgtagtgaagaa caaaaacttataagtgaagaagatttattgaggtcagaggaacagaaatta atttcagaagaggacctttta	4818

Table 4
A list of genes identified as putative interactors for GlRac

Gene	Accession number	Morpholino available?	Comments/function
Kinase	GL50803_17368	CAATCTGTTCGTAG CGATGCATAGA	Endogenous protein not detected
Alpha-annexin	GL50803_103373)	No good oligo	Membrane traffic, cytoskeleton
Sec31	GL50803_2562	Yes	Conventional membrane trafficking
ERP3	GL50803_15204	Yes	Conventional membrane trafficking
Synaptobrevin-like protein	GL50803_14469	Yes	Conventional membrane trafficking
Syntaxin-like protein	GL50803_7309	Yes, but self-complementary	Conventional membrane trafficking
Alpha-SNAP putative	GL50803_17224	Yes	Conventional membrane trafficking
Hypothetical protein	GL50803_6464	Yes, but self-complementary	Membrane trafficking, cytoskeletal protein, Bro1 Alix, Escrt III-binding site
Hypothetical protein	GL50803_15334	Yes, but less affinity (32% GC content)	Vacuolar sorting protein 39 domain 2, cytoskeletal protein
copineI	GL50803_8903	Yes	Membrane trafficking, cytoskeletal protein
LEK1	GL50803_86761	Yes	Cytoskeletal protein
Kinesin-1	GL50803_13825	No good oligo	Cytoskeletal protein
Hypothetical protein, t-SNARE superfamily	GL50803_10013	Yes	Membrane trafficking
Hypothetical protein, Sec1 superfamily	GL50803_16773	Yes	Membrane trafficking
Hypothetical protein	GL50803_7323	Yes	Membrane trafficking
Hypothetical protein	GL50803_2107	Yes	Membrane trafficking, CSK
Hypothetical protein	GL50803_40067	Yes, but self-complementary	Membrane trafficking, cytoskeletal protein

We submitted morpholino design requests in order to use morpholino availability as part of our prioritization scheme for follow-up studies

Table 5
Examples of morpholinos successfully used to knockdown gene expression in *Giardia*

Gene analyzed	Accession number	Morpholino sequence (5′ to 3′)	Protein reduced	Comments/function
Standard control	NA	CCTCTTACCTCAGTTACAATTTATA	NA	General Ctrl
GlActin	GL50803_40817	GGTTGTCGTCTGTCATTTTAC	Yes	5× mispaired CtrlGGTTcTCcTCTGT cATTTaAC [11]
Gl14-3-3	GL50803_6430	CGCGTAAATGCCTCGGCCATAGGTT	Yes	–
GlRac	GL50803_8496	TATCCTCATTTCCTGTACTAGTCAT	Yes	Rho GTPase [12]
Kinase	GL50803_8445	CTGCCAGCTTCTCCGCAATACCCAT	Yes	[13]
Kinase	GL50803_9421	CTAAATAGGTGTGCCATTTCGGTTC	Yes	[13]
Kinase	GL50803_11364	ACGCACGTCTGTGTCGCTCATGTCC	Yes	[13]
Kinase	GL50803_13215	TCTTCTGATACTCGTACAACATCAC	Yes	[13]
Kinase	GL50803_16034	TGTTGATGACATACGGACAAGCCAA	Yes	[13]
GlClathrin	GL50803_102108	CATTAATAGACAAGGCGCTGGCCAT	No	Membrane traffic

will use Actin GL50803_40817. To find the UTR plus the first 25 bp of the coding sequence, type the ORF# in the Gene ID box. Select "Begin at translation start" −10; and end at "Translation Start (ATG)" +25. Hit "Get Sequence." The output is GCAAAGTAAAATGACAGACGACAACCCTGCCATAG.

3. In order to submit the sequence to GeneTools, it is necessary to put parentheses around the start codon at bp 10: GCAAAGTAAA(ATG)ACAGACGACAACCCTGCCATAG.
4. If Gene Tools determines that there is no suitable sequence, determine what the actual UTR is for your gene using 5′ RACE, then you may use more of the UTR. Another, however untested option, is to try the morpholino in another *Giardia* assemblage (GS); the coding region may be different enough to allow morpholinos to be designed in GS when they cannot be designed for a WB strain.

3.2 Antibody or Epitope Tag for Detecting Your Protein of Interest

If you have an antibody that recognizes your protein of interest, you can proceed to Subheading 3.6. If not, you will need to endogenously tag your target gene. *See* Figs. 1, 2, 3, and 4 and **Notes 1–4** for guidance on this process.

3.3 Construct Linearization and Ethanol Precipitation

1. After cloning your gene of interest into the appropriate vector, the construct needs to be linearized in order to integrate into the genome. Complete linearization of the circular construct is crucial. *See* **Note 5**.
2. For linearization set up a 350 μL digestion in a microfuge tube (convenient for subsequent ethanol precipitation).

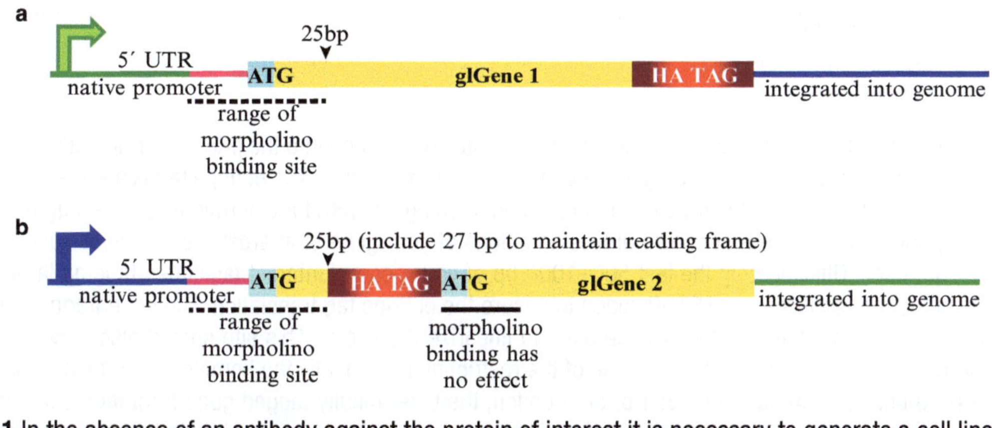

Fig. 1 In the absence of an antibody against the protein of interest it is necessary to generate a cell line containing either a C- or N-terminally tagged copy of your gene of interest (**a** and **b**, respectively). Schematic sequences are color coded: *blue lines* and *yellow boxes* with *black letters* are for vector sequences; *green lines*, and *boxes* with *white letters* are for genomic DNA, *red boxes* with *white letters* for tags. The parts of native promoters that are targets for morpholinos are in *pink*

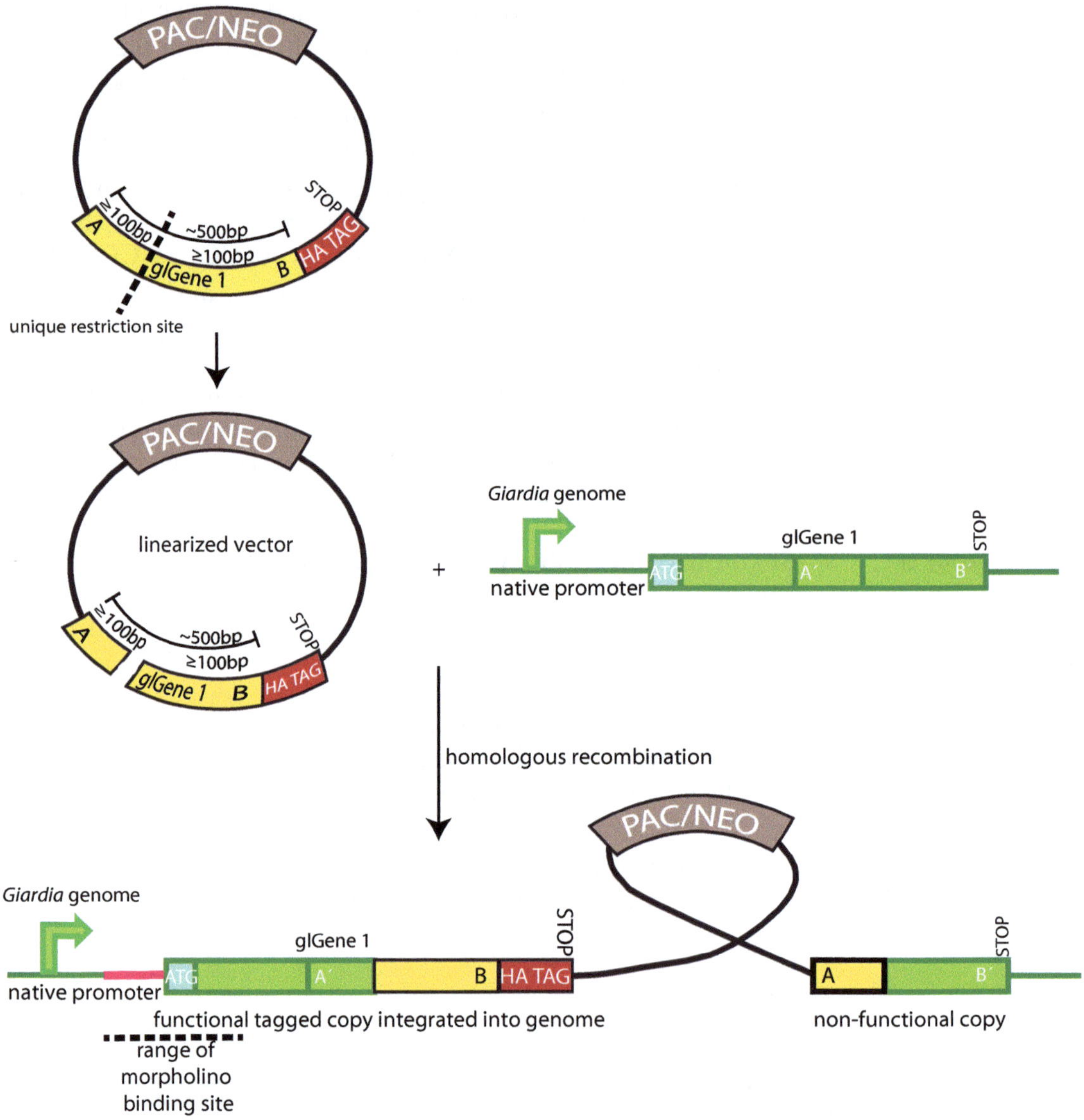

Fig. 2 The cloning of C-terminally tagged genes for morpholino targeting into the pKS vector and integration of pKS construct carrying C-terminally tagged gene into the *Giardia* genome. The epitope tag is necessary for the measurement of knockdown efficacy by immunoblotting. Cloning into pKS for C-terminal epitope tagging is the simplest option and thus the first choice unless homologs of your gene of interest are known to have issues with this approach. Utilizing only the last 500–1000 bp of your gene of interest facilitates tagging large proteins. Remember to omit the gene STOP codon to ensure the epitope tag translation. The gene fragment must contain a unique restriction site that will be used to linearize the vector. This site should also leave at least 90 bp of homologous sequence on either side of the restriction site; ideally the homologous regions would be of 500–1000 bp. By omitting a promoter or start codon, the C-terminally tagged gene fragment is integrated into the genome under the control of its endogenous promoter, resulting in one full length tagged copy and one truncated, nonfunctional copy void of a promoter and start codon. Schematic sequences are color coded: *blue lines* and *yellow boxes* with *black letters* are for vector sequences; *green lines*, and *boxes* with *white letters* are for genomic DNA, *red boxes* with *white letters* for tags. The parts of native promoters that are targets for morpholinos are in *pink*

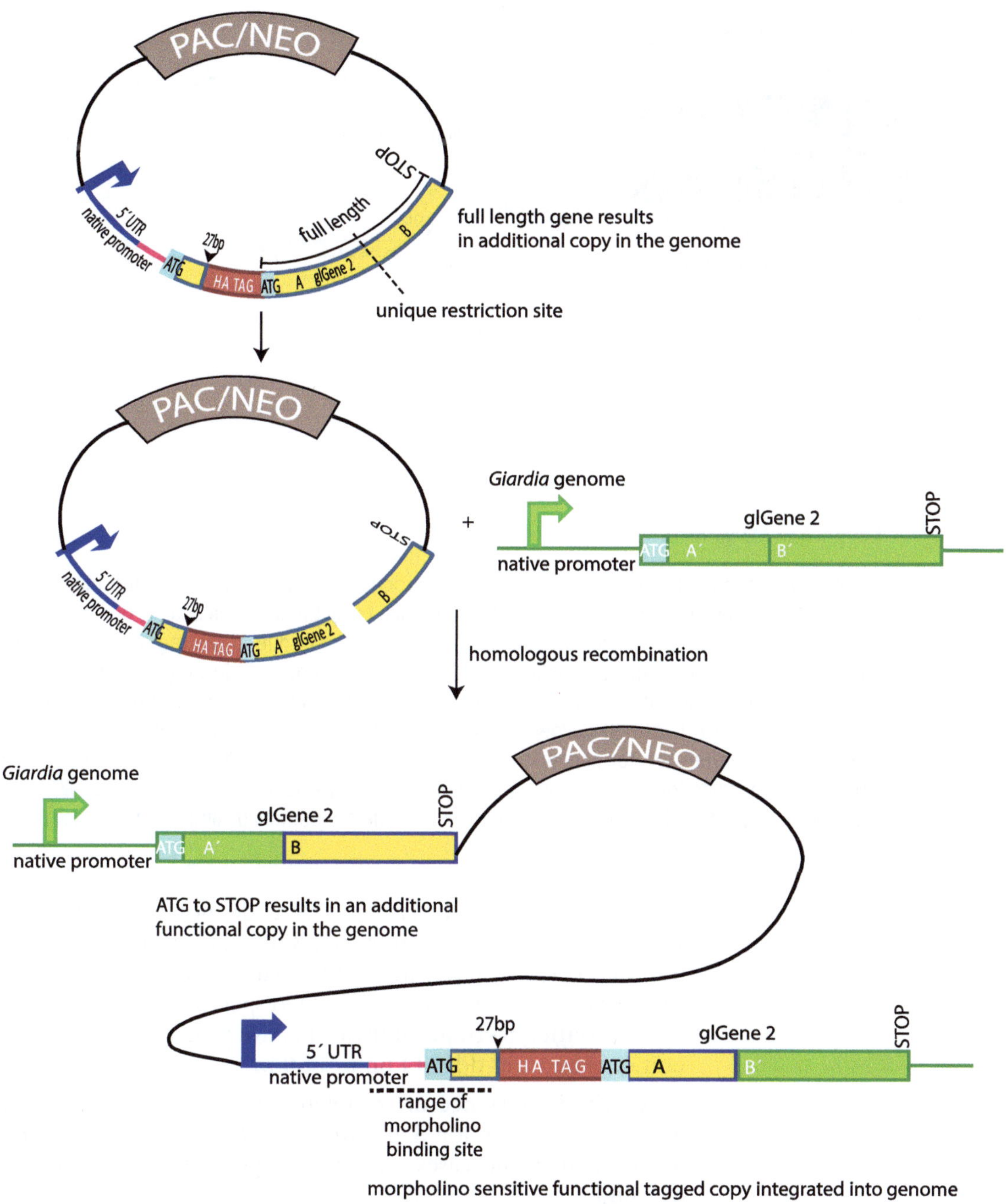

Fig. 3 The cloning of N-terminally tagged genes for morpholino targeting into the pKS vector and integration of pKS construct carrying N-terminally tagged gene into the *Giardia* genome (See [12]). For morpholino-sensitive N-terminal tagging, the 5 ′UTR region should contain the full length native promoter ~100–150 bp. It is crucial to include the first 27 bp of the gene before the tag. As a result, the morpholino oligomer can bind the ATG and surrounding region and block the translation of this transcript just as it will for the untagged copies. The tag is followed by a full length gene copy including ATG. It is also crucial to integrate the construct into the *Giardia* genome. The integration results in a duplication where one gene copy is under the native promoter, and additional tagged copy is under the promoter sequence included in your construct. Schematic sequences are color coded: *blue lines* and *yellow boxes* with *black letters* are for vector sequences; *green lines*, and *boxes* with *white letters* are for genomic DNA, *red boxes* with *white letters* for tags. The parts of native promoters that are targets for morpholinos are in *pink*

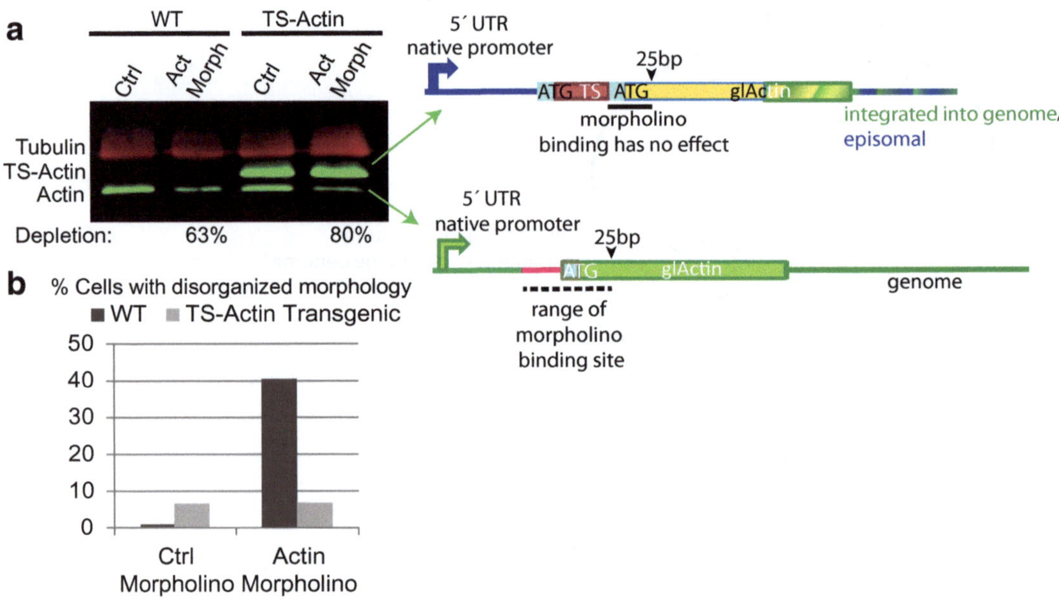

Fig. 4 Conventional, N-terminally tagged genes are morpholino insensitive and can rescue the morpholino disruptive effects. (**a**) Western blot showing actin (*green*) and tubulin (*red*). Endogenous actin is depleted while the N-terminally tagged TS-Actin is morpholino resistant. The resistance is due to the tag providing a new translation initiation sequence. Once the translation of TS-Actin fusion protein is in progress, morpholino-specific binding to the actin sequence has no effect. Ctrl = 5 bp mismatch in this example. Schematic sequences are color coded: *blue lines* and *yellow boxes* with *black letters* are for vector sequences; *green lines*, and *boxes* with *white letters* are for genomic DNA, *red boxes* with *white letters* for tags. The parts of native promoters that are targets for morpholinos are in *pink*. (**b**) Actin depletion results in 40% of WT cells having abnormal organization, the TS-Actin construct rescued this defect demonstrating that the construct is functional. Adapted from [10]

3. Digest 20–100 μg of plasmid DNA, do not use more than 150 μL of your plasmid prep in a 350 μL reaction (if concentration is below 150 ng/μL, it is probably too low to proceed). Mix plasmid DNA, 35 μL of the respective 10× restriction enzyme buffer, 5–10 μL of the respective restriction enzyme (1 unit/μg) and add ddH_2O up to 350 μL.
4. Given the importance of complete linearization, we typically incubate reactions for 3 h with NEB high fidelity enzymes; however, in some cases, complete digestion can take up to overnight.
5. Before proceeding, verify linearization of the plasmid by comparing uncut and cut plasmid with agarose gel electrophoresis.
6. For ethanol precipitation, measure the volume of the DNA sample as some should have been used for verifying linearization.
7. Add 1/10 volume of sodium acetate, pH 5.2 (35 μL in the case the DNA sample volume is 350 μL). Mix well.

8. Add 2.5–3 volumes of cold 100% ethanol (use 1 mL). Mix well.
9. Place on ice or at −20 °C for >20 min.
10. Centrifuge at maximum speed in a microfuge for 10–15 min.
11. Carefully decant supernatant to avoid losing the DNA pellet.
12. Add 1 mL 70% ethanol. Mix. Centrifuge for 5 min at maximum. Carefully decant supernatant and pipette out any remaining liquid carefully to avoid disrupting the pellet.
13. Air dry pellet in a biosafety cabinet to maintain sterility, usually about 5 min.
14. Resuspend the pellet in 50 μL of sterile TE, water, or elution buffer. Ideally, the DNA will be 150–500 ng/μL after ethanol precipitation. Lower concentrations will result in low transformation efficiency.

3.4 Transformation with Tagged Gene for Integration into the Giardia Genome

1. Transfect WB-C6 strain with 5–50 μg of completely linearized vector DNA. The maximum volume of DNA that can be used for transfection is 50 μL. Typically, we use half of the ethanol precipitated DNA.
2. One confluent tube of cells will be enough for up to three electroporations. Place cells on ice for 30′ to detach.
3. Add DNA to a 4 mm cuvette and ice while waiting for cells.
4. After the cells have detached ~30 min, spin down at 500 × *g* for 5 min, discard media, and replace with 1.0 mL of ice cold *Giardia* growth medium.
5. Transfer 300 μL of cells to each cuvette, mix gently by flicking and incubate on ice for 15 min. Immediately before electroporation, make sure to mix the cells again so that they are evenly distributed in the cuvette when the electric pulse is delivered.
6. Electroporate the trophozoites (375 V, 1000 μF, 750 Ω; Gene Pulser Xcell, BioRad, Hercules, CA) and chill for additional 10 min on ice.
7. Resuspend the cells in prewarmed media and incubate.
8. After overnight recovery, begin selection with the appropriate antibiotic. Select the transfected cells by increasing volume of the respective antibiotic every 2–3 days. For example, begin at ~20 μg/mL of puromycin or 200 μg/mL of G418 antibiotic. Final concentrations should be 38.5 μg/mL of puromycin or 385 μg/mL of G418, respectively. This corresponds to 50 μL of typical stock solutions, 10 mg/ml puromyocin and 100 mg/ml G418, in 13 mL.
9. After selection is complete, we maintain cultures with 38.5 μg/mL of puromycin or 385 μg/mL of G418, respectively, except during experiments. It is also advisable create a stock of the cell line for long-term storage at −80 °C.

3.5 Verification of the Integration

Integration of the linearized vector into the genome can be verified by PCR from genomic DNA of the transfected cell line using three primers:

1. Primer 1 should be targeted to a sequence that is unique to the vector such as the epitope tag. Primer 2 should be directed to the genomic DNA outside of the cloned sequence. Primer 3 should be directed against a portion of the coding region included in the vector.
2. With genomic DNA from WT and your transgenic cell line, Primer 1 and 2 can be used to verify integration in the experimental cell line; WT serves as a negative control.
3. Primer 1 and 3 serve as a positive control. This primer set will verify that the plasmid is present in the experimental cell line and should not produce a product when WT genomic DNA is used as a template. Plasmid DNA can be used to validate the primer set.
4. If a band is present for Primer 1 and 3, but not Primer 1 and 2, then resistant cell lines were generated due to incomplete plasmid linearization; transformation with a new plasmid digest should be performed. As an alternate strategy, western blotting can be used. *See* **Note 5** for specific circumstances.

3.6 Preparation of Morpholino Stock Solutions

1. Morpholinos are shipped in 100 or 300 nmol quantities as lyophilized powder. Dissolve the morpholino oligomers by adding 0.1 or 0.3 mL of sterile water, respectively, to achieve 1 mM stock solution that can be kept at RT or frozen.
2. Prior to use, heat the stock solution at 65 °C for 5 min to dissolve oligomer precipitates.

3.7 Electroporation of Morpholino Oligomers

1. Ice a confluent tube of cells for 30 min.
2. During the incubation add 30 μL of morpholino to each cuvette and place on ice to chill.
3. After the cells have detached ~30 min, spin down at 500 × *g* for 5 min and replace media with 1.0 mL of ice cold *Giardia* growth medium.
4. Transfer 300 μL of cells to each cuvette, gently mix and incubate on ice for 15 min. Immediately before electroporation, flick cuvettes to resuspend cells.
5. Electroporate using a GenePulser Xcell, (Bio-Rad) programmed to 375 V, 1000 μF, 750 Ω.
6. Return the cells to ice for 5–10 min and transfer to fresh prewarmed media in a screw top 16 × 120 mm tube.
7. Incubate cells at 37 °C to allow recovery. Typically the cells are used between 8 and 48 h after electroporation for analysis of protein expression or other applications.

3.8 Evaluation of Protein Levels in Control and Knockdown Cells

1. Chill culture tubes on ice for 30 min to detach cells, pellet the cells at 700 × *g*, 5 min and wash once in cold HBS.
2. If you plan to measure protein concentration (alternative later) for even loading, lyse cells in DC lysis buffer, boil for 3 min, and then freeze all but 20 μL of each sample. Use the 20 μL aliquot for measuring protein concentration. Freezing the main sample immediately after boiling protects it from protease degradation. Protein concentration can be measured with Bio-Rad's detergent compatible DC assay kit or equivalent (see manufacturer's instructions).
3. After concentration of each sample is determined, adjust the frozen main samples to the same concentration using 5× sample buffer. Typically, 1 mg/mL is used because 10–30 μL will provide 10–30 μg of extract, our recommended amount for western blotting. Proceed to **step 8**.
4. If protein degradation is a problem, or you want to skip measuring extract concentrations, begin with even numbers of cells.
5. After washing with HBS, resuspend the cells in 300 μL of lysis buffer.
6. Sonicate with a microtip using two cycles of 2 × 1 s at 30% power.
7. Clear the lysate by centrifugation at 10,000 × *g* for 10 min at 4 °C.
8. Boil in sample buffer and run 10–30 μL of samples on SDS PAGE.
9. In either case, blot using Immobilon-FL PVDF membrane (Millipore) following the manufacturer's directions.
10. Block the membrane in blocking solution for 1 h at RT or overnight at 4 °C, either way gently agitating on a horizontal shaker (make sure the proteins on the membrane are facing up and that the solution is evenly spread on the surface of the membrane).
11. Dilute the antibodies to their working concentration in blocking solution (*see* Tables 1 and 2). Pour off the blocking solution from the membrane and replace with diluted antibody solution.
12. Incubate the membrane in primary antibodies solution for 1 h at RT or overnight at 4 °C, gently rotate on a horizontal shaker. Primary antibody solution can be reused for several months if stored with 0.01% sodium azide.
13. Wash in TBST while agitating on horizontal shaker (2× quickly, then 15 min, then 3 × 5 min).

14. Incubate the membrane in secondary antibodies solution for 1 h at RT while agitating.
15. Wash in TBST while agitating (2× quickly, then 5, 10 and 15 min for total of 30 min washing).
16. Multiplex immunoblots can be imaged using a Chemidoc MP (Bio-Rad).
17. If protein levels were quantified, your loading should appear even. If you used the alternative method, it may be necessary to adjust the amount of the samples loaded, according to your loading control and rerun the blot for publication quality results.

4 Notes

1. It is important to endogenously tag the gene of interest in order to maintain endogenous protein levels [9]. Episomal plasmids are problematic because they have variable expression levels due to variance in copy number and integration into sites other than the endogenous locus may result in differential regulation due to differences in chromatin structure. Therefore, we recommend using a pKS vector to epitope tag your gene of interest with a 3×HA, 3×VSVG, or 3×Myc tag (in our order of preference, *see* also Table 3). These vectors (available on request) utilize puromycin, neomycin, or blasticidin resistance and are suitable for tagging up to three genes in a single cell line.
2. The position of the tag either on N- or C- terminus needs to be decided according to the protein's specific features (subcellular localization, presence of signaling peptide, function, domains present). Unfortunately, knowing whether a tag will interfere with the protein localization and/or function requires empirical testing. For example, a C-terminal tag is versatile for cytoplasmic proteins, because it usually does not interfere with localization. On the other hand, membrane proteins harboring domains essential for their association with membranes, such as prenylation CAAX domain, at their C-terminus, need to be tagged at the N-terminus.
3. The cloning of C- or N-terminally tagged genes into the pKS vector, as well as integration into *Giardia* genome, differs slightly (Figs. 1, 2, and 3). Tagging the gene of interest at the N-terminus introduces additional complications; therefore, your first choice should be to tag the C-terminus. Not only is this a simpler cloning strategy, C-terminal cloning only requires you to utilize the last 500–1000 bp of your gene of interest and this facilitates tagging large proteins. Remember to omit the gene STOP codon, so that the downstream epitope tag will be translated. The gene fragment must, however, contain a

unique restriction site that will be used to linearize your vector. The site within your gene of interest should have at least 90 bp of homologous sequence on either side of the restriction site; ideally the homologous regions are 500–1000 bp. By omitting a promoter or start codon, the C-terminally tagged gene fragment is integrated into the genome under the control of its endogenous promoter, resulting in one full-length tagged copy and one truncated, nonfunctional copy void of a promoter and start codon (*see* Fig. 1) [9].

4. For an N-terminally tagged gene, the 5′UTR region should contain the full length native promoter. The integration results in a duplication of one gene copy under the native promoter, and an additional tagged copy is under the promoter sequence included in construct (Fig. 2). An important complication is that N-terminally tagged genes are insensitive to morpholino-mediated block of translation because the N-terminal tag introduces a new start codon and surrounding sequence [10]. This results in N-terminally tagged genes being morpholino insensitive (*see* Fig. 4). For morpholino-sensitive N-terminal tagging, it is necessary to include the first 27 bp of the gene before the tag (*see* Fig. 3). As a result, the morpholino oligomer can bind the ATG and surrounding region, block the translation machinery assembly, and at the same time the knockdown can be measured by immunoblotting. The 5′UTR (promoter) followed by the first 27 bp can be PCR-amplified from *Giardia* genomic DNA as a single fragment and cloned before the tag into the pKS plasmid. The tag is followed by a full length gene copy including ATG (Fig. 3). While a full length coding region is not technically required to generate an N-terminally tagged gene, we prefer to include the entire length because not doing so would generate a second truncated copy of the gene and this could potentially generate unintended dominant negative phenotypes and/or be lethal.
5. In the case of a C-terminal construct without a promoter, incomplete linearization could yield transformants that do not express the tagged copy of the gene. Successful integration can be verified by western blotting. For N-terminal tagged constructs, you will observe a tagged protein in either case, so the only way to verify integration is to use PCR (*see* Subheading 3.5).

Acknowledgements

This work was supported by NIH 1R01AI110708-01A1 grant to A.P., J.K. was further supported by the project NPUI (LO1417) of the Czech Ministry of Education, Youth and Sports during chapter preparation. We thank Kelly Hennessey for editing this protocol.

References

1. Dawson SC, Paredez AR (2013) Alternative cytoskeletal landscapes: cytoskeletal novelty and evolution in basal excavate protists. Curr Opin Cell Biol 25(1):134–141
2. Morrison HG et al (2007) Genomic minimalism in the early diverging intestinal parasite *Giardia lamblia*. Science 317(5846):1921–1926
3. Lalle M (2010) Giardiasis in the post genomic era: treatment, drug resistance and novel therapeutic perspectives. Infect Disord Drug Targets 10(4):283–294
4. Prucca CG et al (2008) Antigenic variation in *Giardia lamblia* is regulated by RNA interference. Nature 456(7223):750–754
5. Rivero MR, Kulakova L, Touz MC (2010) Long double-stranded RNA produces specific gene downregulation in *Giardia lamblia*. J Parasitol 96(4):815–819
6. Carpenter ML, Cande WZ (2009) Using morpholinos for gene knockdown in *Giardia intestinalis*. Eukaryot Cell 8(6):916–919
7. Moulton, J.D. and Y.L. Yan, Using Morpholinos to control gene expression. Curr Protoc Mol Biol , 2008Chapter 26: p. Unit 26.8
8. Eisen JS, Smith JC (2008) Controlling morpholino experiments: don't stop making antisense. Development 135(10):1735–1743
9. Gourguechon S, Cande WZ (2011) Rapid tagging and integration of genes in *Giardia intestinalis*. Eukaryot Cell 10(1):142–145
10. Paredez AR et al (2014) Identification of obscure yet conserved actin-associated proteins in *Giardia lamblia*. Eukaryot Cell 13(6):776–784
11. Paredez, A.R., et al., (2011) An actin cytoskeleton with evolutionarily conserved functions in the absence of canonical actin-binding proteins. Proc Natl Acad Sci U S A 108(15):6151–6156
12. Krtkova, J., et al., Rac Regulates Giardia lamblia Encystation by Coordinating Cyst Wall Protein Trafficking and Secretion. MBio, 2016. 7(4)
13. Hennessey, K.M., et al., Identification and Validation of Small-Gatekeeper Kinases as Drug Ta rgets in Giardia lamblia. PLoS Negl Trop Dis, 2016. 10(11): p. e0005107

Chapter 12

Regulation of Isoform Expression by Blocking Polyadenylation Signal Sequences with Morpholinos

Qiuming Gong and Zhengfeng Zhou

Abstract

Alternative polyadenylation is increasingly being recognized as an important layer of gene regulation. Antisense-mediated modulation of alternative polyadenylation represents an attractive strategy for the regulation of gene expression as well as potential therapeutic applications. In this chapter, we describe methods to upregulate the functional Kv11.1 isoform expression by blocking intronic polyadenylation signal sequences with antisense morpholinos.

Key words Alternative polyadenylation, Morpholino, *KCNH2*, *hERG*, Potassium channel, Long QT syndrome

1 Introduction

Recent high-throughput sequencing studies reveal that 60–70% of human genes undergo alternative polyadenylation, leading to the generation of alternative mRNA transcripts with different coding sequences or variable 3′-untranslated regions (3′-UTRs) [1, 2]. Alternative polyadenylation is increasingly being recognized as an important layer of posttranscriptional gene regulation. Furthermore, aberrant alternative polyadenylation has been associated with a wide variety of human diseases [3–5]. Thus, there is a broad interest in modulating alternative polyadenylation for the regulation of isoform expression as well as potential therapeutic applications.

Antisense strategies have been extensively used in downregulation of gene expression and modulation of splicing events in pre-mRNA processing [6]. However, the use of antisense oligonucleotides that target polyadenylation sites to modulate alternative polyadenylation has only been reported in several cases [7–10]. We recently reported that inhibition of alternative intronic polyadenylation by antisense morpholinos can upregulate the functional Kv11.1 potassium channel expression [9].

Hong M. Moulton and Jon D. Moulton (eds.), *Morpholino Oligomers: Methods and Protocols*, Methods in Molecular Biology, vol. 1565, DOI 10.1007/978-1-4939-6817-6_12, © Springer Science+Business Media LLC 2017

The *KCNH2* (*hERG*) gene encodes the Kv11.1 potassium channel that conducts the rapidly activating delayed rectifier current (I_{Kr}) in the heart [11–14]. I_{Kr} is essential for cardiac action potential repolarization and plays an important role in both inherited and drug-induced forms of long QT syndrome [15, 16]. Alternative polyadenylation of KCNH2 pre-mRNA leads to the generation of two Kv11.1 C-terminal isoforms, the full-length, functional Kv11.1a isoform and the truncated, nonfunctional Kv11.1aUSO isoform [17]. As shown in Fig. 1, Kv11.1a is produced by splicing of intron 9 and polyadenylation at a distal poly(A) site in exon 15, whereas Kv11.1aUSO is generated by polyadenylation at a proximal poly(A) site in intron 9. The alternative polyadenylation of KCNH2 pre-mRNA is regulated in a tissue-specific manner. In the heart, two-thirds of KCNH2 pre-mRNA is polyadenylated at intron 9 to form the nonfunctional Kv11.1aUSO isoform, while in the brain, the levels of Kv11.1a and Kv11.1aUSO are similar [17, 18]. Because the formation of Kv11.1a and Kv11.1aUSO is mutually exclusive, the nonfunctional Kv11.1aUSO is generated at the expense of the functional Kv11.1a isoform. Thus, the relative expression of the C-terminal isoforms plays an important role in the regulation of Kv11.1 channel function. Dysregulation of alternative polyadenylation of KCNH2 pre-mRNA results in isoform switch from Kv11.1a to Kv11.1aUSO and causes long QT syndrome [5]. Given its key role in inherited and drug-induced forms of long QT syndrome, the Kv11.1 channel has become an important target for drug safety screening and therapeutic modulations.

Here we describe methods to upregulate the functional Kv11.1a isoform expression by inhibition of intronic polyadenylation with antisense morpholino oligos. We identify cis-acting elements required for the intron 9 poly(A) signal activity and design a morpholino oligo to target a 25-base sequence of the U/GU-rich downstream elements of *KCNH2* intron 9 poly(A) signal. Inhibition of intron 9 poly(A) signal by antisense morpholinos results in a decrease in the truncated Kv11.1aUSO isoform and a concomitant increase in the full-length, functional Kv11.1a isoform [9]. Our findings suggest that inhibition of *KCNH2* intron 9

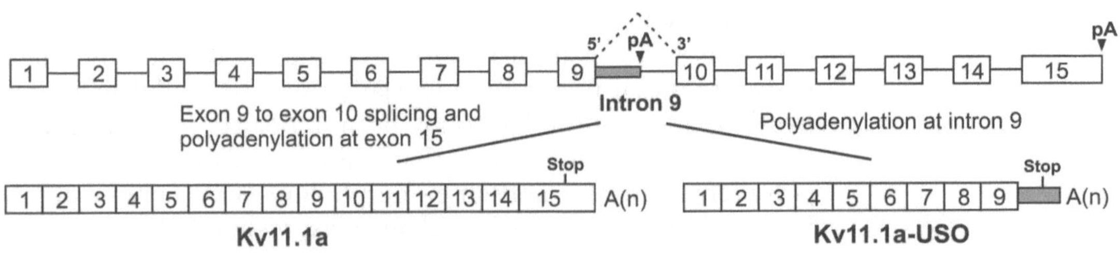

Fig. 1 Splicing of intron 9 leads to the formation of the full-length, functional Kv11.1a channel, whereas polyadenylation at intron 9 results in the production of the C-terminally truncated, nonfunctional Kv11.1aUSO isoform

polyadenylation by the antisense approach may represent a novel therapeutic strategy for long QT syndrome. In addition, antisense-mediated modulation of alternative polyadenylation could be a useful strategy to upregulate the expression of full-length isoforms of other genes that are regulated by alternative intronic polyadenylation.

2 Materials

2.1 Cell Line, Culture Medium, and Transfection

1. Flp-in 293 cells (Life Technologies).
2. Flp recombinase expression vector pOG44 (Life Technologies).
3. The short *KCNH2* gene construct that comprises all *KCNH2* exons, introns, and the poly(A) signal with shortening of the two longest introns, intron 2 (14,974 bp) and intron 5 (4437 bp) to 600 bp (Fig. 2). The short *KCNH2* genes are driven by a CMV promoter. The construct also contains the hygromycin resistance gene with an Flp recombination target site embedded in the 5′ coding region which is obtained from the pcDNA5/FRT vector (Life Technologies).
4. Culture medium: Dulbecco's Modified Eagle Medium (DMEM) supplemented with 10% fetal bovine serum, 100 units/ml penicillin and 100 μg/ml streptomycin.
5. Effectene Transfection Reagent (Qiagen).
6. Hygromycin.
7. Endo-Porter (Gene Tools).
8. Morpholino (Gene Tools).

2.2 RNA Extraction

1. RNeasy Mini Kit from Qiagen.
2. RLN solution: 50 mM Tris–HCl, pH 8.0, 140 mM NaCl, 1.5 mM $MgCl_2$, 0.5% (v/v) Noidet P-40. Just before use, add 1 mM DTT and precool to 4 °C.

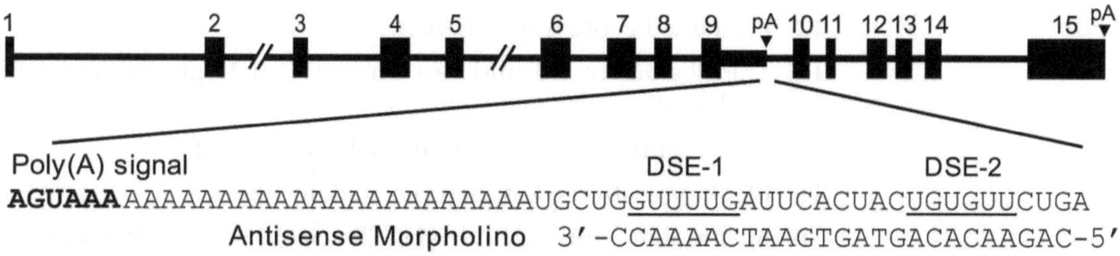

Fig. 2 The diagram of the short *KCNH2* gene construct and schematic representation of the *KCNH2* intron 9 poly(A) signal sequence and antisense morpholino. The length of the short *KCNH2* gene is 15,249 bp with shortening of intron 2 and intron 5 to 600 bp. The noncanonical poly(A) signal (AGUAAA) is shown in *bold* and downstream U/GU-rich elements DSE-1 and DSE-2 are *underlined*

2.3 RNase Protection Assay (RPA)

1. pCRII vector (Life Technologies).
2. Dual Promoter TA Cloning Kit (Life Technologies).
3. MAXIscriptIII in vitro Transcription Kit from Ambion for labeling probes.
4. Biotin 14-CTP.
5. RPA III Ribonuclease Protection Assay Kit from Ambion.
6. 40% Acrylamide gel (acryl:bis-acryl = 19:1).
7. Urea.
8. 10× TBE: 0.9 M Tris Base, 0.9 M Boric Acid and 20 mM EDTA.
9. Positively Charged Nylon membrane.
10. UV crosslinker.
11. BrightStar BioDetect Kit from Ambion for detection of biotin labeled probes.

2.4 SDS-Polyacrylamide Gel Electrophoresis and Western Blot

1. Bio-Rad Mini-Protein Electrophoresis system.
2. Bio-Rad Mini-Protein Transfer unit.
3. Cell lysis buffer: 50 mM Tris–HCl, pH 7.4, 150 mM NaCl, 10 mM EDTA and 1% Triton × 100 (v/v) with proteinase inhibitor cocktail (Roche).
4. Bio-Rad DC Protein Assay Kit.
5. 40% Acrylamide (acryl:bis-acryl = 29:1).
6. 4× running gel buffer: 1.5 M Tris–HCl, pH 8.8, 0.4% SDS.
7. 4× stacking gel buffer: 1 M Tris–HCl, pH 6.8, 0.4% SDS.
8. *N,N,N,N′*-Tetramethylethylenediamine (TEMED).
9. 10% (w/v) ammonium persulfate.
10. 10× Tris–glycine electrophoresis buffer: 0.25 M Tris Base, 1.92 M glycine, 1% (w/v) SDS.
11. 6× loading buffer: 375 mM Tris–HCl pH 6.8, 10% SDS, 30% glycerol, 9% 2-Mercaptoethanol, and 0.03% bromophenol blue.
12. Prestained protein size marker.
13. Transfer buffer: 20 mM Tris Base, 192 mM glycine and 10% methanol.
14. PBS-T buffer: PBS containing 0.2% Tween 20.
15. Blocking buffer: 5% nonfat dry milk in PBS-T.
16. Antibodies: Anti-Kv11.1 (H175, Santa Cruz), horseradish peroxidase (HRP)-conjugated anti-rabbit IgG.
17. ECL reagent.

3 Methods

3.1 Design of Morpholino Oligos

1. Based on our findings that the U/GU-rich downstream elements (DSE-1 and DSE-2) are required for the activation of the *KCNH2* intron 9 poly(A) signal, we design a morpholino oligo 5′-CAGAACACAGTAGTGAATCAAAACC-3′ to target a 25-base sequence of the U/GU-rich downstream elements of *KCNH2* intron 9 poly(A) signal (Fig. 2) (*see* **Note 1**).
2. Two control morpholino oligos are used. One is the invert morpholino oligo with the same sequence but in a reverse orientation, 5′-CCAAAACTAAGTGATGACACAAGAC-3′ and the other is a standard control oligo from Gene Tools, 5′-CCTCTTACCTC AGTTACAATTTATA-3′.
3. The morpholino oligos are made by Gene Tools and are delivered as salt-free lyophilized solids. Stock solutions (1.0 mM) are prepared by adding water and stored at 4 °C.

3.2 Cell Cultures, Transfection, and Morpholino Treatment

1. In order to study the alternative splicing and polyadenylation of the *KCNH2* gene, the short *KCNH2* gene construct is stably transfected into Flp-in 293 cells which contain a single FRT genomic locus, allowing the integration of a single copy of the *KCNH2* gene construct.
2. Flp-in 293 cells are grown in culture medium. When approaching ~80–90% confluency, cells are passed with 0.25% Trypsin.
3. To establish stable transfected cell lines, cells are seeded in a 60 mm dish 1 day prior to transfection. Transfection is performed at ~60% of confluency using Effectene Transfection Reagent with the short *KCNH2* gene construct (0.1 μg) and the Flp recombinase expression vector pOG44 (0.9 μg). After 2 days of transfection, the cells are split into two 100 mm culture dishes and culture medium is replaced with hygromycin B-containing medium at 100 μg/ml. The medium is changed every 3 days until foci can be identified. For polyclonal selection, the hygromycin B-resistant foci are pooled and expended. For monoclonal selection, 5–20 hygromycin B resistant foci are picked and expended (*see* **Note 2**).
4. For RNA extraction and Western blot analysis experiments, the 2–3 × 10^5 cells are seeded in 60 mm culture dishes 1 day prior to morpholino treatment. The Endo-Porter delivery system (Gene Tools) is used to deliver the antisense and control morpholino oligos into the cells (*see* **Note 3**).

3.3 RNA Extraction

1. Completely aspirate cell-culture medium.
2. Add 175 μl of ice cold RLN solution directly to cell culture dish.

3. Detach cells gently using a rubber policeman, and transfer to a microcentrifuge tube. Incubate on ice for 5 min.
4. Centrifuge lysate at 4 °C for 2 min at 300 × *g*. Transfer supernatant to a new tube and discard the pellet.
5. Add 600 μl Buffer RLT to the supernatant and mix.
6. Add 430 μl 100% ethanol to the lysate.
7. Continue the RNA isolation following the instruction of RNeasy Mini Kit provided by the manufacturer (*see* **Note 4**).
8. Measure the RNA concentration using a NanoDrop UV spectrophotometer.

3.4 RNase Protection Assay

1. The RNA probe spanning region of *KCNH2* exon 9 and 10 containing 309-nt is amplified by PCR and cloned into the pCRII vector using Dual Promoter TA cloning Kit (*see* **Note 5**).
2. The pCRII vector containing the Kv11.1 probe is linearized by restriction enzyme XbaI and then used as the template for antisense probe synthesis (*see* **Note 6**).
3. The antisense probe is transcribed and labeled nonisotopically with SP6 RNA polymerase in the presence of biotin-CTP using MAXIscriptIII in vitro Transcription Kit. The total length of the probe is 440 nt, containing 309 nt of *KCNH2* and sequences from the pCRII vector at both ends (*see* **Note 7**).
4. Following the instruction of RPA III ribonuclease Protection Assay Kit, 20 μg RNA is precipitated by adding 1/10 volume of 5 M NH_4OAc and 2.5 volume of 100% ethanol, and incubated on ice for 30 min (*see* **Note 8**).
5. The RNA is pelleted by centrifuging at 10,000 × *g* for 20 min at 4 °C, and washed with 70% ethanol. Let pellet air dry for 5 min.
6. Add 600 pg of probe to the pellets, and resuspend the pellets in 10 μl of Hybridization buffer III.
7. Heat samples to 90–95 °C for 3 min to denature the RNA.
8. Incubate tubes at 42 °C overnight for hybridization.
9. Add 150 μl of 1:200 diluted RNase solution and incubate 30 min at 37 °C.
10. Add 225 μl of RNase inactivation solution and incubate 15 min at −20 °C.
11. Centrifuge 15 min in microcentrifuge at maximum speed to pellet the precipitated products.
12. Remove all supernatant and wash the pellets with 70% ethanol.
13. Resuspend the pellets in 8 μl of loading buffer and incubate 3 min at 95 °C to denature the RNA.

14. Continue to follow the manufacturer's instruction to prepare 5% acrylamide and 8 M urea gel with 40% Acrylamide (acryl:bis-acryl = 19:1) and 10× TBE buffer using Bio-Rad Mini-Protein Electrophoresis system.
15. Load samples on the gel and run the gel in 1× TBE buffer.
16. Transfer the gel to a positively charged nylon membrane by electroblotting using the Bio-Rad Mini-Protein transfer unit, and UV crosslink nucleic acids to the membrane.
17. Detect the biotin labeled probe by following the instruction of the BrightStar BioDetect Kit (Ambion).

3.5 SDS-Polyacrylamide Gel Electrophoresis and Western Blot

1. Wash cells with PBS.
2. Add 100–200 μl of ice cold lysis buffer.
3. Scrape cells and transfer to a microcentrifuge tube.
4. Incubate on ice for 30 min.
5. Centrifuge at 10,000 × *g* for 20 min at 4 °C.
6. Collect the supernatant and measure the protein concentration with the Bio-Rad DC Protein assay kit.
7. Prepare 7.5% separation gel and 4% stacking gel with 4× running gel buffer and 4× stacking gel buffer, respectively, and 40% Acrylamide (acryl:bis-acry = 29:1), using the Bio-Rad Mini-Protein Electrophoresis system.
8. Prepare the cell lysates containing 30–50 μg protein by adding 6× loading buffer.
9. Denature the cell lysates for 5 min at 85 °C.
10. Load the samples on the gel and load a well with a prestained protein size marker.
11. Run the gel at 30 mA (two gels at 60 mA) in 1× Tris-glycine electrophoresis buffer.
12. Prepare the PVDF membrane by wetting it in 100% methanol for 5 s, transferring into distilled H_2O for 2 min, and then equilibrating in transfer buffer for 5 min.
13. Set up "transfer sandwich" by following the Bio-Rad Mini Transfer instructions.
14. Insert the sandwich into the transfer tank.
15. Run at 100 V for 1 h in transfer buffer.
16. Incubate the membrane blot in blocking buffer for 1 h at room temperature.
17. Incubate with primary antibody (Anti-Kv11.1, H175, Santa Cruz) in blocking buffer overnight at 4 °C (*see* **Note 9**).
18. Wash the membrane four times, 10 min each time at room temperature with PBS-T buffer.

19. Incubate the blot with secondary antibody (HRP-anti-rabbit-IgG) in blocking buffer for 1 h at room temperature.
20. Wash the membrane four times, 10 min each time at room temperature with PBS-T buffer.
21. Prepare the ECL reagent following the manufacturer's instructions.
22. Expose the membrane to a chemiluminescence detection film.

4 Notes

1. The morpholino oligo is 25-bases long as recommended by the manufacturer. It is important to follow the guidelines of having a high enough GC content (40–60%) and not having a stretch of four or more contiguous G.
2. Transfection condition can be optimized to obtain the highest transfection efficiency and low toxicity by using different transfection reagents, adjusting the cell density and varying the amount or ratio of transfection reagent and DNAs.
3. Delivering morpholino oligos by Endo-Porter should be optimized for each cell type. Endo-Porter concentration can be varied from 2 to 10 μM and the concentration of morpholino oligos can be varied from 2 to 15 μM. Cells can be assayed from 24 to 72 h after morpholino treatment.
4. We recommend to isolate cytoplasmic RNA with the RNeasy Mini Kit.
5. Expression of two Kv11.1 isoforms can be detected simultaneously in a given sample by RPA with one probe which protects a 309 nt fragment of the Kv11.1a isoform and a 210 nt fragment of the Kv11.1aUSO isoform. The relative abundance of the mRNA levels of the two isoforms can be quantified using ImageJ software and adjusted for the number of biotin-labeled cytidines in each protected fragment.
6. The orientation of the probe in the pCRII vector should be checked by sequencing to ensure that the antisense probe is synthesized.
7. Since the prematurely terminated transcription products in probes will cause background bands in RPA, we recommend performing denaturing polyacrylamide gel electrophoresis to purify the full-length probe.
8. The signal-to-noise ratio in the RPA should be optimized by varying the dilution of RNase, and the amount of probe and RNA used. For quantitative detection of mRNA, it is essential that the probe is present in molar excess over the target

mRNA. Yeast RNA can be used as a control for the complete digestion of the probes by RNase.

9. Different dilutions of antibodies can be tested for optimal condition. Incubation of primary antibody in Western blot can also be done for 1.5 h at room temperature. Because the H-175 anti-Kv11.1 antibody recognizes an epitope in the N-terminus of Kv11.1a and Kv11.1aUSO, the expression of both Kv11.1 isoforms can be detected in one blot. The relative protein levels of the two Kv11.1 isoforms can be quantified using ImageJ software.

Acknowledgments

This study was supported in part by NIH grant HL68854.

References

1. Shi Y (2012) Alternative polyadenylation: new insights from global analyses. RNA 18:2105–2117
2. Tian B, Manley JL (2013) Alternative cleavage and polyadenylation: the long and short of it. Trends Biochem Sci 38:312–320
3. Elkon R, Ugalde AP, Agami R (2013) Alternative cleavage and polyadenylation: extent, regulation and function. Nat Rev Genet 14:496–506
4. Hollerer I, Grund K, Hentze MW, Kulozik AE (2014) mRNA 3′end processing: A tale of the tail reaches the clinic. EMBO Mol Med 6:16–26
5. Gong Q, Stump MR, Deng V, Zhang L, Zhou Z (2014) Identification of Kv11.1 isoform switch as a novel pathogenic mechanism of long QT syndrome. Circ Cardiovasc Genet 76:26–32
6. Bennett CF, Swayze EE (2010) RNA targeting therapeutics: molecular mechanisms of antisense oligonucleotides as a therapeutic platform. Annu Rev Pharmacol Toxicol 50:259–293
7. Vickers TA, Wyatt JR, Burckin T, Bennett CF, Freier SM (2001) Fully modified 2′ MOE oligonucleotides redirect polyadenylation. Nucleic Acids Res 29:1293–1299
8. Vorlová S, Rocco G, Lefave CV, Jodelka FM, Hess K, Hastings ML, Henke E, Cartegni L (2011) Induction of antagonistic soluble decoy receptor tyrosine kinases by intronic polyA activation. Mol Cell 43:927–939
9. Gong Q, Stump MR, Zhou Z (2014) Upregulation of functional Kv11.1 isoform expression by inhibition of intronic polyadenylation with antisense morpholino oligonucleotides. J Mol Cell Cardiol 76:26–32
10. Raz V, Buijze H, Raz Y, Verwey N, Anvar SY, Aartsma-Rus A, van der Maarel SM (2014) A novel feed-forward loop between ARIH2 E3-ligase and PABPN1 regulates aging-associated muscle degeneration. Am J Pathol 184:1119–1131
11. Warmke JW, Ganetzky B (1994) A family of potassium channel genes related to eag in Drosophila and mammals. Proc Natl Acad Sci USA 91:3438–3442
12. Sanguinetti MC, Jiang C, Curran ME, Keating MT (1995) A mechanistic link between an inherited and an acquired cardiac arrhythmia: HERG encodes the IKr potassium channel. Cell 81:299–307
13. Trudeau MC, Warmke JW, Ganetzky B, Robertson GA (1995) HERG, a human inward rectifier in the voltage-gated potassium channel family. Science 269:92–95
14. Zhou Z, Gong Q, Ye B, Fan Z, Makielski JC, Robertson GA, January CT (1998) Properties of HERG channels stably expressed in HEK 293 cells studied at physiological temperature. Biophys J 74:230–241
15. Curran ME, Splawski I, Timothy KW, Vincent GM, Green ED, Keating MT (1995) A molecular basis for cardiac arrhythmia: HERG mutations cause long QT syndrome. Cell 80:795–803
16. Kannankeril P, Roden DM, Darbar D (2010) Drug-induced long QT syndrome. Pharmacol Rev 62:760–781

17. Gong Q, Stump MR, Dunn AR, Deng V, Zhou Z (2010) Alternative splicing and polyadenylation contribute to the generation of hERG1 C-terminal isoforms. J Biol Chem 285:32233–32241

18. Kupershmidt S, Snyders DJ, Raes A, Roden DM (1998) A K^+ channel splice variant common in human heart lacks a C-terminal domain required for expression of rapidly activating delayed rectifier current. J Biol Chem 273:27231–27235

Chapter 13

Targeting Functional Noncoding RNAs

Madzia P. Crossley and Torsten Krude

Abstract

Noncoding RNAs have essential biochemical functions in different areas of cellular metabolism, including protein synthesis, RNA splicing, protein secretion, and DNA replication. We have successfully used Morpholino antisense oligonucleotides for the functional inactivation of small noncoding RNAs required for DNA replication (Y RNAs in vertebrates and stem-bulge RNAs in nematodes). Here we discuss specific issues of targeting functional noncoding RNAs for inactivation by Morpholino antisense oligonucleotides. We present protocols for the design, preparation, and efficacy controls of Morpholino antisense oligonucleotides, as well as brief descriptions for their delivery into vertebrate and nematode embryos.

Key words Noncoding RNAs, Functional inhibition, Morpholino design, Noncoding RNA domains, Functional assay to test Morpholino efficacy and specificity, *Xenopus laevis*, *Danio rerio*, *Caenorhabditis elegans*, Morpholino delivery, Dominant-negative inhibition

1 Introduction

Noncoding RNAs are involved in a multitude of cellular processes, including epigenetic regulation; RNA processing; transcriptional, posttranscriptional, and translational control; the DNA damage response; chromosomal DNA replication; and many others (reviewed in ref. [1]). However, the function of many noncoding RNAs remains unknown [1]. Loss-of-function experiments are essential to establish the functions of noncoding RNAs, but these approaches are not always straightforward. In some cases, noncoding RNAs may be refractory to RNA interference [2]. Additionally, functionally redundant noncoding RNAs may be present in the organism of interest, making it difficult to achieve sufficient loss of function through knockdown or knockout approaches to observe clear phenotypes. Morpholino antisense oligonucleotides (Morpholinos) may serve as useful alternative tools to study noncoding RNA function. We have used Morpholinos to functionally inhibit noncoding Y RNAs and stem-bulge RNAs (sbRNAs) in vertebrates and nematodes, respectively, thereby establishing an

Hong M. Moulton and Jon D. Moulton (eds.), *Morpholino Oligomers: Methods and Protocols*, Methods in Molecular Biology, vol. 1565, DOI 10.1007/978-1-4939-6817-6_13, © Springer Science+Business Media LLC 2017

essential and conserved function for these noncoding RNAs in DNA replication, cell proliferation, and early embryonic development [3–5].

The approach to use Morpholinos to target noncoding RNAs differs from the conventional approach to target protein-coding mRNAs. In order to inhibit expression of candidate proteins, specific hybridization of Morpholinos with the complementary mRNA sequence is employed to block the assembly or the progression of a ribosome on the target mRNA. In this approach, the Morpholino therefore acts indirectly with regard to the target and any effect depends on the downstream pathway of protein synthesis. In contrast, many noncoding RNAs have a clear direct function, for instance, rRNAs acting as scaffolds for the assembly of a functional ribosome, U snRNAs as cofactors for site-specific mRNA splicing, 7SL RNAs as an essential component of the signal recognition particle complex for protein secretion, or Y and sbRNAs as essential DNA replication factors. The targeting of such functional noncoding RNAs by Morpholinos therefore acts directly, most likely by disrupting the local RNA secondary structures essential for function, or by interfering with the assembly of functional ribonucleoprotein complexes.

Many functional noncoding RNAs are relatively short, for instance, Y RNAs and stem-bulge RNAs are only about a hundred nucleotides long [5]. This reduces drastically the availability of target sites for the hybridization with a Morpholino, compared with much longer mRNAs. Furthermore, these functional noncoding RNAs often have complex secondary structures comprising of double-stranded RNA domains, which cover a large proportion of the entire RNA. These features limit the identification of suitable target sites amenable for the synthesis of antisense Morpholinos. In this chapter, we will therefore consider these limitations and discuss technical ways to resolve them in detail.

Noncoding RNAs are often comprised of families of functionally redundant individual RNA homologs. For instance, there are up to four different Y RNA genes in vertebrates, and tens of different sbRNAs in nematodes [5]. Each of these genes is transcribed from a single locus by RNA polymerase III, leading to relatively high copy numbers in the cell. The four different Y RNA gene lineages have arisen during vertebrate evolution through gene duplication events [6, 7]. Furthermore, sbRNA genes are organized as clusters of homolog types in nematode genomes, which also suggests an evolution through complex gene duplication events [8]. We have found that Y RNAs and sbRNAs are functionally redundant for the initiation of chromosomal DNA replication in vertebrates and nematodes [4, 9, 10]. The reason for this functional redundancy is the presence of a conserved structural motif that is essential and sufficient for their function in DNA replication [4, 10, 11]. It comprises of a short, inherently instable double-stranded RNA helix flanked either side by G–C clamps, which isolate it from adjacent single-stranded loops or bulges [11]. Importantly, additional structural domains present in these short RNAs are involved in other metabolic activities, so that it becomes important to consider the noncoding RNAs as composite entities with a modular organization [5].

An antisense approach to target these functionally redundant noncoding RNAs would seem complex as all individual RNA homologs in the organisms would need to be inactivated in order to obtain a knock-down phenotype. Indeed, this is precisely the case when individual Y RNAs are inactivated through biochemical degradation in human cell extracts by DNA antisense-mediated RNAseH digestion [9]. However, we found that this is not true for Morpholino-mediated inactivation of functionally redundant Y or sbRNAs [3, 4]. Targeting one active Y RNA in vertebrate cells (or one sbRNA in *C. elegans*) resulted in a dominant phenotype despite the presence of functionally redundant RNAs that were not targeted by the Morpholino used [3, 4]. In control experiments, an addition of functionally redundant nontargeted RNAs to the inactive Morpholino–RNA complexes did not rescue the specific inhibition of Y or sbRNA function in vitro [3, 4]. Therefore, Morpholinos target functionally redundant RNA families in a dominant-negative manner. It is very likely that this dominant-negative effect arises because Morpholino–noncoding RNA complexes are functionally inactive, but can still interact via additional domains with crucial proteins or nucleic acids to form an inactive complex that poisons the metabolic pathway [3]. This dominant-negative effect provides us now with a powerful tool to study the biological function of these noncoding RNAs by antisense Morpholinos.

In this chapter, we provide methods for the design of Morpholinos to target functional noncoding RNAs, for determination of the efficacy and specificity of these Morpholinos, for noncoding RNA-targeted Morpholino delivery in *Xenopus laevis*, *Danio rerio*, and *Caenorhabditis elegans*, and for interpretation of the resulting phenotypes.

2 Materials

All aqueous solutions are made up in ultrapure water.

2.1 Morpholino Resuspension and Storage

1. Resuspend Morpholinos according to the manufacturer's instructions—for Gene Tools, LLC, this recommendation is 1 mM stock solutions in ultrapure water.
2. Aliquot Morpholino stock solutions and store at −20 °C.

2.2 Preparation of Morpholino Working Solutions

1. Heat block set to 65 °C.
2. Bench top centrifuge.

2.3 Delivery In Vivo by Microinjection

1. Microinjection facilities suitable for target organism. The details for maintaining populations of the organism (we have used *Xenopus laevis*, *Danio rerio*, and *Caenorhabditis elegans*) and the manipulation of specialist instrumentation are highly diverse and beyond the scope of this chapter. They should therefore be sourced and be available locally.

3 Methods

3.1 Design of Morpholinos for the Functional Inhibition of Noncoding RNAs

It is important to carefully consider the Morpholino design when targeting noncoding RNAs. Considerations may vary between different noncoding RNAs. Below, we provide an approach that is generally applicable to different classes of functional noncoding RNAs.

1. Obtain the nucleotide sequences of the target noncoding RNAs.
2. Use the nucleotide sequences of the noncoding RNAs to generate predicted RNA secondary structures using the Mfold v.3.6 folding algorithm (web server at: http://unafold.rna.albany.edu) under default conditions [12, 13] or RNAfold (web server at http://rna.tbi.univie.ac.at).
3. Identify the functional RNA domain (predetermined experimentally) and adjacent secondary structure motifs within the predicted RNA secondary structures.
4. Contact the Morpholino manufacturer (e.g., GeneTools, LLC; web server at http://www.gene-tools.com) and provide the primary nucleotide sequence of either the full-length noncoding RNA (for small noncoding RNAs < 200 nucleotides in length) or the functional RNA domain and adjacent sequences (for long noncoding RNAs). The manufacturer will then return a list of antisense Morpholinos which are chemically feasible for synthesis. For small noncoding RNAs, this list may be limited to 1–3 possible Morpholino sequences per RNA molecule.
5. We have previously used Morpholinos to successfully target a double-stranded, nine base-pair motif within noncoding vertebrate Y RNAs and nematode sbRNAs [3, 4]. In these cases, effective Morpholinos were designed to hybridize to a single-stranded loop or bulge directly adjacent to the functional double-stranded domain and also extend into the 5′ or 3′ strand of the target functional domain (*see* Fig. 1 for the complementary

RNA binding sites of these Morpholinos). If the target functional domain within the noncoding RNA is double stranded, including a region within the Morpholino that could bind an adjacent single-stranded RNA motif may increase the likelihood of favorable binding to, and hence functional inhibition of, the target RNA domain. Select the Morpholinos for synthesis using these considerations and also those that are within the manufacturer's chemical feasibility and stability constraints.

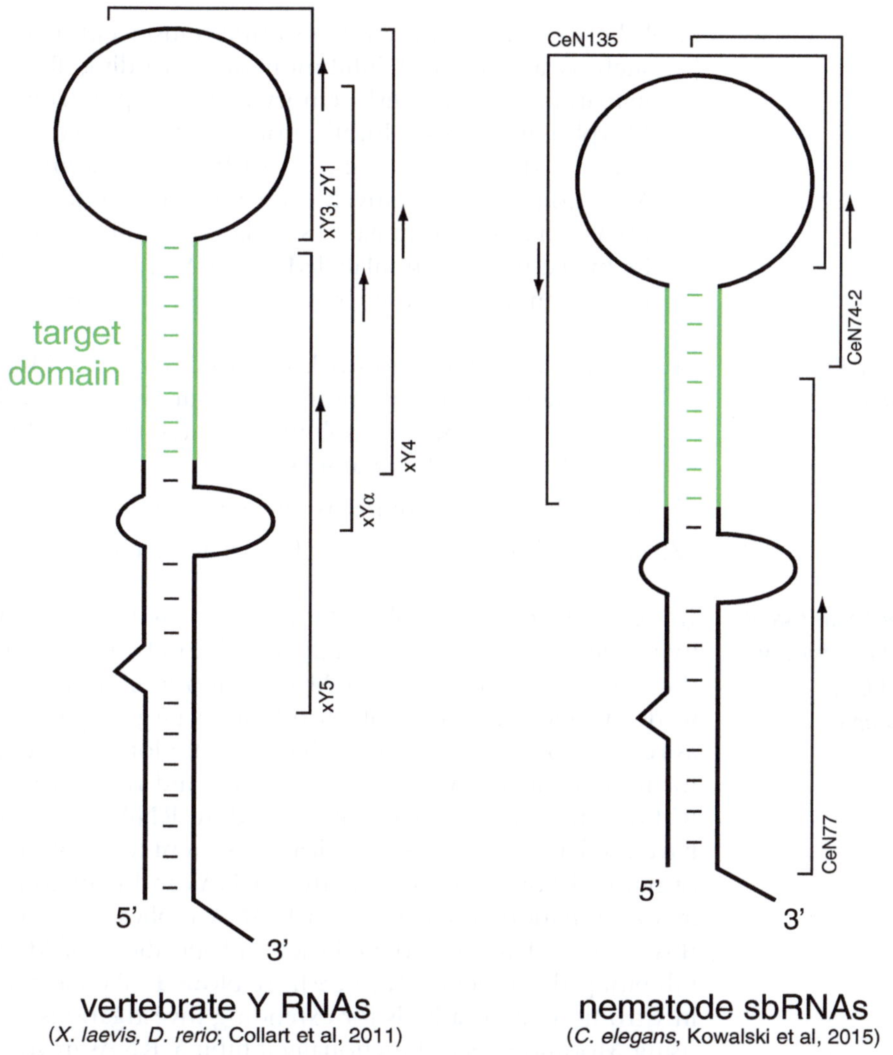

Fig. 1 Targeting noncoding RNAs by Morpholino antisense oligonucleotides. Schematic representations of canonical RNA structures are shown for: (**a**) vertebrate Y RNAs and (**b**) nematode stem-bulge RNAs (sbRNAs). The target domain for inactivation of noncoding RNA function is highlighted in green. Squared brackets indicate the positions of effective Morpholinos that that were shown to inactivate these noncoding RNAs in vitro and in vivo [3, 4]. Positions of effective Morpholinos either partially overlap with or are adjacent to the target domain. Note that a portion of each effective Morpholino hybridizes also with a single-stranded domain adjacent to the target domain in the target RNA

Where possible, select multiple antisense Morpholinos designed to hybridize to the target noncoding RNA domain.

6. For control experiments, purchase a control Morpholino, such as the standard control Morpholino that targets a human beta-globin intron mutation that causes beta-thalassemia. This Morpholino causes little phenotypic effect in any system tested (except human beta-thalassemic hematopoietic cells) and is widely used as a negative control (http://www.gene-tools.com).
7. Where feasible in experiments using translucent tissues in vivo, such as a functional inhibition of noncoding RNAs during nematode or zebrafish embryogenesis, purchase modified Morpholinos whose localization can be tracked during and after delivery, e.g., carboxyfluorescein-conjugated Morpholinos. Alternatively, and more cost effectively, a carboxyfluorescein-conjugated standard control Morpholino may be mixed with the nonlabeled target Morpholino and thereby used as an in vivo tracer (ref. [4], *see* Subheading 3.4.3).

3.2 Morpholino Preparation

1. Using the Morpholino stock solution, dilute the Morpholinos to the desired concentration in ultrapure water, heat for 7 min at 65 °C, and pellet insoluble residues by centrifugation at 16,000 × *g*, for 30 min at 4 °C.
2. Keep Morpholino solutions on ice until required.
3. Prepare fresh Morpholino solutions for each use.

3.3 Functional Assay to Determine Efficacy and Specificity of Morpholinos

It is essential to have a robust biochemical assay for the functionality of the target noncoding RNA, which can be used to determine the efficacy and specificity of Morpholino inhibition. Assays used to determine the efficacy of Morpholinos targeting mRNAs, such as readout of protein synthesis by immunoblot, are not applicable for noncoding RNAs. The exact nature and readout of the assay will depend on the functional noncoding RNAs under study. We have used a cell-free DNA replication system to assay the function of purified noncoding vertebrate Y RNAs and nematode sbRNAs in the initiation of chromosomal DNA replication, and adapted this assay to determine the efficacy and specificity of Morpholinos inhibiting this function [3, 4, 9], *see* **Note 1**. We have used these in vitro findings as a basis for extending investigations in vivo and using Morpholinos to functionally inhibit Y RNAs in *X. laevis* and *D. rerio* [3] and sbRNAs in *C. elegans* [4].

The functional assay should be used to determine the efficacy and specificity of Morpholinos targeting noncoding RNAs in the following ways:

1. Titrate each Morpholino targeting an individual noncoding RNA, as well as the control Morpholino, to identify an effi-

cacious dose. Addition of the target Morpholino should significantly inhibit noncoding RNA function, while the control Morpholino should have no significant effect. If testing multiple Morpholinos against the same target noncoding RNA, select those Morpholinos which display the most robust inhibitory effects for further testing. For the inhibition of sbRNAs in vitro, Morpholino concentrations of 500–700 nM are efficacious [4].

2. Assess the specificity of the Morpholinos by testing whether each Morpholino selected only inhibits its target noncoding RNA in vitro. Design mutant versions of the noncoding RNAs in which the overall structures and essential nucleotide sequences are maintained, while the target sequences of the Morpholinos are different. These mutant RNAs should be functionally active but remain unaffected by the addition of the specific Morpholino (or control Morpholino) [3].

3.4 Morpholino Delivery

After selecting those Morpholinos which show specific functional inhibition of their target noncoding RNA in vitro, the Morpholino may be delivered in vivo. Below, we provide the methods used to deliver Morpholinos into vertebrate and nematode embryos, including Morpholino doses that were used to inhibit Y RNA and sbRNA function.

1. Thaw the Morpholino stock solutions and dilute the Morpholinos to the desired concentration in ultrapure water, heat for 7 min at 65 °C and pellet insoluble residues by centrifugation at 16,000 × *g*, for 30 min at 4 °C.
2. Keep Morpholino solutions on ice until required and prepare fresh Morpholino solutions for each round of delivery on separate days.

3.4.1 Delivery into *X. laevis*

1. Obtain *X. laevis* embryos by artificial fertilization and stage them as described previously [14].
2. Inject *X. laevis* embryos at the one-cell stage with 50–100 ng of Morpholino per embryo [3].

3.4.2 Delivery into *D. rerio*

1. Obtain *D. rerio* embryos through natural spawning and stage them as described previously [15].
2. Inject *D. rerio* embryos in the yolk at the one-cell stage with 10 ng of Morpholino per embryo [3].

3.4.3 Delivery into *C. elegans*

1. If studying the function of noncoding RNAs during nematode development, carboxyfluorescein-conjugated Morpholinos may be used in order to trace delivery and incorporation into the progeny developing *in* and *ex utero*. Alternatively, carboxyfluorescein-conjugated standard control Morpholino

at 10–50 nM may be mixed with the target Morpholino to serve as a tracer [4].

2. Inject Morpholinos at 100–300 nM, bilaterally into the syncytial gonads of healthy, nonstarved young adult worms [16].

3.5 Scoring Phenotypes and Data Interpretation

1. Successful delivery of the Morpholinos using fluorescently labeled Morpholinos should be determined where possible to eliminate scoring false negatives, especially where delivery is technically challenging.
2. As with the functional assays described in Subheading 3.3, it is advisable to test dose-dependent effects of Morpholino inhibition on the phenotype of interest by titrating the Morpholino concentration in the injection mix.
3. If targeting multiple noncoding RNAs, additive effects of Morpholino inhibition may be determined by delivery of multiple Morpholinos in one injection mix targeting different noncoding RNAs together [3].
4. Experimentally determined noncoding RNA expression level data may be used to inform the interpretation of the time of onset or severity of different phenotypes arising from the Morpholino inhibition in vivo, and also to inform comparison of phenotypes arising from the inhibition of multiple noncoding RNAs. In *X. laevis*, the more abundant the targeted noncoding Y RNA, the earlier the onset of the phenotype (embryonic lethality) after Morpholino injection [3].
5. Morpholinos inhibit noncoding Y RNA and sbRNA function in a dominant-negative manner [3, 4]. In functional assays of DNA replication, addition of nontargeted but otherwise functionally redundant Y RNA or sbRNAs from within the same noncoding RNA families cannot rescue the inhibition of function effected by the specific, complementary Morpholino [3, 4]. Likewise the phenotypic effects of Morpholinos targeting Y RNAs in vivo cannot be rescued by coinjection of the specific Morpholino with nontargeted Y RNAs. Noncoding Y RNAs and sbRNAs may therefore form stable dominant-negative inhibitor complexes, which cannot simply be overcome by an excess of functionally redundant noncoding RNAs [3, 4].

4 Notes

1. Cell-free DNA replication system

   We are referring to this system here only as an example, with the sole intention to highlight the necessity to have a functional assay available to directly test the functionality of the noncoding RNA under study. The initiation of DNA replication in vitro is the functional assay used in our laboratory to study Y RNAs and sbRNAs.

In this system, late G1-phase template nuclei from mammalian cells initiate semiconservative chromosomal DNA replication upon incubation in a cytosolic extract from proliferating human cells, which contains DNA replication factors and endogenous Y RNAs [9, 17–19]. In order to test for the function of exogenous noncoding RNAs in this system, the endogenous Y RNAs are removed from the cell extract by biochemical fractionation [9]. This yields two protein fractions, containing all essential soluble DNA replication proteins, but lacking the endogenous Y RNAs. Exogenous RNAs are then added to the two protein fractions and the proportions of replicating nuclei are scored by confocal immunofluorescence microscopy [9, 10].

Technical and procedural details for this particular system can be found in a chapter of a previous book in this series [20].

Acknowledgements

This work was supported by grants from Cancer Research UK, the Medical Research Council, the Wellcome Trust and more recently, from the Biotechnology and Biological Sciences Research Council (BBSRC doctoral training grant DTG (BB/F016581/1) and grant BB/K013378/1).

References

1. Morris KV, Mattick JS (2014) The rise of regulatory RNA. Nat Rev Genet 15:423–437
2. Ploner A, Ploner C, Lukasser M, Niederegger H, Huttenhofer A (2009) Methodological obstacles in knocking down small noncoding RNAs. RNA 15:1797–1804
3. Collart C, Christov CP, Smith JC, Krude T (2011) The midblastula transition defines the onset of Y RNA-dependent DNA replication in *Xenopus laevis*. Mol Cell Biol 31:3857–3870
4. Kowalski MP, Baylis HA, Krude T (2015) Non-coding stem-bulge RNAs are required for cell proliferation and embryonic development in *C. elegans*. J Cell Sci 128:2118–2129
5. Kowalski MP, Krude T (2015) Functional roles of non-coding Y RNAs. Int J Biochem Cell Biol 66:20–29
6. Mosig A, Guofeng M, Stadler BMR, Stadler PF (2007) Evolution of the vertebrate Y RNA cluster. Theory Biosci 129:9–14
7. Perreault J, Perreault JP, Boire G (2007) The Ro associated Y RNAs in metazoans: evolution and diversification. Mol Biol Evol 24:1678–1689
8. Boria I, Gruber AR, Tanzer A, Bernhart SH, Lorenz R, Mueller MM, Hofacker IL, Stadler PF (2010) Nematode sbRNAs: homologs of vertebrate Y RNAs. J Mol Evol 70:346–358
9. Christov CP, Gardiner TJ, Szüts D, Krude T (2006) Functional requirement of noncoding Y RNAs for human chromosomal DNA replication. Mol Cell Biol 26:6993–7004
10. Gardiner TJ, Christov CP, Langley AR, Krude T (2009) A conserved motif of vertebrate Y RNAs essential for chromosomal DNA replication. RNA 15:1375–1385
11. Wang I, Kowalski MP, Langley AR, Rodriguez R, Balasubramanian S, Hsu ST, Krude T (2014) Nucleotide contributions to the structural integrity and DNA replication initiation

activity of noncoding y RNA. Biochemistry 53:5848–5863

12. Mathews DH, Sabina J, Zuker M, Turner DH (1999) Expanded sequence dependence of thermodynamic parameters improves prediction of RNA secondary structure. J Mol Biol 288:911–940
13. Zuker M (2003) Mfold web server for nucleic acid folding and hybridization prediction. Nucleic Acids Res 31:3406–3415
14. Nieuwkoop PD, Faber J (1975) Normal table of *Xenopus laevis* (Daudin): A systematical and chronological survey of the development from the fertilized egg till the end of metamorphosis, 2nd edn. North-Holland Pub. Co., Amsterdam
15. Westerfield M (1993) The zebrafish book: A guide for the laboratory use of zebrafish (*Brachydanio rerio*). University of Oregon Press, Eugene
16. Mello C, Fire A (1995) DNA transformation. Methods Cell Biol 48:451–482
17. Krude T (2000) Initiation of human DNA replication in vitro using nuclei from cells arrested at an initiation-competent state. J Biol Chem 275:13699–13707
18. Szüts D, Christov C, Kitching L, Krude T (2005) Distinct populations of human PCNA are required for initiation of chromosomal DNA replication and concurrent DNA repair. Exp Cell Res 311:240–250
19. Szüts D, Kitching L, Christov C, Budd A, Peak-Chew S, Krude T (2003) RPA is an initiation factor for human chromosomal DNA replication. Nucleic Acids Res 31:1725–1734
20. Marheineke K, Goldar A, Krude T, Hyrien O (2009) Use of DNA combing to study DNA replication in xenopus and human cell-free systems. Methods Mol Biol 521:575–603

Chapter 14

Use of Morpholino Oligomers for Pretargeting

Guozheng Liu

Abstract

Differing from the conventional direct-targeting strategy in which a probe or payload is directly loaded onto a targeting molecule that binds to the native target, pretargeting is an improved targeting strategy. It converts the native target to an artificial target specific for a secondary targeting molecule loaded with the probe or payload (effector). The effector is small and does not accumulate in normal tissues, which accelerates the targeting process and generates high target to nontarget ratios. DNA/cDNA analogs can serve as the recognition pair, i.e., the artificial target and the secondary targeting effector. Morpholino oligomers are so far the most investigated and the most successful DNA/cDNA analog recognition pairs for pretargeting. Herein, we describe the pretargeting principles, the pretargeting strategy using Morpholino oligomers, and the preclinical success so far achieved.

Key words Pretargeting, Morpholino oligomers, Semiempirical pretargeting model, Antibody, Antibody modification, Radiolabeling of Morpholino oligomers, In Vitro and In Vivo targeting evaluation, Preclinical study

1 Introduction

This chapter describes the principles of pretargeting using Morpholino oligomers (shortened as Morpholinos), wherein a pretargeting molecule attached with a Morpholino (MORF) is administrated to target the organ of interest, followed after a defined time by an effector that consists of a complementary Morpholino (cMORF) and the effecting probe or therapeutic payload. The cMORF effector as a secondary targeting molecule binds the MORF-pretargeting molecule and is thus retained in the target organ. The unbound cMORF effector in normal tissues is rapidly excreted. We will provide an overview of the pretargeting strategy; quantitatively describe the relationships of target accumulation with the dosages, timing parameters, and the pharmacological properties of pretargeting systems; demonstrate the construction strategies of a pretargeting system; and briefly describe some preclinical results. At the end, to exemplify the construction of a pretargeting system, we will provide the materials and methods to

Hong M. Moulton and Jon D. Moulton (eds.), *Morpholino Oligomers: Methods and Protocols*, Methods in Molecular Biology, vol. 1565, DOI 10.1007/978-1-4939-6817-6_14, © Springer Science+Business Media LLC 2017

prepare the pretargeting agent and the effectors for a two-step Morpholino pretargeting.

1.1 Overview of Pretargeting

Pretargeting as a modified magic bullet concept has a long history, though less than that of direct targeting in which the payload is directly attached to the targeting molecule (the magic bullet concept) [1–3]. Its focus has been on tumor imaging and therapy. Although fewer investigations are on pretargeting than on direct targeting, the enthusiasm for the pretargeting strategy continues to increase. A pretargeting strategy starts with the injection of a primary targeting agent (pretargeting agent) and, after the pretargeting is essentially complete and the pretargeting agent in the circulation is cleared, a second injection of the effector is administrated. The effector exclusively binds the pretargeting agent in the targeted organ and is excreted rapidly elsewhere to provide a very low normal tissue background.

This pretargeting strategy has several advantages. Widely appreciated is the advantage for radiotherapy. The delayed injection of radioactivity avoids the radiation exposure to normal tissues during the long pretargeting phase of the pretargeting agent. This is especially true when antibody is used as the pretargeting agent, as its targeting and clearance kinetics is slow. If a therapeutic nuclide were directly attached to the pretargeting agent, it would impose a high radiation dose to normal tissues. Another advantage of the delayed injection is that it enables the use of short half-life diagnostic nuclides such as fluorine-18 along with the antibodies for imaging. A less known advantage is the sequestration of the pretargeting agent internalized in normal tissues [4]. If directly radiolabeling the pretargeting agent, the normal tissue background would be higher even at the time equivalent when the effector is injected [5]. Finally, another advantage so far not systematically investigated is the flexibility in modulation of the effector structure for improved tumor accumulation [6]. Modulation of the direct targeting molecules may be constrained by the targeting moiety.

Pretargeting can be mediated by different recognition systems and its nature is to convert the native target in the target tissue to an artificial target specific for the effector. The antibody/hapten and the (strept)avidin/biotin systems are the two most investigated recognition families for pretargeting [2, 3]. The (strept) avidin/biotin family encounters several difficulties. Avidin cannot be used to modify the pretargeting agent, because a liver receptor immediately captures molecules bearing the avidin and essentially deprives of the opportunity for the pretargeting agents to reach the target tissue. Streptavidin is an option but there is an immunogenicity issue. Attachment of biotin to the pretargeting agents is another option but it involves the same issues because avidin or streptavidin has to be part of the effector structure for recognition.

In addition, streptavidin as an effector leaves a higher normal tissue background. The antibody/hapten is comparatively more successful and currently under clinical trial [7]. The technical difficulties are reported to be related to quantitatively understanding the pretargeting process [8].

The DNA/cDNA recognition systems are relatively new [9]. Natural DNA structures are not stable in vivo, so synthetic analogs are used [10]. Morpholinos are so far the most successful DNA analogs for pretargeting due to their in vivo stability, water solubility, the flexible choice of chain length and sequences, and the favorable pharmacokinetics of the cytosine-free species [11]. The Diels–Alder reaction may be the latest found recognition system [12], but to what extent it is advantageous is yet to be evaluated. Bioorthogonality is thought to be one of the advantage, at least in theory, but not all the interactions for other pretargeting mechanisms are endogenous. Also, there is no evidence that the short Morpholino sequences chosen for pretargeting enter cells to bind the nucleic acids. Even if that would happen to some extent, the issue could be addressed by use of the L-form oligomers [13, 14].

The disadvantage of pretargeting is the multiple injections that complicate the targeting procedure. Although the complexity can be addressed based on the pretargeting principles, the multiple injections certainly involve a higher cost and do not favor the FDA approval. Nevertheless, it is almost impossible for direct antibody targeting to achieve tumor eradication or address some imaging challenge such as pancreatic beta cell imaging.

1.2 Pretargeting Principles

A pretargeting system includes at least two injections, the pretargeting agent and the effector. To further improve normal tissue background, another injection of a clearing agent may be added in between. The biodistribution of the labeled effector defines the final pretargeting results. A pretargeting regimen is determined by the dosage and timing parameters. Part of the complexity of the pretargeting strategy lies in the optimization of these parameters. However, we have established an optimization guideline based on our investigation on the pretargeting principles [15].

The nature of pretargeting is a conversion of native target to artificial target. The targeting kinetics of the effector dictates its accumulation in the target organ just as that of a direct targeting agent does. If there are a sufficient number of target sites, the overall target accumulation would be the integration of the accumulation increment (dA) in each indefinitely small time interval (dt). If the effector is overdosaged, the accumulated percentage would be smaller, because the effector transported to the site after target saturation would not stay. If not overdosaged, the percent target accumulation is at the highest value, namely, at the maximum percent target accumulation (MPTA), assuming there is no subsequent dissociation. The MPTA in percent of injected dosage per gram of tissue (%ID/g) can be mathematically described as [16, 17]:

$$MPTA(\%ID/g) = F \times f \times W^{-1} \times E \times \int_{t=0}^{t=\infty} C(\%ID/g)_{\text{blood}} \times dt = F \times f \times W^{-1} \times E \times AUC_{\text{blood}}$$

where F, f, and W are the cardiac output, the fraction of blood getting to the target tissue, and tissue weight, respectively. E is the tissue trapping fraction, and AUC_{blood} is the area under the blood curve.

Take a MORF-Ab pretargeted tumor as an example. Fig. 1 shows the absolute tumor accumulation of the cMORF effector in ng/g (A) and the percent tumor accumulation in %ID/g (B) from one of our documented reports [16]. The solid lines are theoretical without considering dissociation, while the dashed lines represent the observed. They are very close because of the strong affinity of the effector to the artificial target. This observation has been confirmed in another two occasions [18, 19].

A complication of the target conversion in a pretargeting process is the residual artificial target sites left in the circulation and normal tissues. They will bind the effector to generate a higher normal tissue background. Except for the organs excreting the effector, our observation is that the background levels (to be accurate, correct for the minimal level free effector) are in equilibrium with the blood concentration. Table 1 lists the concentrations of a cMORF effector in MORF-Ab %ID/g in pretargeted mice along

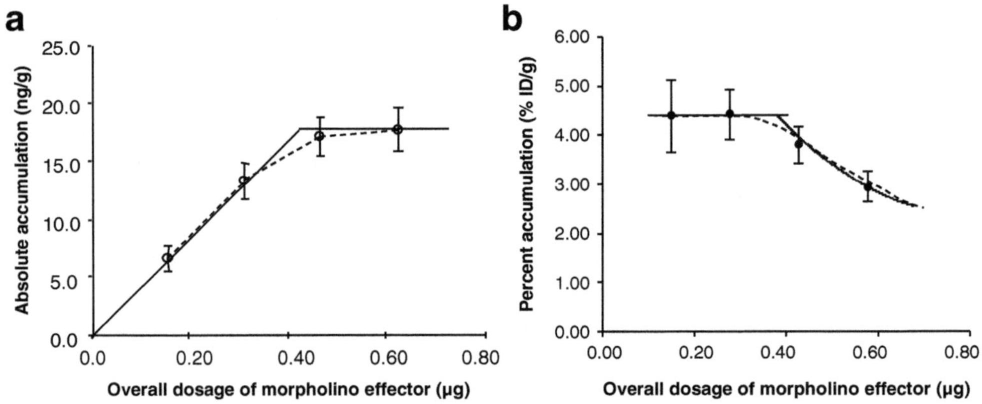

Fig. 1 The absolute tumor accumulation and percent tumor accumulation in a study where increasing number of Morpholino effector injections were given to mice in groups that are pretargeted with MORF-Ab. The *x*-axis represents the overall dosage each group of mice received. The *solid lines* make a prediction for the tumor accumulations: At dosage below saturation of the artificial target sites: *Percent tumor accumulation = MPTA;* At dosage above saturation of the artificial target sites: *Percent tumor accumulation = (MPTA × saturation dosage)/injected dosage*

Table 1
Percent accumulations (%ID/g) of a ^{99m}Tc labeled Morpholino effector in mouse blood and several normal organs at 3 h post the effector administration. Organ-to-blood ratios have been corrected for the contributions of free ^{99m}Tc-labeled effector

Organs	Free [a]	Study 1[a]	Study 2[a]	Study 3[a]	Study 4[a]	Study 5[b]	Study 7[b]	Average
Blood	0.04	5.44±1.07	5.20±0.48	3.34±1.03	3.10±0.46	1.27±0.17	0.93±0.08	
Liver	0.27	1.90±0.24	1.93±0.18	1.33±0.11	1.23±0.15	0.60±0.05	0.54±0.03	
Liver/blood		0.30	0.32	0.32	0.31	0.27	0.30	0.31
Spleen	0.17	1.09±0.26	0.94±0.22	0.70±0.12	0.63±0.08	0.33±0.06	0.29±0.05	
Spleen/blood		0.17	0.15	0.16	0.15	0.13	0.13	0.15
Lung	0.12	2.04±0.42	1.95±0.19	1.37±0.24	1.20±0.30	0.74±0.16	0.41±0.07	
Lung/blood		0.36	0.35	0.48	0.35	0.50	0.33	0.40
Heart	0.06	1.23±0.18	1.05±0.07	0.75±0.07	0.71±0.20	0.30±0.07	0.22±0.03	
Heart/blood		0.22	0.19	0.21	0.21	0.20	0.18	0.20
Muscle	0.03	0.49±0.08	0.45±0.04	0.42±0.09	0.36±0.04	0.19±0.02	0.13±0.02	
Muscle/blood		0.09	0.08	0.12	0.11	0.13	0.11	0.11

[a]Taken from ref. [20].
[b]Taken from ref. [16].

with that in nonpretargeted control mice. As seen, although the concentration can be very different in different studies, the organ-to-blood concentration ratios are essentially the same. This suggests the background effector levels at the time when free effector is cleared can be predicted from the blood effector level.

In the Morpholino pretargeting strategy, kidneys are the main organs to clear the free Morpholino effector. The concentration level is dependent on the pharmacokinetics of the ^{99m}Tc-labeled Morpholino effector. It is a constant in a given study but varies considerably among different studies, probably due to mouse age and the kidney function at the time. Liver also clears the Morpholino effector, but only to the extent of about 2%. At 3 h when the free Morpholino effector is completely excreted, the molecules have been transported from the liver to the intestines. That is why in Table 1, the free effector concentration level is also minimal in the liver.

The equilibrium of the accessible antibody between organs and blood explains the constant organ-to-blood ratios. Unlike tumor accumulation, the effector levels in these normal tissues are normally the result of saturating of accessible antibody, unless the dosage of the pretargeting agent is extremely large or the effector dosage is extremely small. The quantitative relationship between the artificial target number and the effector level in an organ is:

$$\text{Percent accumulation} = (\text{number of artificial targets} / \text{number of injected effector}) \times 100\%$$

As shown in one of our studies (Fig. 2), where the dosage of the pretargeting antibody is increased while the effector dosage is kept constant [4, 18], the effector accumulation in tumor plateaus (reached the maximum) much earlier than the effector levels in normal tissues. This earlier plateau signals earlier arrival at the condition for MPTA, i.e., the faster saturation of the targets in tumor. The sufficient blood flow to normal tissues as compared to tumor and the less antibody concentration in normal tissues should have both contributed to the easier saturation of the targets in normal tissues. As seen, when using a large antibody dosage, the normal tissue background is high and the tumor-to-nontarget tissue ratios are low. We note these conditions are extremely suboptimal and are used herein to demonstrate the principles of pretargeting. In an application rather than for this particular investigation, a dosage in proximity to the optimal should be used to provide the MPTA and highest tumor-to-normal tissue ratios. The optimal dosage is at the sharp point of the solid lines in Fig. 1 where the artificial target sites in the tumor are just to be saturated. It may be worth noting that back in 2003 we erroneously explained the plateau of tumor accumulation (Fig. 2) was due to target saturation instead of delivery limitation [4].

Understanding these pretargeting principles accelerates the development of pretargeting systems, as these examples illustrate. There was a concept of "amplification pretargeting" that is now clear

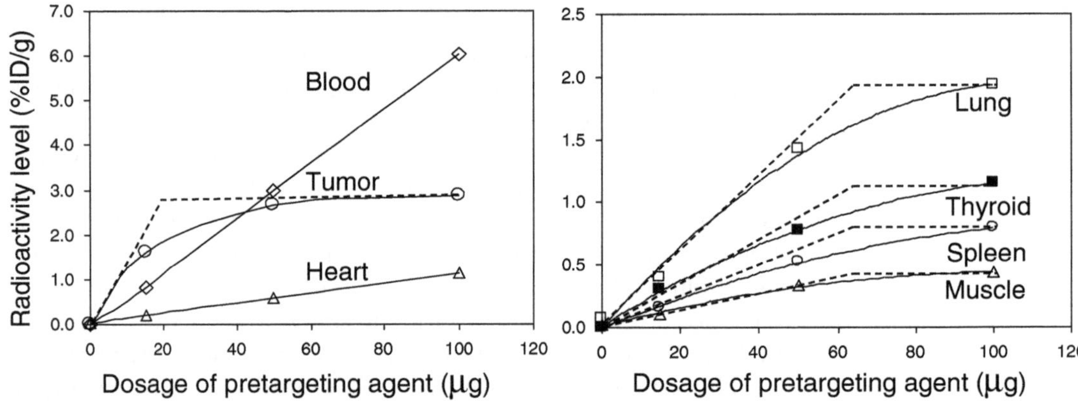

Fig. 2 Radiolabeled effector level at 3 h in pretargeted mice bearing LS174T tumors under the conditions of a varying dosage of a Morpholino-modified antibody MN14, 2 day pretargeting interval, 0.15 μg of the Morpholino complementary to the Morpholino attached to the antibody and labeled with ^{99m}Tc

to be equivalent to the effect of using a larger dosage of pretargeting agent, though more efficiently. Unlike previously perceived, the increased number of artificial target sites would not improve the MPTA (refer to Fig. 1b) although it does favor the absolute tumor accumulation (Fig. 1a). In other words, amplification of the target sites shifts the saturation point in Fig. 1 to the right, but does not improve the percent tumor accumulation. At a fixed nonsaturating effector dosage, a larger dosage of the pretargeting agent or amplification of the target sites may slightly shorten the distance between the observed and the theoretical solid line due to reduced dissociation, but the effect would be weak and not obvious. If the affinity between the recognition pairs is sufficiently strong, the observed line and the solid line would almost superimpose. Also, for tumor accumulation choosing a different pretargeting antibody is similar to employing an amplification mechanism. Any effort to change the MPTA by varying the pretargeting agent would be ineffective, though it may have an effect on the normal tissue background [21].

1.3 Construction of a Morpholino Pretargeting System

1.3.1 Choice of a Pretargeting Agent

The pretargeting agent has to be able to bind a native target specific for the diseases or conditions of interest. So far, the pretargeting strategy has been applied mainly to tumor therapy and imaging. The technology is thus being advanced on the basis of tumor model. However, as mentioned, pretargeting may apply to any diseases or conditions in general. It would have a place wherever high target-to-normal tissue ratios are required, e.g., in pancreatic beta cell imaging [5, 22]. Beta cell mass is a fundamental indicator for the progression of diabetes. However, beta cell imaging requires high endocrine to nonendocrine ratios, because beta cell mass constitutes only about 1–2% of the total pancreatic mass and the current imaging physics cannot differentiate the islets from the nonislet (exocrine) tissues. In order to achieve sufficient specific signal intensity, the beta cells have to achieve an accumulation 99 times higher than that of the exocrine pancreatic tissues.

The Morpholino pretargeting requires the pretargeting molecule be modified by a Morpholino recognition group. In addition, after binding to the target, the pretargeting agent has to be accessible for the Morpholino effector, i.e., to stay on the cell surface. Thus, it is reasonable that the pretargeting strategy is based on the use of noninternalizing antibodies. Nevertheless, any cell-binding antibody would internalize, including those typical noninternalizing antibodies [19]. The reality is, if the noninternalized portion of the cell-bound antibody is considerable, the pretargeting would be successful [23], especially for imaging that does not require a large number of the artificial target sites. The key issue is whether high target-to-normal tissue ratios of the artificial target site for the effector can be sufficiently high.

Intact antibody is often chosen to construct the pretargeting agent, but bear in mind any tumor targeting agents could be used as long as the pretargeting strategy can improve the target-to-normal tissue ratios. The pretargeting strategy was originally intended to address the slow kinetics of large molecules, but there have been studies using antibody fragments and minibodies [24–26]. Justification of choosing smaller targeting molecules for the pretargeting agent lies in the fact that they reduce the pretargeting interval and the pretargeting agent that has been internalized, metabolized, and/or trapped in normal organs can be sequestered [4]. By shortening the pretargeting interval, the small pretargeting agents may obviate the need of a clearing agent.

1.3.2 Choice of a Morpholino Effector

Because the DNA/cDNA interaction is to provide the mutual recognition between the pretargeting agent and the effector, any pair of mutually complementary Morpholinos would theoretically function. However, we found that the oligomers including cytosine bases would elevate their kidney accumulation although the background in other normal organs remains similarly clearable [11, 20]. Figure 3 shows how kidney accumulation increases with the number of cytosines and it can go to as high as about 80%ID. Cytosine should therefore be excluded from the Morpholino effector, though cytosine is allowed to be in the Morpholino attached to the pretargeting agent as in this case it is not labeled with the probe or toxic therapeutic group.

The chain length and the sequence together dictate the affinity of a Morpholino to its complement. As said in the principle section, affinity is an important factor for the success of a pretargeting system. If it is not sufficiently strong, the Morpholino effector may still bind its complement on the pretargeting agent but will dissociate. We have developed a method to readily attach an amine-derivatized Morpholino to a serum protein to evaluate its in vivo binding affinity for its complement [27]. Figure 4 shows the biodistributions of a labeled 18-mer Morpholino in 5 groups of mice, 4 of which have been pretreated with serum protein–

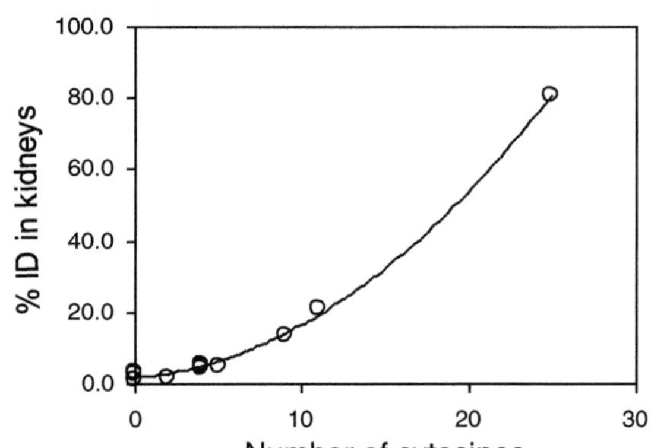

Fig. 3 Kidney accumulation at 3 h after injection of ^{99m}Tc-labeled Morpholinos (12–25 mer) that contains different number of cytosines

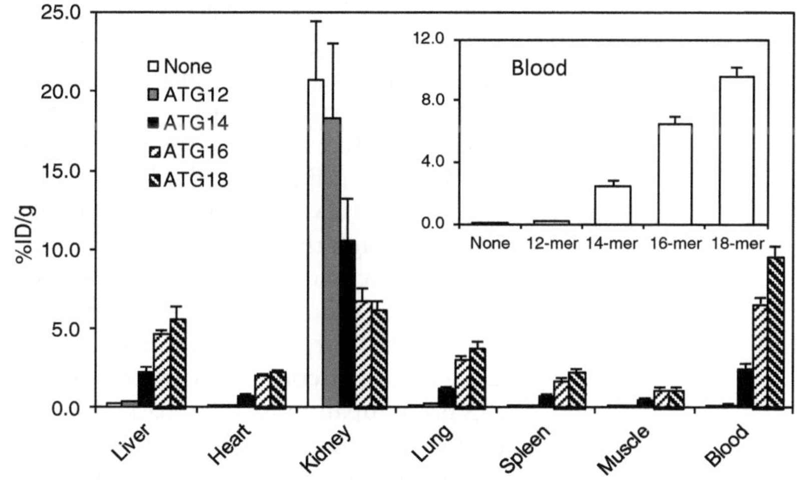

Fig. 4 Biodistribution at 3 h of a radiolabeled 18-mer Morpholino in mice pre-injected with its C-free complements at different lengths (ATG-12 mer to 18 mer) conjugated to serum protein

Morpholino complement conjugates at 4 different lengths (ATG-12 to 18). At 3 h postinjection, the labeled Morpholino effector is confirmed to be bound to the Morpholino–serum protein conjugate, though to a different extent. The 18-mer cytocine-free Morpholino (ATG18) is currently being chosen to be labeled instead and employed for our tumor pretargeting investigations.

We note, in tumor-bearing mice pretargeted with Morpholino-conjugated antitumor antibodies, we observed a lower blood level with the 15-mer Morpholino effector than the chosen 18-mer Morpholino (though not very significant) and the tumor accumulation at 3 h is comparable [28]. However, our aim is at therapy and the higher affinity should provide better toxicity retention in tumor, which is why we adopted the 18-mer Morpholino as the effector.

1.3.3 Choice of a Clearing Agent

Pretargeting is a strategy rapidly to provide high target-to-nontarget (T/NT) ratios. For imaging it allows for the use of short half-life nuclides and for therapy it reduces toxicity exposure to normal tissues. The price for the improved T/NT ratios is the two injections and the wait time (pretargeting interval) prior to the effector injection. The T/NT ratios can be further improved by an additional injection of a clearing agent. Without the clearing agent, there are higher levels of the circulating pretargeting agent that will lead to higher level of the effector in normal tissues and blood. That is the why small pretargeting agents are drawing increasing attention [24–26].

The use of a clearing agent adds another injection to a pretargeting procedure. The increased complexity in part explains why a clearing agent was not included in most clinical trials. More injections increase the cost. Nevertheless, the question is whether it is worthwhile. The reality is that directly attaching an effecting group to a targeting agent either small or large has failed in many applications including tumor therapy due to the low T/NT ratios. Pretargeting with a clearing agent can provide a high tumor accumulation with extremely low normal tissue background [6].

The real difficulty is the lack of an ideal clearing agent. There have been relatively fewer investigations, although the research continues through the past decades. The majority of the clearance studies employ a mechanism in which the clearing agent shares the same recognition group with the effector. Thus, a considerable portion of the artificial target sites created for the effector is preoccupied. Under normal conditions where the effector is not overdosaged, this preoccupation is often masked by the principle that the percent tumor accumulation (%ID or %ID/g) is a constant that is only dependent on the properties of the effector instead of the number of the target sites [29]. The number of the available targets may later be found to be insufficient due to target number reduction [30]. Currently, we are developing a mechanism using separate groups to recognize the clearing agent and the effector [31–33].

1.4 Preclinical Morpholino Pretargeting

Thus far Morpholino pretargeting utilizes antibodies and is in the preclinical stage. There is a wide dosage range for the antibodies and we normally choose 15–30 μg of the Morpholino-modified antibody. In a conventional two-step procedure, we usually choose 2 days as the pretargeting interval, although the longer it is, the lower the normal tissue background will be. The dosage of the Morpholino effector is dependent on three factors: its own pharmacokinetics, the dosage and the pharmacokinetics of the pretargeting antibody, and the Morpholino recognition groups per antibody. When using a CC49 antibody, the optimal dosage in mice for the 18-mer Morpholino effector selected from Fig. 4 can be calculated from the optimal dosage ratio of (Morpholino effector/Morpholino-antibody) = 3.1 measured previously [34]. The imaging should be performed at a time when the free cMORF is essentially excreted (after 1 h). For therapy, this is not a parameter.

The conventional two-step procedure has been applied to tumor imaging and therapy [15–22, 33–37]. Figure 5 shows the images of mice bearing tumors in the thigh that are pretargeted 2 days earlier with an 18-mer Morpholino-CC49 antibody. The imaging time is at 3 h after the labeled 18-mer Morpholino effector is injected. The kidneys as the excreting organ have the highest background. The background in other normal tissues is low and is dependent on the nuclide to some extent, higher with ^{111}In than with ^{99m}Tc and ^{188}Re. We have also performed tumor pretargeting with ^{90}Y at a tracer level [38]. No image was available because it is a pure beta emitter. Using ^{188}Re for therapy we have observed tumor remittance although with recurrence [34].

We have extended the Morpholino pretargeting to other areas such as pancreatic beta cell imaging in connection with

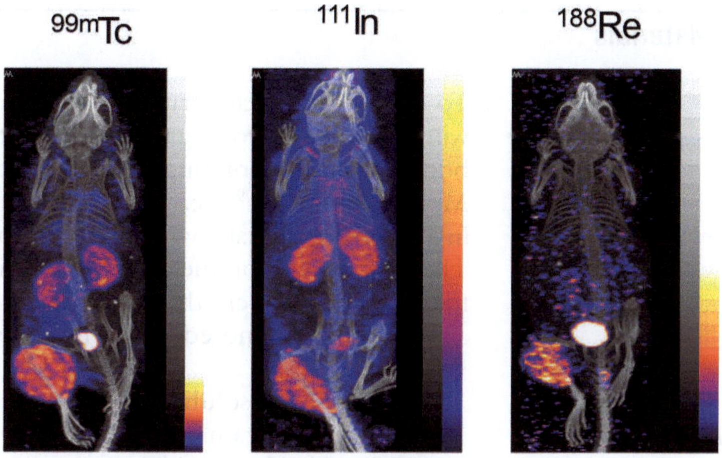

Fig. 5 The fused posterior NanoSPECT/CT images of the mice that were pretargeted 2 days earlier by Morpholino-conjugated antibody CC49 (MORF-CC49), 3 h after injection of the 18-mer complementary Morpholino (cMORF) labeled ^{99m}Tc, ^{111}In, or ^{188}Re. The tumor is in the left hind leg

diabetes [5, 22]. Human islets implanted in immunodeficient mice can be discerned using an anti-human islet antibody but we have not imaged the native mouse islets due to the lack of anti-mouse islet antibody. Although pretargeting greatly improved the target-to-nontarget ratios, like any other islet targeting technologies, Morpholino-based pretargeting has not reached the required endocrine-to-exocrine ratios of over 100. Thus, further improvement is still needed.

1.5 Prospect of Morpholino Pretargeting

Although previously fewer researchers were engaged in the pretargeting area, pretargeting strategy is useful, as it is difficult or almost impossible for the direct targeting to reach the required high target accumulation and target-to-nontarget ratios required for tumor therapy and some imaging applications. Pretargeting is becoming an increasingly active area and is currently being tested in humans. Pretargeting systems are based on several recognition mechanisms and Morpholino oligomers provide a family of the recognition pairs. The pretargeting principles indicate different recognition mechanisms make no fundamental difference as long as the free effector is rapidly targeting, rapidly excreted, and not retained in the normal tissues. Nevertheless, the recognition mechanism could make a difference in modulating the labeled effector to provide higher target accumulation and, coupled with the use of a clearing agent, to provide extremely high target-to-nontarget ratios. The Morpholino pretargeting favorably allows varying the chain length, sequence, and chemical modifications. The flexibility will also help to better understand the tumor accumulation principles and advance the pretargeting technology for targeted systemic therapy or human disease imaging.

2 Materials

To exemplify the construction of a pretargeting system, below we will describe the conjugation of a Morpholino to the CC49 antibody as the pretargeting agent and labeling the complementary Morpholino with ^{99m}Tc, ^{188}Re, ^{111}In, and ^{90}Y as the effectors for use in the conventional two-step Morpholino pretargeting protocol. The Morpholino on the antibody and its complement in the effector are 18-mer in length and are amine-derivatized at the 3′-equivalent terminal (denoted as MORF18-NH_2 and cMORF18-NH_2).

2.1 Conjugate the MORF18-NH_2 to an Antibody

1. 4FB DNF solution: dissolve 1.88–3.76 mg of 4FB (4-formylbenzamide) into 0.10 mL DMF (*N*, *N*-dimethylformamide).
2. 0.2 M pH 8.0 HEPES (4-(2-hydroxyethyl)-1-piperazineethanesulfonic acid)
3. MORF18-NH_2: the 3′-primary amine derivatized 18-mer Morpholino from Gene-Tools with a sequence of 5′-TCTTCTACTTCACAACTA.
4. MORF HEPES solution: dissolve 3 mg of MORF18-NH_2 into 0.95 mL of 0.2 M pH 8.0 HEPES.
5. PD-10 column: an open column filled with Sephadex-G25 (NeoRx Corp).
6. The pH 5 MES buffer: 0.1 M pH 5.0 MES (2-(*N*-morpholino) ethane-sulfonic acid).
7. CC49: a murine antiTAG-72 IgG antibody prepared by Strategic Biosolutions using the gift CC49 hybridoma from Dr. Jeff Scholm (NCI, NIH, Bethesda, MD, USA).
8. S-HyNic DMF/PB solution: S-HyNic (succinimidyl-6-hydrazino-nicotinamide) at a concentration of 80 mM in DMF/10 mM pH 6 PB (v/v=1/2).
9. Sephadex-G100 column: a glass Econo-column filled with Sepahadex-G100 resin from Bio-Rad Laboratories.
10. Measure the number of MORF18s per CC49: The size-exclusion HPLC system includes a Superose-12 HR10/30 column with an optimal separation range between 1×103 and 3×105 Da (Amersham Pharmacia Biotech, Piscataway, NJ) using 0.10 M pH 7.2 phosphate buffer as eluant at a flow rate of 0.6 mL/min, an inline radioactivity detector, and a Waters 2487 dual wavelength absorbance detector (Milford, MA). Routinely measure the recovery of radioactivity and confirm it is over 90% prior to the measurement. Use the radioactivity channel to calculate the gpm, while set the two UV channel to 280 nm sensitive for Ab and 265 nm sensitive for MORF18 for peak identification.

2.2 Label the cMORF18-NH_2 as the Effector

1. cMORF18-NH_2: the 18-mer Morpholino complementary to the MORF18 on the pretargeting antibody CC49 is derivatized at the 3′ equivalent terminal with primary amine.
2. P4 column: a glass Econo-column filled with P4 resin from Bio-Rad Laboratories.
3. Ammonium acetate buffer: 0.25 M pH 5.2 NH_4OAc.
4. Tartrate buffer: 50 μg/μL Sodium Tartrate Dihydrate in a buffer of 0.5 M Na_2HCO_3, 0.25 M NH_4OAc, 0.175 M NH_3, pH 9.2.

2.3 Conjugate the cMORF18-NH_2 with MAG_3

1. S-acetyl NHS-MAG_3: NHS activated S-acetyl mercaptoacetyl-triglycine, house made and chemically characterized.
2. Tin Solution #1: 10 μg/μL $SnCl_2{\cdot}2H_2O$ and 1 μg/μL Sodium·Ascorbate in 10 mM HCl.

2.4 ^{99m}Tc Labeling

1. ^{99m}Tc-pertechnetate from a ^{99}Mo-^{99m}Tc generator.
2. Tin solution #2: 4 μg/μL $SnCl_2{\cdot}2H_2O$ and 1 μg/μL Sodium·Ascorbate in 10 mM HCl.

2.5 ^{188}Re Labeling

1. ^{188}Re perrhenate from a ^{188}W-^{188}Re generator.
2. Tin solution #3: 10 μg/μL $SnCl_2{\cdot}2H_2O$ and 1 μg/μL of Sodium Ascorbate in 20 mM HCl.

2.6 Conjugate p-SCN-Bn-DOTA to the cMORF18-NH_2

1. p-SCN-Bn-DOTA: A bifunctional DOTA chelator from Macrocyclics (Dallas, TX).
2. Disodium carbonate buffer: 0.5 M Na_2CO_3-$NaHCO_3$ buffer at pH 9.8.

2.7 ^{111}In or ^{90}Y Radiolabeling

$^{111}In\ Cl_3$ or $^{90}YCl_3$ in 50 mM HCl.

3 Methods

3.1 Conjugating MORF18-NH_2 to an Antibody (See Note 1)

1. **Modify the MORF18-NH2 with 4FB**:
   - Mix 4FB DNF and MORF HEPES solutions at the MORF:4FB molar ratios of 1:10–20.
   - Incubate the mixture for 2 h at room temperature.
   - Purify the reaction mixture over a PD-10 column using the pH 5 MES buffer as the eluent.
   - Pool the peak 4FB-MORF18 fractions and store in a refrigerator.
2. **Modify CC49 antibody with S-HyNic:**
   - Mix 8 mg (1.5 mL) of CC49 in 20 mM pH 7.2 PBS with 27 μL of S-HyNic DMF/PB solution (*see* **Note 2**).

- Incubate 24 h at 4 °C.
- Incubate 0.5 h at RT.
- Purify the reaction mixture using a PD-10 column with the pH 5 MES buffer as the eluent.
- Pool the peak fractions as the CC49-HyNic product.

3. **Attach the MORF18-NH2 to the CC49 antibody:**
   - Mix the CC49-HyNic with the 4FB-MORF18.
   - Let the mixture react at 4 °C for 20 h.
   - Remove the MORF18-NH_2 not bound to the antibody using a 1×50 cm Sephadex-G100 column and the 50 mM pH 7.0 PB as the eluent.
   - Pool the peak fractions as the MORF18-CC49 conjugate.
4. **Measure the number of MORF18s per CC49:**
   - Sample 4 aliquots of the MORF18-CC49 conjugate.
   - Mix each with a known but elevated excess of ^{99m}Tc-labeled cMORF18 (see later for preparation).
   - Analyze each by HPLC. If the free labeled cMORF18 is in excess, measure the area of the radioactivity peak on the MORF18-Ab and that off the antibody. Be sure the off/on ratio is between 2 and 4; if not, disregard it and make another sample
   - Calculate the gpm from the area ratio, the known amount of Ab, and the know amount of cMORF18, based on the assumption that the one MORF18 binds one cMORF18.

3.2 Label a Morpholino as the Effector

We have attached optical probes and radiometals (^{99m}Tc, ^{188}Re, ^{90}Y, and ^{111}In) as the effecting group to the cMORF-NH_2 Morpholinos. The attachment of the optical groups is straightforward [39], but radiometal attachment includes chelator conjugation and the subsequent site-specific radiolabeling. The ^{99m}Tc and $^{186, 188}$Re labeling chemistry using MAG_3 chelator (S-acetylmercaptoacetyltriglycine) has been well established. High labeling efficiency for ^{90}Y and ^{111}In can be achieved using DOTA and DTPA chelators, but specific radioactivity for ^{90}Y is yet to be improved for radiotherapy.

1. **Conjugate the cMORF18-NH2 with MAG3** (*See* **Note 3**):
   - Dissolve 1.5 mg of the cMORF18-NH_2 in 0.2 M, pH 8.0 HEPES to a concentration of 0.5 mg/100 μL.
   - Transfer the cMORF18-NH_2 solution to a vial containing the S-acetyl NHS-MAG_3 powder at a MAG_3/amine molar ratio of 10–25.

- After 2 h incubation at room temperature, purify the reaction mixture over a P4 column (0.7 × 20 cm) using the ammonium acetate buffer as the eluant.
- Pool the peak fractions with the OD at 265 nm greater than 1.5.
- Mix the pooled conjugate solution with the tartrate buffer and a fresh Tin Solution #1 at a volume ratio of conjugate solution/tartrate buffer/tin solution of 15/5/1.
- Heat for 20 min to dissociate the metastable conjugate.
- Purify over a P4 column (1.0×50 cm) using the ammonium acetate buffer as the eluant.
- Pool the peak fractions with OD values greater than 1.5 as the MAG_3-cMORF18 product solution and store it at −20 °C (*see* **Note 4**).

2. **99mTc labeling protocol**:
   - Add 50 μL of the ^{99m}Tc-pertechnetate eluant from the ^{99}Mo-^{99m}Tc generator into a combined solution of 30 μL (preferably no less than 0.5 μg) of the MAG_3-cMORF18 product solution, 10 μL of the tartrate·buffer, and 3 μL of Tin solution #2.
   - Heat the solution at 100 °C for 20 min.
3. **188Re labeling protocol**:
   - Add 50 μL of ^{188}Re perrhenate to a combined solution of 30 μL of the MAG_3-cMORF18 product solution containing more than 0.2 μg but preferably more than 0.5 μg, 10 μL of Na_2tartrate·$2H_2O$ water solution (50 μg/μL) and 20 μL of Tin Solution #3.
   - Heat the solution in a boiling water bath for 1 h.
4. **Conjugate the cMORF18-NH2 with p-SCN-Bn-DOTA** (*See* **Note 5**):
   - Dissolve 2.5 mg of p-SCN-Bn-DOTA in 0.25 mL of the disodium carbonate buffer and mix with 2.3 mg of the lyophilized cMORF18-NH_2.
   - Incubate overnight at room temperature.
   - Add 1 mL of the ammonium acetate buffer.
   - Heat the conjugation solution at 100 °C for about 3 h to destroy the metastable conjugates.
   - Load the solution onto a 1 × 50 cm P4 column.
   - Elute with the ammonium acetate buffer.
   - Pool three peak fractions (about 0.5 mL in total) as the DOTA-cMORF18 product solution.

5. **111In and 90Y radiolabeling protocol**:
   - For radiolabeling, add 1–3 μL of $^{111}InCl_3$ or $^{90}YCl_3$ solution in 50 mM HCl to 10–30 μL of the purified DOTA- cMORF18 product solution.
   - Heating at 100 °C for 10 min (*see* **Note 6**).

4 Notes

1. The pretargeting molecules modified with Morpholinos are so far limited to antibodies. The free carboxylate groups and amine groups are useful for the modifications. Our earlier conjugation of Morpholinos to the carboxylic groups was successful [4, 20, 28, 35], but with poorer gpm (< 0.3 groups of Morpholino oligomers per antibody molecule) and in one case even failed due to the loss of cell binding affinity. Since then, a commercially available method, the "Hydralink" method, utilizing the amine groups was utilized [36, 40]:

   Following the provider's instruction, we succeeded in attaching the Morpholinos to antibody with higher conjugation efficiency compared to the previous conjugation to carboxylic groups [41]. The conjugation protocol was not fully optimized. However, a typical conjugation protocol described in this chapter provides a product with 0.5–1.5 Morpholino groups per antibody while preserving the tumor cell binding affinity.
2. The molar ratio of S-HyNic:CC49 antibody =15:1.
3. Previously investigators encountered a low labeling efficiency problem in labeling biomolecules with ^{99m}Tc and $^{186, 188}Re$ using NHS (*N*-hydroxysuccinimide) activated S-acetyl MAG_3, although the ^{99m}Tc labeling of non-NHS-activated free MAG_3 (betiatide) was routinely performed in the clinic as a kidney imaging agent. In the clinical kit, a benzoyl group was used to protect the sulfhydryl group of the non-NHS-activated MAG_3 that needs heating to facilitate the release of the sulfhydryl for coordination. An acetyl group in place of the benzoyl group was proposed to protect the sulfhydryl group for labeling at room temperature [42]. Nevertheless, the labeling efficiency was low (60–80%). We applied the NHS-activated S-acetyl MAG_3 to the labeling of Morpholinos and found that heating was still necessary, in addition to the observed low labeling efficiency [43]. The labeled impurity was free ^{99m}Tc-MAG_3. To identify the source of impurity, we performed repeated purifications and labeled the eluated product in each purification cycle to determine whether incomplete separation was responsible. It turned out that during the labeling process some MAG_3 chelators were shed off the Morpholino conjugate. We therefore incorporated a nonradioactive fake-labeling step to

dissociate the metastable conjugate followed by its removal prior to the radiolabeling [43]. The presence of tin (Sn^{2+}) in this step was necessary, which was found a couple of years later [20, 44], probably due to the deposit of some tin oxide precipitate on the top of the purification column or due to the water quality (presence of trace metals) at that time [43]. The MAG_3 modified Morpholino was successfully labeled with either ^{99m}Tc or ^{188}Re [44]. However, the labeling protocols were not exactly the same is that the $^{188}ReO_4^-$ was more difficult to be reduced than the $^{99m}TcO_4^-$.

4. The conjugation protocol is now well established to provide a high labeling efficiency (>90%) with both ^{188}Re and ^{99m}Tc [44]. The MAG_3-cMORF18 preparation can be used for at least a year. We have developed another conjugation protocol in which the first column purification is replaced by a centrifuge separation [45]. The chemistry is essentially the same and the product is equally effective.
5. The radiolabeling of Morpholino with ^{111}In and ^{90}Y using p-SCN-Bn-DOTA as the chelator experienced a heat-instability issue, similar to that for the labeling with ^{99m}Tc using MAG_3 chelator [46]. In the beginning when labeling with ^{111}In at room temperature, we did not observe this issue. This heat instability was revealed in the ^{90}Y labeling of the DOTA–Morpholino conjugate where heating was required. We resolved this problem based on our experience with the metastable impurity in the MAG_3–Morpholino conjugate. A heating prior to the labeling readily destroyed the heat-sensitive species and released the chelators metastably attached to the Morpholino. The modified conjugation protocol provides a high labeling efficiency of over 90%.
6. The specific radioactivity for ^{90}Y labeled Morpholino cannot be very high, possibly due to radiolysis [38].

References

1. Pressman D, Korngold L (1953) The in vivo localization of anti-Wagner osteogenic-sarcoma antibodies. Cancer 6:619–612
2. Reardan DT, Meares CF, Goodwin DA et al (1985) Antibodies against metal chelates. Nature 316(6025):265–268
3. Hnatowich DJ, Virzi F, Rusckowski M (1987) Investigations of avidin and biotin for imaging applications. J Nucl Med 28:1294–1302
4. Liu G, Liu C, Zhang S, He J, Liu N, Gupta S, Rusckowski M, Hnatowich DJ (2003) Investigations of technetium-99 m morpholino pretargeting in mice. Nucl Med Commun 24:697–705
5. Liu G, Dou S, Akalin A, Rusckowski M, Streeter PR, Shultz LD, Greiner DL (2012) Pretargeting vs direct targeting of human betalox5 islet cells subcutaneously implanted in mice using an anti-human islet cell antibody. Nucl Med Biol 39:645–651
6. Liu G (2016) A clearance step will become increasingly crucial for pretargeted tumor therapy when tumor accumulation is improved. J Cancer Clin Trials 1:2
7. Goldenberg DM, Chang CH, Rossi EA, JW MB (2012) Sharkey RM Pretargeted molecular imaging and radioimmunotherapy. Theranostics 2:523–540

8. van de Watering FC, Rijpkema M, Robillard M, Oyen WJ, Boerman OC (2014) Pretargeted imaging and radioimmunotherapy of cancer using antibodies and bioorthogonal chemistry. Front Med (Lausanne) 1:44
9. Kuijpers WH, Bos ES, Kaspersen FM, Veeneman GH, van Boeckel CA (1993) Specific recognition of antibody-oligonucleotide conjugates by radiolabeled antisense nucleotides: a novel approach for two-step radioimmunotherapy of cancer. Bioconjug Chem 4:94–102
10. Hnatowich DJ (1997) Pharmacokinetics considerations in the development of oligomers as radiopharmaceuticals. Q J Nucl Med 41:91–100
11. Liu G, He J, Zhang S, Liu C, Rusckowski M, Hnatowich DJ (2002) Cytosine residues influence kidney accumulations of ^{99m}Tc-labeled morpholino oligomers. Antisense Nucleic Acid Drug Dev 12:393–398
12. Rossin R, Verkerk PR, van den Bosch SM, Vulders RCM, Verel I, Lub J, MS R (2010) In vivo chemistry for pretargeted tumor imaging in live mice. Angew Chem Int Ed 49:3375–3378
13. Förster C, Bergmann R, Schubert M, Walther M, Pietzsch J, Vonhoff S, Klussmann S, Pietzsch H-J, Steinbach J (2010) Radiolabeled L-oligonucleotides with tunable pharmacokinetics—a suitable complementary system for pretargeting technologies. Nucl Med Biol 37:706
14. Vater A, Klussmann S (2015) Turning mirror-image oligonucleotides into drugs: the evolution of Spiegelmer® therapeutics. Drug Discov Today 20:147–155
15. Liu G, Hnatowich DJ (2008c) A semiempirical model of tumor pretargeting. Bioconjug Chem 19:2095–2104
16. Liu G, He J, Dou S, Gupta S, Rusckowski M, Hnatowich DJ (2005) Further investigations of morpholino pretargeting in mice—establishing quantitative relations in tumor. Eur J Nucl Med Mol Imaging 32:1115–1123
17. Liu G (2013) Rules of thumb for maximum percent tumor accumulation. Nucl Med Biol 40:865–867
18. Liu G, Dou S, He J, Liu X, Rusckowski M, Hnatowich DJ (2007) Predicting the biodistribution of radiolabeled cMORF effector in MORF-pretargeted mice. Eur J Nucl Med Mol Imaging 34:237–246
19. Liu G, Dou S, Pretorius PH et al (2010) Tumor pretargeting in mice using MORF conjugated CC49 antibody and radiolabeled complimentary cMORF effector. Q J Nucl Med Mol Imaging 54:333–340
20. Liu G, He J, Dou S, Gupta S, Vanderheyden J-L, Rusckowski M, Hnatowich DJ (2004) Pretargeting in tumored mice with radiolabeled morpholino oligomer showing low kidney uptake. Eur J Nucl Med Mol Imaging 31:417–424
21. Liu G, Dou S, Rusckowski M, Hnatowich DJ (2008) An experimental and theoretical evaluation of the influence of pretargeting antibody on the tumor accumulation of effector. Mol Cancer Ther 7:1025–1032
22. Liu G, Dou S, Cheng D, Leif J, Rusckowski M, Streeter PR, Shultz LD, Hnatowich D, Greiner D (2011) Human islet cell MORF/cMORF pretargeting in a xenogeneic murine transplant model. Mol Pharm 8:767–773
23. Sharkey RM, van Rij CM, Karacay H, Rossi EA, Frielink C, Regino C, Cardillo TM, McBride WJ, Chang CH, Boerman OC, Goldenberg DM (2012) A new Tri-Fab bispecific antibody for pretargeting Trop-2-expressing epithelial cancers. J Nucl Med 53:1625–1632
24. Behr TM, Becker WS, Sharkey RM, Juweid ME, Dunn RM, Bair HJ, Wolf FG, Goldenberg DM (1996) Reduction of renal uptake of monoclonal antibody fragments by amino acid infusion. J Nucl Med 37:829–833
25. Honarvar H, Westerlund K, Altai M, Sandström M, Orlova A, Tolmachev V, Karlström AE (2016) Feasibility of affibody molecule-based PNA-mediated radionuclide pretargeting of malignant tumors. Theranostics 6:93–103
26. Altai M, Perols A, Tsourma M, Mitran B, Honarvar H, Robillard M, Rossin R, Ten Hoeve W, Lubberink M, Orlova A, Karlström AE, Tolmachev V (2016) Feasibility of affibody-based bioorthogonal chemistry-mediated radionuclide pretargeting. J Nucl Med 57:431–436
27. Dou S, Liu G (2016) A method to characterize in vivo binding of morpholinos for drug design. J Drug Discov Develop and Deliv 3:1018
28. Liu G, Zhang S, He J, Liu N, Gupta S, Rusckowski M, Hnatowich DJ (2002) The influence of chain length and base sequence on the pharmacokinetic behavior of ^{99m}Tc-morpholinos in mice. Q J Nucl Med 46:233–243
29. Axworthy DB, Reno JM, Hylarides MD, Mallett RW, Theodore LJ, Gustavson LM, Su F, Hobson LJ, Beaumier PL, Fritzberg AR (2000) Cure of human carcinoma xenografts by a single dose of pretargeted yttrium-90 with negligible toxicity. Proc Natl Acad Sci U S A 97:1802–1807
30. Cheal SM, Yoo B, Boughdad S, Punzalan B, Yang G, Dilhas A, Torchon G, Pu J, Axworthy

DB, Zanzonico P, Ouerfelli O, Larson SM (2014) Evaluation of glycodendron and synthetically modified dextran clearing agents for multistep targeting of radioisotopes for molecular imaging and radioimmunotherapy. Mol Pharm 11:400–416
31. Dou S, Virostko J, Greiner DL, Powers AC, Liu G (2015) Quantitative correlation of in vivo properties with in vitro assay results: the in vitro binding of a biotin-DNA analogue modifier with streptavidin predicts the in vivo avidin-induced clearability of the analogue-modified antibody. Mol Pharm 12:3097–3103
32. Dou S, Virostko J, Rusckowski M, Greiner DL, Powers AC, Liu G (2014) Differentiation between temporary and real non-clearabilityof biotinylated IgG antibody by avidin in mice. Front Pharmacol 5:172
33. Liu G, Dou S, Chen X, Chen L, Liu X, Rusckowski M, Hnatowich DJ (2010) Adding a clearing agent to pretargeting does not lower the tumor accumulation of the effector as predicted. Cancer Biother Radiopharm 25:757–762
34. Liu G, Dou S, Baker S, Akalin A, Cheng D, Chen L, Rusckowski M, Hnatowich DJ (2010) A preclinical ^{188}Re tumor therapeutic investigation using MORF/cMORF pretargeting and an antiTAG-72 antibody CC49. Cancer Biol Ther 10:767–774
35. Liu G, Mang'era K, Liu N, Gupta S, Rusckowski M, Hnatowich DJ (2002) Tumor pretargeting in mice using technetium-99 m labeled morpholinos, a DNA analog. J Nucl Med 43:384–391
36. Liu G, Dou S, Mardirossian G, He J, Zhang S, Liu X, Rusckowski M, Hnatowich DJ (2006) Successful radiotherapy of tumor in pretargeted mice by ^{188}Re-radiolabeled phosphorodiamidate morpholino oligomer, a synthetic DNA analogue. Clin Cancer Res 12:4958–4964
37. Liu G, Cheng D, Dou S, Chen X, Liang M, Pretorius H, Rusckowski M, Hnatowich DJ (2009) Replacing ^{99m}Tc with ^{111}In improves MORF/cMORF pretargeting by reducing intestinal accumulation. Mol Imaging Biol 11:303–307
38. Liu G, Dou S, Liu Y, Wang Y, Rusckowski M, Hnatowich DJ (2011) ^{90}Y labeled phosphorodiamidate morpholino oligomer for pretargeting radiotherapy. Bioconjug Chem 22:2539–2545
39. He J, Rusckowski M, Wang Y, Dou S, Liu X, Zhang S, Liu G, Hnatowich DJ (2007) Optical pretargeting of tumor with fluorescent MORF oligomers. Mol Imaging Biol 9:17–23
40. Liu G, Dou S, Yin D, Squires S, Liu X, Wang Y, Rusckowski M, Hnatowich DJ (2007) A novel pretargeting method for measuring antibody internalization in tumor cells. Cancer Biother Radiopharm 22:33–39
41. He J, Liu G, Dou S, Gupta S, Rusckowski M, Hnatowich DJ (2007) An improved method for covalently conjugating morpholino oligomers to anticancer antibodies. Bioconjug Chem 18:983–988
42. Winnard P Jr, Chang F, Rusckowski M, Mardirossian G, Hnatowich DJ (1997) Preparation and use of NHS-MAG3 for technetium-99 m labeling of DNA. Nucl Med Biol 24:425–432
43. Liu G, Zhang S, He J, Zhu Z, Rusckowski M, Hnatowich DJ (2002) Improving the labeling of S-acetyl NHS-MAG_3 conjugated morpholino oligomers. Bioconjug Chem 13:893–897
44. Liu G, Dou S, He J, Yin D, Gupta S, Zhang S, Wang Y, Rusckowski M, Hnatowich DJ (2006) Radiolabeling of MAG_3-morpholino oligomers with ^{188}Re at high labeling efficiency and specific radioactivity for tumor pretargeting. Appl Radiat Isot 64:971–978
45. Wang Y, Liu G, Hnatowich DJ (2006) Methods for MAG_3 conjugation and ^{99m}Tc radiolabeling of biomolecules. Nat Protoc 1:1477–1480
46. Liu G, Dou S, Liu Y, Chen L, Cheng D, Greiner D, Rusckowski M, Hnatowich DJ (2011) Unexpected side products in the conjugation of an amine-derivatized morpholino oligomer with p-isothiocyanate benzyl DTPA and their removal. Nucl Med Biol 38:159–163

Chapter 15

Diagnostic Applications of Morpholinos and Label-Free Electrochemical Detection of Nucleic Acids

Rastislav Levicky, Ursula Koniges, and Napoleon Tercero

Abstract

Diagnostic applications of morpholinos take advantage of their unique properties including backbone charge neutrality, a weak impact of ionic strength on their hybridization behavior, and their resistance to enzymatic degradation. This chapter overviews how these properties have advanced transduction and other capabilities useful for the analysis of nucleic acids. In many cases, the benefits stem from electrostatic mechanisms; for example, use of low ionic strengths improves sensitivity of detection while decreasing background signals because only the nucleic acid analyte is charged. While most literature reports focus on in vitro assays in buffer, morpholinos have been also used for biodistribution measurements of species such as fungal rRNA and miRNA. After reviewing the diagnostic applications of morpholinos, the chapter describes preparation of morpholino monolayers on metal supports for electrochemical diagnostics and the procedure for performing label-free detection of DNA from changes in surface capacitance.

Key words Biosensor, Surface, Interface, Microarray, FISH, Fluorescence, Imaging

1 Introduction

Morpholinos bring a number of key advantages to diagnostic applications. Compared to other uncharged DNA analogs, such as peptide nucleic acids (PNA) and methylphosphonates [1], morpholinos offer flexibility with regard to oligomer sequence and length combined with an optimal balance of binding affinity. The binding affinity is well suited to diagnostics using oligomer lengths of around 20 nucleotides. These lengths provide good sequence specificity and affinity to bind analyte nucleic acids (targets) at the fM to nM concentrations of interest without the affinity being so high as to induce strong background signals from shorter and/or mismatched sequences [2]. Like other nucleic acid analogs morpholinos possess excellent resistance to enzymatic degradation.

Diagnostic applications of morpholinos (MOs) were initiated nearly two decades ago. The first example devised a purification strategy in which nucleic acids of interest were hybridized with

Hong M. Moulton and Jon D. Moulton (eds.), *Morpholino Oligomers: Methods and Protocols*, Methods in Molecular Biology, vol. 1565, DOI 10.1007/978-1-4939-6817-6_15, © Springer Science+Business Media LLC 2017

complementary MOs in solution, followed by solid phase capture of the hybrids [3]. This early study already highlighted certain key advantages bestowed by the charge-neutral nature of morpholinos. For example, it was recognized that hybridization between a charge-neutral MO and a nucleic acid has a much reduced dependence on ionic strength, compared to base pairing between two nucleic acids. This insensitivity to salt concentration can be used to devise assays in which low ionic strengths destabilize secondary structure in a nucleic acid analyte and thus improve its availability for detection. This general strategy was exploited in morpholino microarrays [4] as well as nanoparticle-based diagnostics in which thermally induced aggregation of MO-modified gold nanoparticles was used to identify analytes of interest [5, 6]. The nanoparticle aggregation strategy was recently combined with asymmetric PCR to develop a single nucleotide polymorphism assay for warfarin sensitivity [7]. These studies found ionic strengths in the 10–50 mM range as optimal for realizing good diagnostic signals while suppressing analyte secondary structure. The lower limit arises because, in surface hybridization such as encountered in microarrays, overly low ionic strengths decrease signals because of electrostatic penalties to accumulation of charged nucleic acid analyte on the array [4].

Perhaps the broadest impact of morpholinos on diagnostic research has been through advancement of mechanisms of transduction, especially ones based on electrostatic and conductivity principles. Tercero and coworkers demonstrated label-free detection of hybridization between immobilized morpholino probes and solution DNA targets by monitoring changes in interfacial capacitance [8, 9]. The observed capacitive signals were attributed to changes in the near surface ionic strength and dielectric constant that accompanied accumulation of charged DNA strands. The signals were several times higher when uncharged morpholino probes were used instead of DNA probes, illustrating the benefit of restricting electrostatic interactions to just the analyte. Gao et al. took advantage of the accumulation of analyte surface charge in a different way by using it to electrostatically immobilize, once hybridization was complete, cationic polymers [10, 11]. The final extent of hybridization could be detected using amperometric methods in combination with a redox-active, positively charged polymer for the readout step [11]. Alternately, detection could be accomplished through impedance measurements and an insulating polymer as the contrast agent [10]. A somewhat related approach by another group of investigators used Zr^{4+} cations to "label" hybridized morpholino films through association with the bound DNA, followed by complexation of the associated Zr^{4+} with carboxylate rhodamine in a second step [12]. This method provided linear fluorescent readout from 1 pM to 1 nM concentration of analyte. The detection limit was further lowered to 10 fM by switching to hybridization on microspheres instead of a planar support and implementing posthybridization labeling with streptavidin/alkaline phosphatase, coupled to a phosphatase-catalyzed reaction cascade for fluorescent detection [13].

Several approaches have relied on unique properties of nanomaterials [14–17]. The narrow pores of porous anodic alumina nanostructures, on the order of a few tens of nm in diameter, were modified with morpholino probes. Because of the narrowness of these pores, subsequent hybridization effectively fills the pores with charged nucleic acids together with mobile countercharge needed to maintain electroneutrality. These changes can be readily measured, especially at lower ionic strengths, to monitor the progress of hybridization. Various approaches have been used, including measurements of the AC impedance across the nanopores [14] and monitoring the through-pore diffusion of charged electroactive species, such as ferricyanide anions, whose partitioning into the pores is affected by the amount of hybridized nucleic acid [15]. Alternately, Gao et al. immobilized MO–DNA heteroduplexes on nanopore walls; if the sample was sufficiently complementary an exchange reaction occurred that transferred the heteroduplex DNA to the solution sample, leading to increase in ferricyanide diffusion [16]. Other nanostructures explored for morpholino diagnostic schemes include nanowires, as reported by Zhang et al., who functionalized silicon nanowires with morpholino probes and detected binding of charged nucleic acids from changes in nanowire conductance [17]. Binding of negatively charged DNA analyte leads to repulsive field effects within the nanowire that lower its conductance. Approximately a 100 fM sensitivity was reported.

The compatibility of morpholinos with hybridization under low ionic strengths enables additional capabilities. For example, Martins et al. exploited low ionic strength buffers to concentrate charged DNA analyte using an electrokinetic ion concentration polarization strategy, reaching 1000-fold increases in analyte concentration within several minutes and accelerating rates of DNA hybridization to immobilized MO probes by over 50-fold [18]. Zhao and coworkers combined morpholino-modified gold nanoparticles with pH-sensitive folding of i-motif DNA to demonstrate plasmonic pH sensors capable of functioning in low ionic strength environments [19].

The synthetic origins of morpholinos endow them with excellent resistance to enzymatic degradation, a consideration that becomes especially relevant when morpholinos are introduced into live cells or tissue matrices. A number of investigators have exploited this advantage for diagnostic purposes. Liong et al. used morpholino linkers to bestow DNase resistance in their approach to magnetic immunolabeling of cells [20]. Cao and coworkers demonstrated that the DNA strand in a MO–DNA duplex is protected against DNase I cleavage in surface hybridization applications [21]. Burki et al. took this notion a step further by developing an assay for morpholino concentrations in tissues and sera based on nuclease digestion of unhybridized DNA probes immobilized on well plates. In this approach, only MO–DNA hybrids remained undigested and are subsequently detected through an alkaline phosphatase-mediated mechanism that leads to a fluorescent product [22].

Morpholinos have also found significant use in biodistribution measurements. They can be modified with radiolabels and antibodies for PET and SPECT tomography imaging, as demonstrated by Cheng et al. in the context of studying drug distributions in mice [23]. Radiolabeled morpholinos have also been used to detect presence of fungal rRNA as a diagnostic intervention, based on direct hybridization of the morpholino with the rRNA sequences of interest in the lungs of infected mice [24]. Most recently, fluorescein-modified morpholinos were used for in situ hybridization with small RNA in tissue samples to reveal the RNA content of synaptic vesicles, in combination with alkaline phosphatase-based signal amplification [25]. This experimental protocol was previously developed and demonstrated for determining miRNA distributions [26]. The same study demonstrated that morpholino probes are a highly competitive alternative to locked nucleic acid (LNA) probes for in situ hybridization diagnostics. Diagnostic assays have been also developed for the "reverse problem" of quantifying the concentration of morpholino oligomers in tissue and cell samples, using in situ hybridization and fluorescent LNA probes for the morpholino sequences of interest [27].

Having overviewed some of the ways that morpholinos have and continue to contribute to diagnostic applications, the remainder of this chapter will describe a protocol for preparation of morpholino films on electrodes, and of using such films for monitoring binding of nucleic acids through determination of the changes in interfacial capacitance. The protocol foremost derives from studies reported in references [8, 9].

2 Materials

Prepare all solutions using ultrapure water with a resistivity of 18.2 MΩ cm at 25 °C. Use analytical grade reagents and follow all waste disposal regulations when discarding waste materials.

2.1 Components for Oligonucleotide Labeling and Purification

1. HPLC system with a binary mobile phase.
2. Reversed-phase column, e.g., Clarity Oligo-RP column, 100 × 4.6 mm, 3 μm particle size (Phenomenex, Torrance, CA, USA).
3. Vacuum centrifuge, e.g., vacufuge (Eppendorf, Hamburg, Germany).
4. HPLC aqueous mobile phase: triethylamine/hexafluoroisopropanol buffer (TEA-HFIP); 4.5 mM TEA, 100 mM HFIP, pH 8.0.

5. HPLC organic mobile phase: methanol.
6. Morpholino probe solution: 0.25 nmol of morpholino probes (Gene Tools, Philomath, OR, USA) in 1 ml of deionized water. Store at 0 °C until use.
7. Electroactive labels for determining absolute coverage of probes and targets, if desired. These are available commercially and can be amino-reactive such as ferrocene mono-carboxylic acid *N*-hydroxysuccinimide ester (FcCA-NHS) as well as sulfhydryl-reactive such as *N*-(2-ferrocene-ethyl) maleimide (FEM).
8. Disulfide deprotection buffer: 10 mM Tris–HCl, pH 8.0, 10 mM dithiothreitol, 1 mM EDTA. Use solution immediately after preparation.
9. Probe morpholino labeling buffer: 0.5 M sodium carbonate buffer, pH 9.0.
10. DNA target: commercially synthesized and HPLC purified (Eurofins MWG Operon, Louisville, KY, USA).
11. Oligonucleotide purification columns: NAP-10 purification column (GE Healthcare, Chicago, IL, USA) and oligonucleotide purification cartridge (Applied Biosystems, Foster City, CA).
12. Target DNA labeling buffer: 150 mM potassium phosphate buffer, pH 8.0.

2.2 Components for Preparation and Modification of Gold Electrodes

1. Electrochemical etching solution: 0.5 M H_2SO_4.
2. Solution for determination of electrode roughness: 0.1 M NaF.
3. Mechanical polishing solution: 1 μm diamond particle suspension (Bioanalytical Systems, Inc., West Lafayette, IN, USA).
4. Electrode polishing pad (Bioanalytical Systems, Inc., West Lafayette, IN, USA).
5. Passivation solution: 1 mM mercaptopropanol (MCP, 95% purity; Aldrich).

2.3 Materials for Electrochemical Measurement

1. Electrochemical workstation for measuring AC impedance and cyclic voltammetry with a three-electrode cell with ports for an Au working electrode (*see* following **step 3**), a platinum wire counter electrode, and a reference electrode. A typical reference electrode is Ag/AgCl/saturated KCl; 0.197 V vs NHE [normal hydrogen electrode] at 25 °C (*see* **Note 1**). All quoted potentials are expressed relative to this reference.
2. Working electrodes: typically 1.6 mm diameter polycrystalline gold disk electrodes (Bioanalytical Systems, Inc., West Lafayette, IN, USA).

3. DNA target solution: 25 nM complementary DNA target, commercially synthesized and HPLC purified (Eurofins MWG Operon, Louisville, KY, USA) and diluted in deionized water.
4. Hybridization buffer: sodium phosphate buffer, pH 7.0 (*see* **Note 2**).

3 Methods

3.1 Labeling and Purification of Morpholino and Nucleic Acid Oligonucleotides

(*only required if absolute coverages are to be determined*)

1. Label morpholino oligos at the end not used for immobilization with an electroactive label, such as ferrocene. Standard bioconjugate methods are used.

   Example: Combine 150-fold stoichiometric excess of ferrocene mono-carboxylic acid *N*-hydroxysuccinimide ester (FcCA-NHS) with 0.3 mM solution of morpholino probe in 150 mM pH 9.0 sodium carbonate buffer for 1 h at room temperature. Remove unreacted FcCA-NHS using a NAP-10 purification column. Complete purification of recovered morpholino oligos using reverse-phase HPLC with a 20 min linear gradient of 12–100% methanol in triethylamine/hexafluoroisopropanol buffer (4.5 mM TEA, 100 mM HFIP, pH 8.0), followed by 5 min at 100% methanol.
2. Label nucleic acid analyte with an electroactive label. Standard bioconjugate methods apply.

   Example: for amino-modified nucleic acid and amino-reactive electroactive tags, the sample procedure in **step 1** earlier can be followed but using an HPLC gradient for nucleic acids (*see* later). For disulfide-modified nucleic acid and thiol-reactive labels, proceed as follows. Deprotect the disulfide for 2 h in a solution of 10 mM dithiothreitol (DTT), 10 mM Tris–HCl, 1 mM EDTA, pH 8.0. Remove excess DTT on a NAP-10 column. React the recovered solution for 1 h with 50-fold molar excess of a thiol-reactive tag such as *N*-(2-ferrocene-ethyl) maleimide (FEM) in 150 mM potassium phosphate buffer, pH 8.0. Purify the resultant product using a 12–60% gradient of methanol in HFIP-TEA spread over 22 min.
3. Store labeled morpholino and DNA conjugates dry at −20 °C until use.

3.2 Electrode Preparation and Modification

1. Clean a gold disk working electrode by mechanical polishing with 1 μm diamond suspension on an electrode polishing pad; once done rinse well with methanol followed by deionized water. Electrode should have a mirror-like finish. While keeping the electrode covered by a water droplet to prevent contamination from adsorbates in air, transfer it to 0.5 M H_2SO_4 and

perform 60 cycles of cyclic voltammetry electropolishing by cycling from 0.24 to 1.54 V and back to 0.24 V at 0.1 V/s. Electropolishing further cleans the gold through alternate oxidation and reduction cycles. When done, rinse electrode thoroughly with deionized water and transfer, while covered with a water droplet, into 0.1 M NaF for determination of roughness as in **step 2** of Subheading 3.2, or else directly into probe deposition solution as in **step 3** of Subheading 3.2.

2. If desired, determine electrode roughness r and the true area $A = rA_g$ where A_g is the geometric area of the electrode, by measuring the electrode capacitance at −0.8 V in 0.1 M NaF solution. Set up the experiment as in **step 1** of Subheading 3.3. Measure the total electrode capacitance using AC impedance at a bias of −0.8 V, using a superimposed potential modulation of 25 Hz and 5 mV rms amplitude. From the measured in-phase and out-of-phase currents, calculate the electrode capacitance as in Subheading 3.4, making sure to use μF as the units. Divide the electrode capacitance by 22 $\mu F/cm^2$ to arrive at the true electrode area A in cm^2. Roughness factor r can be calculated from A and A_g, $r = A/A_g$.
3. After roughness determination, rinse the electrode with deionized water and, without drying, immerse it in 0.25 μM probe deposition solution. A probe solution concentration range of 0.25–0.75 μM is recommended (*see* **Note 3**).
4. After immobilization, rinse electrode surface with deionized water, then passivate the electrode surface with 1 mM mercaptopropanol solution for 150 min (*see* **Note 4**).

3.3 Electrochemical Measurements

1. Set up the electrochemical workstation with a three-electrode cell filled with the hybridization solution and connect the Au working electrode, a platinum wire counter-electrode, and an Ag/AgCl/saturated KCl (or other suitable) reference electrode.
2. To determine capacitance C_d, make an AC impedance measurement. Typical procedure: Set the DC bias to −0.1 V, at which sensitivity is optimal. On top of this bias, apply a sinusoidal potential modulation of frequency f (*see* **Note 5**) and 5 mV rms amplitude. From the measured in-phase and out-of-phase currents, calculate C_d as in Subheading 3.4. Changes in C_d as a function of time are used to monitor the progress of hybridization. To more accurately estimate C_d, correct for baseline drift by extrapolating drift from points measured prior to addition of target. Alternately, add 1 mM mercaptopropanol to the hybridization solution to decrease baseline drift by better maintaining electrode passivation, or use a second working electrode with a nonhybridizing probe sequence to serve as baseline reference.

3. To determine absolute probe and target coverages, make a cyclic voltammetry measurement over a potential range that spans the electroactive region of the tags used. Typical procedure for ferrocene labels: scan the potential at 20 V/s from 0 to 0.6 V and back to 0 V. The high scanning speed decreases likelihood of ferrocene degradation. Calculate probe and/or target coverage from the measured current vs potential voltammogram as in Subheading 3.5.

3.4 Derivation of Differential Capacitance (C_d) from AC Impedance Data

1. Calculate the out-of-phase impedance $Z'' = -V_{AC}I_O/(I_I^2 + I_O^2)$, where V_{AC} is amplitude of the imposed sinusoidal potential modulation and I_O and I_I are the measured amplitudes of the out-of-phase and in-phase components of the current, respectively.
2. Calculate the differential capacitance from $Cd = 1/(2\pi f|Z''|)$, where f is the frequency of the imposed potential modulation and the vertical bars "| |" signify absolute value. The capacitance can be expressed per area by dividing C_d by the total area A (**step 2** of Subheading 3.2).

3.5 Derivation of Target and Probe Coverages from Cyclic Voltammetry Data

1. Integrate areas of the peaks in the CV voltammogram that correspond to the probe and/or the target tag. The area is given by the total current minus baseline current. The baseline current under the peak can often be well estimated by a sum of a linear and a stretched exponential function.
2. Calculate the corresponding probe or target coverage Γ from the area P of the integrated peak using $\Gamma = P/(eA\nu)$ where $e = 1.602 \times 10^{-19}$ C is the elementary charge, A is the measured electrode area (**step 2** of Subheading 3.2), and ν is the scan rate used for measurement. Confirm dimensional consistency by checking all units carefully.

4 Notes

1. Reference electrodes can be bought commercially or fabricated by enclosing a 5–6 cm piece of 0.5 mm diameter silver wire coated with AgCl within a glass sleeve with a porous glass junction and filled with a saturated KCl solution.
2. A recommended concentration range for the pH 7.0 sodium phosphate hybridization buffer is 0.001–0.5 M (molarity of phosphate).
3. The probe coverage that results from the described protocol is typically between 2×10^{12} and 5×10^{12} probes/cm^2.

4. Mercaptopropanol passivation improves reproducibility of baselines in electrochemical experiments and suppresses non-specific adsorption of nucleic acid analyte.
5. The AC frequency should correspond to a phase angle of about 45° for the chosen cell geometry and hybridization buffer. Caution should be exercised to avoid any secondary capacitive charging that may arise at high frequencies.

References

1. Micklefield J (2001) Backbone modification of nucleic acids: synthesis, structure and therapeutic applications. Curr Med Chem 8:1157–1179
2. Summerton JE (2004) Morpholinos and PNAs compared. Lett Pept Sci 10:215–236
3. Wages JM, Wages GM, Matthews P, Weller D, Summerton J (1997) Affinity purification of RNA: sequence-specific capture by nonionic morpholino probes. Biotechniques 23:1116–1120
4. Qiao W, Kalachikov S, Liu Y, Levicky R (2013) Charge-neutral morpholino microarrays for nucleic acid analysis. Anal Biochem 434:207–214
5. Zu Y, Ting AL, Yi G, Gao Z (2011) Sequence-selective recognition of nucleic acids under extremely low salt conditions by using nanoparticle probes. Anal Chem 83:4090–4094
6. Zu Y, Ting AL, Gao Z (2011) Visualizing low-level point mutations: enzyme-like selectivity offered by nanoparticle probes. Small 7:306–310
7. Zu Y, Tan M-H, Chowbay B, Lee SC, Yap H, Lee MTM, Lu L-S, Chang C-P, Ying JY (2014) Nanoprobe-based genetic testing. Nano Today 9:166–171
8. Tercero N, Wang K, Gong P, Levicky R (2009) Morpholino monolayers: preparation and label-free DNA analysis by surface hybridization. J Am Chem Soc 131:4953–4961
9. Tercero N, Wang K, Levicky R (2010) Capacitive monitoring of morpholino-DNA surface hybridization: experimental and theoretical analysis. Langmuir 26:14351–14358
10. Gao Z, Deng H, Shen W, Ren Y (2013) A label-free biosensor for electrochemical detection of femtomolar microRNAs. Anal Chem 85:1624–1630
11. Gao Z, Ting BP (2009) A DNA biosensor based on a morpholino oligomer coated indium-tin oxide electrode and a cationic redox polymer. Analyst 134:952–957
12. Hu W, Hu Q, Li L, Kong J, Zhang X (2015) Detection of sequence-specific DNA with a morpholino-functionalized silicon chip. Anal Methods 7:2406–2412
13. Hu W, Ning Y, Li L, Kong J, Zhang X (2015) Highly sensitive detection of sequence-specific DNA with morpholino-functionalized magnetic microspheres. Anal Methods 7:6712–6717
14. Wang X, Smirnov S (2009) Label-free DNA sensor based on surface charge modulated ionic conductance. ACS Nano 3:1004–1010
15. Li SJ, Li J, Wang K, Wang C, Xu JJ, Chen HY, Xia XH, Huo Q (2010) A nanochannel array-based electrochemical device for quantitative label-free DNA analysis. ACS Nano 4:6417–6424
16. Gao H-L, Wang M, Wu Z-Q, Wang C, Wang K, Xia X-H (2015) Morpholino-functionalized nanochannel array for label-free single nucleotide polymorphisms detection. Anal Chem 87:3936–3941
17. Zhang GJ, Luo ZH, Huang MJ, Tay GK, Lim EJ (2010) Morpholino-functionalized silicon nanowire biosensor for sequence-specific label-free detection of DNA. Biosens Bioelectron 25:2447–2453
18. Martins D, Levicky R, Song Y-A (2015) Enhancing the speed of morpholino-DNA biosensor by electrokinetic concentration of DNA in a microfluidic chip. Biosens Bioelectron 72:87–94
19. Zhao Y, Cao L, Ouyang J, Wang M, Wang K, Xia XH (2013) Reversible plasmonic probe sensitive for pH in micro/nanospaces based on i-motif-modulated morpholino-gold nanoparticle assembly. Anal Chem 85:1053–1057
20. Liong M, Tassa C, Shaw SY, Lee H, Weissleder R (2011) Multiplexed magnetic labeling amplification using oligonucleotide hybridization. Adv Mater 23:H254–H257
21. Cao L, Zhao Y, Ji LN, Zhang J, Wang K, Xia XH (2013) The enhanced enzymolysis resistance of surface-immobilized DNA caused by hybridizing with morpholino. Electroanalysis 25:1074–1079
22. Burki U, Keane J, Blain A, O'Donovan L, Gait MJ, Laval SH, Straub V (2015) Development and application of an ultrasensitive hybridization-based ELISA method for the determination of

peptide-conjugated phosphorodiamidate morpholino oligonucleotides. Nucleic Acid Ther 25:275–284

23. Cheng D, Wang Y, Liu X, Pretorius PH, Liang M, Rusckowski M, Hnatowich DJ (2010) A comparison of ^{18}F PET and ^{99m}Tc SPECT imaging in phantoms and in tumored mice. Bioconjug Chem 21:1565–1570
24. Wang Y, Chen L, Liu X, Cheng D, Liu G, Liu Y, Dou S, Hnatowich DJ, Rusckowski M (2013) Detection of Aspergillus fumigatus pulmonary fungal infections in mice with 99mTc-labeled MORF oligomers targeting ribosomal RNA. Nucl Med Biol 40:89–96
25. Li H, Wu C, Aramayo R, Sachs MS, Harlow ML (2015) Synaptic vesicles contain small ribonucleic acids (sRNAs) including transfer RNA fragments (trfRNA) and microRNAs (miRNA). Sci Rep 5:14918
26. Lagendijk AK, Moulton JD, Bakkers J (2012) Revealing details: whole mount microRNA in situ hybridization protocol for zebrafish embryos and adult tissues. Biol Open 1:566–569
27. Schnell FJ, Crumley SL, Mourich DV, Iversen PL (2013) Development of novel bioanalytical methods to determine the effective concentrations of phosphorodiamidate morpholino oligomers in tissues and cells. Biores Open Access 2:61–66

Chapter 16

Intranasal Delivery of Peptide-Morpholinos to Knockdown Influenza Host Factors in Mice

Ricardo Rajsbaum

Abstract

Influenza viruses replicate primarily in the lung tissue of different host species. For efficient replication the virus utilizes host factors that are expressed in target cells. Cell-penetrating peptide-conjugated Morpholino oligomers (PPMOs) designed to target viral proteins have shown promising results as potential antiviral drugs in tissue culture and animal models. However, since viruses tend to have high rates of mutations, targeting viral proteins may result in viral escape mutants. An alternative approach to inhibit virus replication with PPMOs is to target host factors that are required for virus replication. Delivery of PPMO through the intranasal route has been shown to be effective in knockdown of host factors or microbial genes leading to protection against respiratory pathogens and reduced microbial burden. In addition, protective host innate antiviral immune responses in the lung can be studied by knockdown of immune signaling factors using PPMOs. Here we describe a successful approach using PPMOs to knockdown either proviral or antiviral host factors leading to changes in influenza virus replication in the lungs of mice, providing a tool to investigate immune responses and host–virus interactions in vivo.

Key words Influenza virus, TRIM E3-ubiquitin ligase, Intranasal administration, Lung tissue, PPMO, Knockdown of host proteins, Innate immune response, TRIM6, UBR4, Antiviral activity, Proviral factors

1 Introduction

Although so far the use of antisense phosphorodiamidate morpholino oligomers (PMO) for in vivo applications remains limited, it can have important applications in research as well as therapeutics. In fact, PMOs for treatment of Duchenne muscular dystrophy have been in clinical trials [1, 2] with one PMO approved as a drug by the US FDA. Multiples studies using animal models have shown promising results using cell penetrating peptide-conjugated PMO (PPMO) to target viral proteins resulting in reduced virus replication [3]. Recent studies have shown that PPMOs designed against Marburg virus, a highly pathogenic member of the Filoviridae family that causes severe hemorrhagic fever, is effective in nonhuman primates and no measurable safety concerns or adverse effects were

Hong M. Moulton and Jon D. Moulton (eds.), *Morpholino Oligomers: Methods and Protocols*, Methods in Molecular Biology, vol. 1565, DOI 10.1007/978-1-4939-6817-6_16,

© Springer Science+Business Media LLC 2017

observed in humans [4]. Despite increasing evidence that PPMOs can be used in vivo, some researchers remain skeptical of PPMO use in clinical settings because of safety concerns, limited knockdown efficiencies, and the potential of generating immune responses against the PPMO or inducing immune activation. So far, the use of PPMOs to inhibit virus replication has been more focused on targeting viral proteins; however, recent studies have shown that PPMO-mediated knock down of host factors in the lung of mice can be extremely useful to study host–pathogen interactions, especially in short-term studies. These studies have shown great success in avoiding immune activation and have resulted in significant phenotypes as well as efficient levels of protein knockdowns [5, 6]. Therefore, while delivery of PPMOs to other tissues/organs may be more difficult, it appears that intranasal administration of PPMOs provides an excellent route to study the function of lung host factors.

PPMOs can be designed to target sites on mRNA sequences to block translation of viral proteins and have been shown to be effective in vivo by effectively reducing pathogenicity and/or inhibiting replication of different viruses including poliovirus, coxsackievirus B3, dengue virus, West Nile virus, Venezuelan equine encephalitis virus, respiratory syncytial virus, Ebola virus, and influenza A virus [3]. Specifically to intranasal administration, PPMO targeting microbial genes have been shown to be effective to reduce bacterial infections by *Acinetobacter lwoffii* and *Acinetobacter baumannii* [7], or respiratory viruses including Influenza [8, 9], Respiratory Syncytial virus [10], and combined intranasal with subcutaneous PPMO treatment against lethal Venezuelan equine encephalitis virus (VEEV) infection [11].

However, since viruses have high rates of mutations, targeting virus genes with antisense sequences has the disadvantage of potentially selecting for viral escape mutants. In contrast, host factors can be targeted resulting in changes in virus replication depending on the function of the host protein targeted. For example, knockdown of the TRIM6 E3-ubiquitin, which is an important component of the host innate immune system and the type-I IFN antiviral response, results in reduced induction of antiviral IFN-stimulated genes and consequently increased virus replication [5] (*see* Fig. 2). In contrast, knockdown in the lungs with PPMOs targeting UBR4, a host E3-ligase important for promoting influenza virus replication, resulted in decreased virus load [6]. Here we present a relatively simple and convenient approach to knockdown lung proteins to study their function in the context of Influenza virus infection. The same methodology can be used for PPMOs designed to target viral RNA, providing additional experimental control to monitor knockdown efficiency. As an example we show here the efficient knockdown of TRIM6 in the lungs using intranasal administration of PPMOs. Knockdown of TRIM6 resulted in approximately 50% reduction in TRIM6 mRNA levels by quantitative RT-PCR (Fig. 2a) and protein levels by immunoblot (Fig. 2b). The reduction in TRIM6 expression by PPMO results in almost tenfold increase in Influenza virus titers (Fig. 2c). Consistent with a role of TRIM6 in type-I Interferon (IFN-I)-mediated responses [5], induction of the IFN-stimulated gene IFI44 was reduced in influenza-infected mice treated with TRIM6-specific PPMO as compared to control PPMO (Fig. 2c). The following is the detailed protocol used to achieve TRIM6 knockdown in the lungs of mice.

2 Materials

1. Previously designed and tested antisense PPMO targeting your specific gene. Nontargeting control (ensure that it does not has sequence complementarity with virus sequences to avoid targeting virus genes) (*see* **Notes 1–3**). As an example, the following are the oligo sequences used in our TRIM6 experiment. TRIM6-targeting PMO sequence: AAGCTTAGGACCG ACCTGGTACTCC (specifically targets the exon2-intron 2 junction of TRIM6 pre-mRNA). Nontargeting control sequence: CCTCTTACCTCAGTTACAATTTATA.
2. Influenza Virus PR8 strain (A/Puerto Rico/8/1934 H1N1 [PR8]).
3. Six- to 8-week-old female BALB/c mice purchased from Jackson Laboratories (or your own sources). Three to five mice per group.
4. Anesthetic (Ketamine and Xylazine, Ketamine or isoflurane).
5. Phosphate-buffered saline (PBS).
6. Syringes for anesthesia administration.
7. Dissection tools (scissors and forceps).
8. Micropipette and sterile tips (20–100 μl).
9. Microcentrifuge tubes (2 ml), with screw cap.

3 Methods

The experimental design for knockdown of host factors in the lung should take in consideration the experimental outcome that is desired to be investigated. For example, if the investigator desires to determine the lethal dose (LD_{50}) of a virus in PPMO-treated mice, groups of at least five sex- and age-matched, immunocompetent mice will need to be infected with different doses of virus and health scoring, survival, and body weights should be monitored on a daily basis for the duration of the experiment. For this kind of experiment, the amount of PPMO and the frequency of treatments (before or after infection) should be tested experimentally. Here we will describe a methodology that is effective for analysis of mRNA expression, protein quantification, and virus titration in the lungs of influenza-infected mice for a short period of time (2 days postinfection) (Fig. 1). This methodology can be used to study early events that occur after virus infection.

1. Dissolve lyophilized PPMOs in sterile PBS or water (DNAase and RNAase free), at a concentration of 100 μg in 40 μl (2.5 μg/ul) (*see* **Notes 1–3**). The final volume to be administered intranasally should not exceed 40 μl.
2. Anaesthetize mice prior to inoculation by intraperitoneal injection of 100 μl Ketamine HCl/Xylazine mix (Ketamine 90–150 mg/kg and Xylazine at 7.5–16 mg/kg) (*see* **Notes 4** and **5**). For a mouse of approx. 0.02 kg, the required concentration is 2 mg of ketamine and 0.15 mg xylazine in sterile water.
3. When the mouse is under anesthesia hold the mouse and using a micropipette release slowly (drop by drop) no more than 40 μl of PPMO solution into the nostrils (20 μl in each nostril) (*see* **Note 6**). Adjust the rate of release so as to allow the mouse

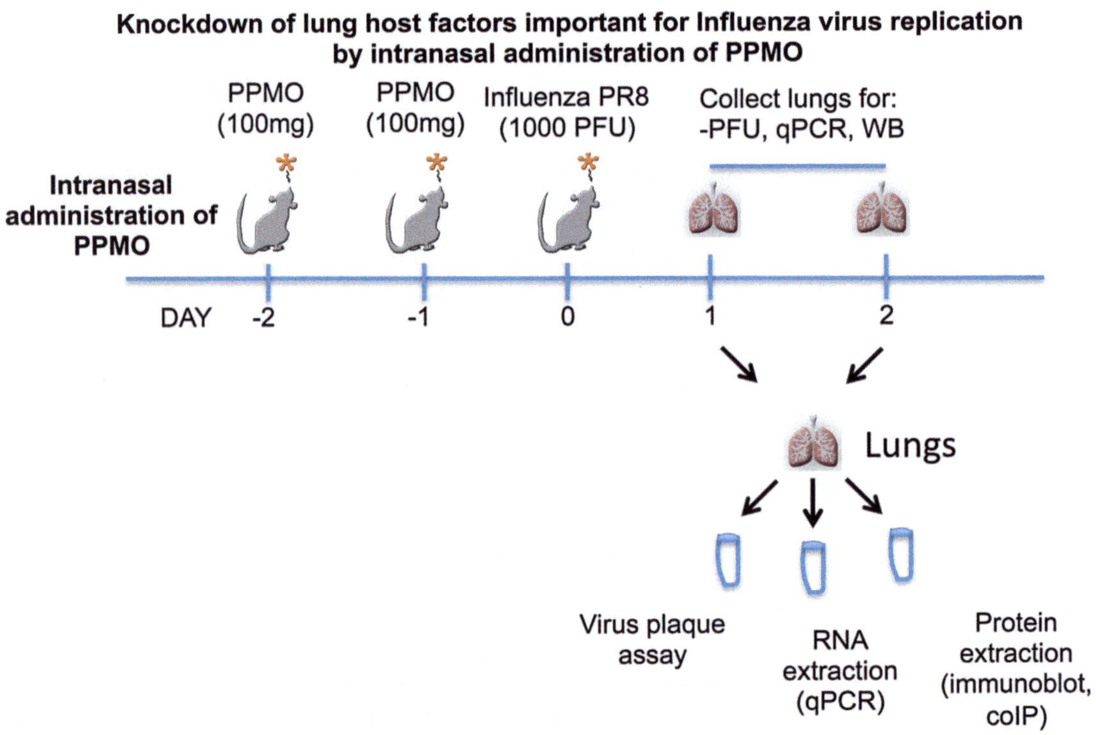

Fig. 1 Schematics of the protocol used for intranasal administration of PPMO and Influenza virus infection. Modified from [5], with permission from Elsevier

to inhale the oligo solution without causing the mouse to choke or interfere with breathing and allow the mouse to inhale the solution without forming bubbles.

4. Hold the mouse for a few seconds until breathing returns to normal.
5. Allow the mice to recover in a warm chamber and return to the cage.
6. After 24 h of the first treatment, repeat the PPMO administration by following **steps 2–5**.
7. After 24 h past the second treatment (48 h total), proceed with Influenza infection. First prepare the Influenza inoculum by diluting the desired amount of virus in sterile PBS (10–1000 PFU in a maximum of 40 μl PBS). Always include a mock-treated control.
8. If pathogenicity will be assessed, measure the body weight of each mouse before infection (*see* **Note 7**). The weight of mice at this time point will be normalized to 100% to calculate percentage of weight loss during the experiment.
9. Anaesthetize mice prior to inoculation as described in **step 2**.
10. Inoculate the virus (40 μl) as described in **steps 3–5**.
11. To assess the effects of protein knockdown in the lungs of infected mice, sacrifice groups of mice (three mice per group are usually enough; groups are infected and treated, infected and untreated, uninfected and treated, and uninfected and untreated). At 24 h postinfection, euthanize the first group of animals and collect lungs using standard procedures. Isolated lungs can be cut in three pieces for mRNA extraction, protein assays, or determination of virus titers (for results *see* Fig. 2) (*see* **Notes 8–11**). Repeat at day 2 or as long as necessary.

4 Notes

1. The major source of problems for in vivo knockdown using PPMO is the efficiency of protein knockdown. This will depend on the stability of the host protein intended to be targeted, the delivery of the PPMO, and the PPMO design. The concentration of PPMO may need to be adjusted depending on the efficiency of knockdown and delivery. This has to be empirically tested. Although the knockdown efficiency, PPMO concentrations, and antisense DNA sequences should be tested in murine cell lines prior to administration into mice, ultimately the PPMO concentration and knockdown efficiency have to be optimized and validated in vivo. Always use a nontargeting PPMO control. In addition, to ensure that the PPMOs do not

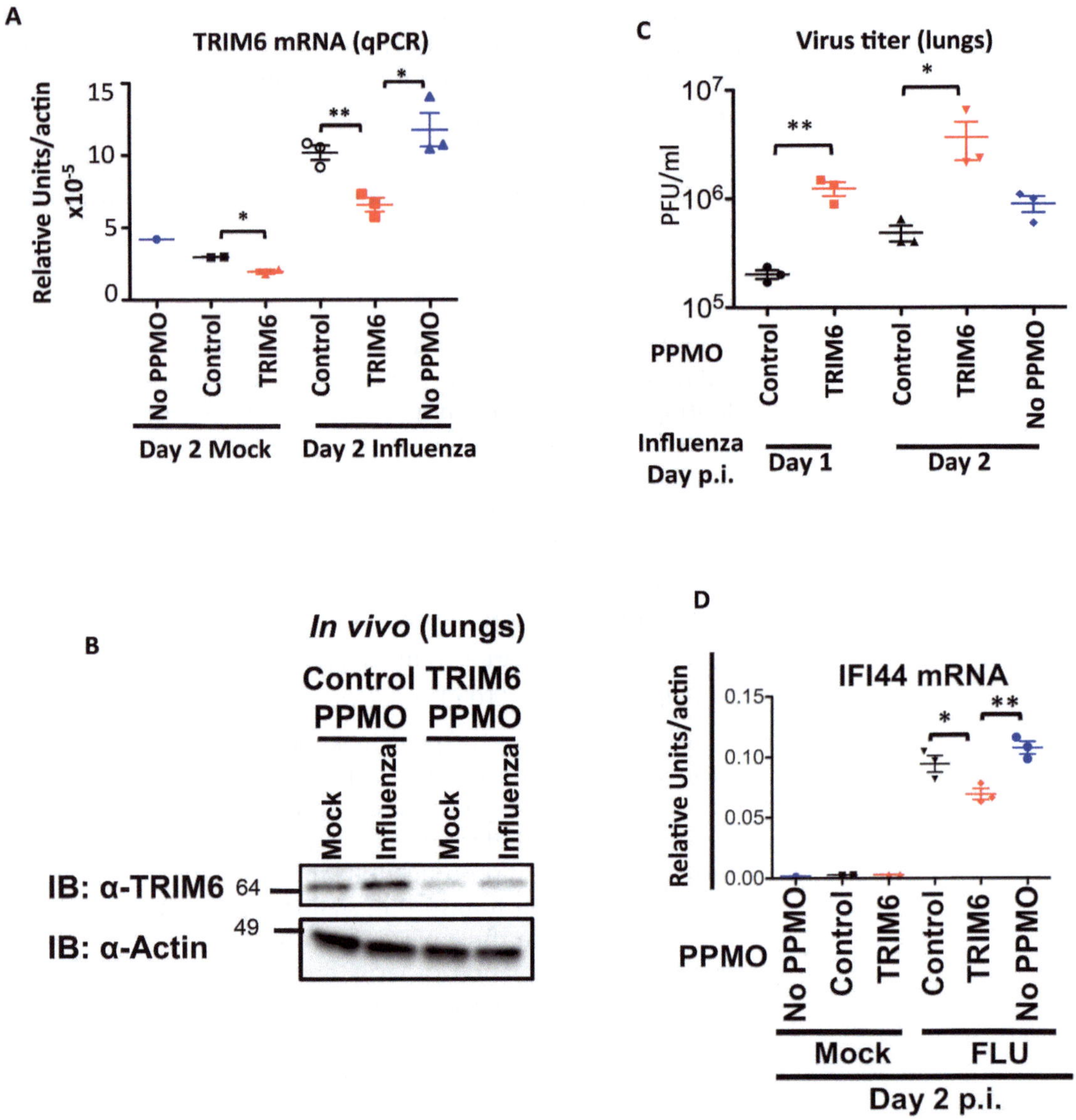

Fig. 2 TRIM6 knockdown by PPMO in the lungs of influenza-infected mice. Mice were treated i.n. with TRIM6-specific or nontargeting control PPMO at 48 h prior to infection. Treatment was repeated 24 h later. Mice were then infected i.n. with 1000 plaque-forming units (PFU) of influenza virus PR8 strain. At 24 and 48 h postinfection, lungs were sliced in three pieces for mRNA quantification by qPCR (**a**, **d**), analysis of TRIM6 protein levels by immunoblot (**b**), or virus quantification by plaque assays (**c**). The different samples were homogenized using a Beadblaster (Denville Scientific). Then samples were cleared by microcentrifugation and supernatants were used for downstream applications. For qPCR analysis, total RNA was isolated using RNeasy kit (Qiagen). DNAse digestion was performed using Turbo DNase (Ambion). Reverse transcription was performed using the high capacity cDNA reverse transcription kit (Applied Biosystems). (**a**, **d**) Real-time qPCR was performed in 384-well plates in triplicate using SYBR green I master mix (Roche) in a Roche LightCycler 480. Relative mRNA values were calculated using the $\Delta\Delta$Ct method using β-actin as housekeeping gene and plotted as relative values or fold induction of mock-treated samples. Figure modified from [5], with permission from Elsevier

cause immune activation in the lungs, always include a group of mice treated with PBS (Mock treatment [no PPMO treatment]).

2. Ensure that the antisense sequences are designed specifically to target the gene of interest for the appropriate mouse genetic line, as some polymorphisms exist.
3. It is important to note that since PPMO use a specific antisense sequence to target gene expression, the use of PPMO can lead to off-target effects. Therefore, whenever possible, it is a good practice to use at least two different PPMO sequences to confirm phenotypes or perform rescue experiments in tissue culture systems (knockdown protein of interest in cell lines, combined later with overexpression of the same protein from RNA lacking the PPMO target).
4. Anesthesia: Depending on the weight of the mouse, the amount of anesthesia can vary from Ketamine 90–150 mg/kg and Xylazine at 7.5–16 mg/kg.
5. If desired, Isoflurane by inhalation (≤5% to effect) can also be used to anesthetize mice.
6. The volume for intranasal administration should not exceed 40 μl. Excess volume or rapid administration will induce suffocation and death.
7. To monitor the effects of the virus, measure the body weight of each mouse at the same time every day for the duration of the experiment. Plot the data as percentages of the initial weight (*see* Fig. 2).
8. To assess the effects of knockdown in downstream applications, it is important to have working solutions ready with appropriate inhibitors, at 4 °C, to reduce the possibility of protein/RNA/virus degradation. Lung tissue can be stored at −80 °C in the appropriate buffer, but once thawed, the tissue should be homogenized and processed as soon as possible. The accuracy of detection of RNA/protein will depend on how well the samples have been homogenized, extracted, and conserved in proper buffers and protease or RNAase inhibitors.
9. For RNA applications, Qiagen RNAlater buffer can be used to stabilize the RNA, although storage in RLT buffer from the RNeasy kit yields enough RNA for cDNA synthesis and Real-Time PCR quantification. Samples should be kept on ice and stored at −80 °C as soon as possible. Process samples immediately after thawing, following manufacturer instructions. RNA can also be extracted using Trizol (results shown in Fig. 2a and d). Proceed with cDNA synthesis and Real-Time PCR [12].
10. For protein applications (immunoprecipitation and/or immunoblots), lungs can be homogenized in PBS-containing proteasome inhibitors (we use a Beadblaster, Denville Scientific).

After homogenization, clear samples by centrifugation at maximum speed in a microcentrifuge, and mix the supernatant with RIPA buffer followed by protein quantification (BCA or Bradford), before proceeding to immunoblot or immunoprecipitation (results shown in Fig. 2b).

11. For virus titer determination, homogenize the slice of lung tissue in 800 μl of PBS containing 0.3% BSA using a Beadblaster (Denville Scientific, or other methods), by two rounds of mechanical treatment for 10 s each at 6.5 m/s. Remove tissue debris by low-speed centrifugation and determine virus titers in supernatants by standard plaque assays [13] (results shown in Fig. 2c).

Acknowledgements

Work in the Rajsbaum lab (UTMB) is supported by the Institute for Human Infections and Immunity (IHII), and the Department of Microbiology and Immunology at the University of Texas Medical Branch (UTMB), Galveston. We also would like to thank Dr. Adolfo Garcia-Sastre (Icahn School of Medicine at Mount Sinai, NY) and his lab, where the original work was performed, as well as Dr. Hong Moulton (Oregon State University) for the design, synthesis, and general advice of the PPMOs.

References

1. Moulton HM (2013) In vivo delivery of morpholino oligos by cell-penetrating peptides. Curr Pharm Des 19(16):2963–2969
2. Yin H, Boisguerin P, Moulton HM, Betts C, Seow Y, Boutilier J, Wang Q, Walsh A, Lebleu B, Wood MJ (2013) Context dependent effects of chimeric peptide morpholino conjugates contribute to dystrophin exon-skipping efficiency. Mol Ther Nucleic Acids 2:e124. doi:10.1038/mtna.2013.51
3. Stein DA (2008) Inhibition of RNA virus infections with peptide-conjugated morpholino oligomers. Curr Pharm Des 14(25):2619–2634
4. Heald AE, Charleston JS, Iversen PL, Warren TK, Saoud JB, Al-Ibrahim M, Wells J, Warfield KL, Swenson DL, Welch LS, Sazani P, Wong M, Berry D, Kaye EM, Bavari S (2015) AVI-7288 for Marburg virus in nonhuman primates and humans. N Engl J Med 373(4):339–348. doi:10.1056/NEJMoa1410345
5. Rajsbaum R, Versteeg GA, Schmid S, Maestre AM, Belicha-Villanueva A, Martinez-Romero C, Patel JR, Morrison J, Pisanelli G, Miorin L, Laurent-Rolle M, Moulton HM, Stein DA, Fernandez-Sesma A, tenOever BR, Garcia-Sastre A (2014) Unanchored K48-linked polyubiquitin synthesized by the E3-ubiquitin ligase TRIM6 stimulates the interferon-IKKepsilon kinase-mediated antiviral response. Immunity 40(6):880–895
6. Tripathi S, Pohl MO, Zhou Y, Rodriguez-Frandsen A, Wang G, Stein DA, Moulton HM, DeJesus P, Che J, Mulder LC, Yanguez E, Andenmatten D, Pache L, Manicassamy B, Albrecht RA, Gonzalez MG, Nguyen Q, Brass A, Elledge S, White M, Shapira S, Hacohen N, Karlas A, Meyer TF, Shales M, Gatorano A, Johnson JR, Jang G, Johnson T, Verschueren E, Sanders D, Krogan N, Shaw M, Konig R, Stertz S, Garcia-Sastre A, Chanda SK (2015) Meta- and orthogonal integration of influenza "OMICs" data defines a role for UBR4 in virus budding. Cell Host Microbe 18(6):723–735
7. Geller BL, Marshall-Batty K, Schnell FJ, McKnight MM, Iversen PL, Greenberg DE (2013) Gene-silencing antisense oligomers

inhibit acinetobacter growth in vitro and in vivo. J Infect Dis 208(10):1553–1560

8. Gabriel G, Nordmann A, Stein DA, Iversen PL, Klenk HD (2008) Morpholino oligomers targeting the PB1 and NP genes enhance the survival of mice infected with highly pathogenic influenza A H7N7 virus. J Gen Virol 89(Pt 4):939–948
9. Lupfer C, Stein DA, Mourich DV, Tepper SE, Iversen PL, Pastey M (2008) Inhibition of influenza A H3N8 virus infections in mice by morpholino oligomers. Arch Virol 153(5):929–937
10. Lai SH, Stein DA, Guerrero-Plata A, Liao SL, Ivanciuc T, Hong C, Iversen PL, Casola A, Garofalo RP (2008) Inhibition of respiratory syncytial virus infections with morpholino oligomers in cell cultures and in mice. Mol Ther 16(6):1120–1128
11. Paessler S, Rijnbrand R, Stein DA, Ni H, Yun NE, Dziuba N, Borisevich V, Seregin A, Ma Y, Blouch R, Iversen PL, Zacks MA (2008) Inhibition of alphavirus infection in cell culture and in mice with antisense morpholino oligomers. Virology 376(2):357–370
12. Livak KJ, Schmittgen TD (2001) Analysis of relative gene expression data using real-time quantitative PCR and the 2(−Delta Delta C(T)) Method. Methods 25(4):402–408
13. Szretter KJ, Balish AL, Katz JM (2006) Influenza: propagation, quantification, and storage. Curr Protoc Microbiol. Chapter 15:Unit 15G 11

Chapter 17

Systemic Delivery of Morpholinos to Skip Multiple Exons in a Dog Model of Duchenne Muscular Dystrophy

Rika Maruyama, Yusuke Echigoya, Oana Caluseriu, Yoshitsugu Aoki, Shin'ichi Takeda, and Toshifumi Yokota

Abstract

Exon-skipping therapy is an emerging approach that uses synthetic DNA-like molecules called antisense oligonucleotides (AONs) to splice out frame-disrupting parts of mRNA, restore the reading frame, and produce truncated yet functional proteins. Multiple exon skipping utilizing a cocktail of AONs can theoretically treat 80–90% of patients with Duchenne muscular dystrophy (DMD). The success of multiple exon skipping by the systemic delivery of a cocktail of AONs called phosphorodiamidate morpholino oligomers (PMOs) in a DMD dog model has made a significant impact on the development of therapeutics for DMD, leading to clinical trials of PMO-based drugs. Here, we describe the systemic delivery of a cocktail of PMOs to skip multiple exons in dystrophic dogs and the evaluation of the efficacies and toxicity in vivo.

Key words Morpholinos (phosphorodiamidate morpholino oligomers), Dystrophic dogs, Canine X-linked muscular dystrophy (CXMD), Duchenne/Becker muscular dystrophies (DMD/BMD), Exon skipping, Dystrophin

1 Introduction

Phosphorodiamidate morpholino oligomers (PMOs) are antisense oligonucleotides characterized by charge-neutral phosphorodiamidate linkages and six-membered morpholine rings that have been widely used for studies of gene function in nonmammalian animal models, such as zebra fish, frogs, and sea urchins [1]. It has been recently shown that PMOs are a powerful tool to modulate the splicing of pre-mRNA and restore the expression of mutated genes such as *dystrophin* in mouse and dog models of Duchenne muscular dystrophy (DMD) in vivo [2–5]. Based on these results, antisense drugs based on PMOs are currently in late-stage clinical trials [6] and one, eteplirsen, has been approved by the US FDA to treat humans with some DMD mutations.

DMD is one of the most common lethal genetic disorders, caused by the lack of dystrophin protein [7]. In DMD, frameshift

Hong M. Moulton and Jon D. Moulton (eds.), *Morpholino Oligomers: Methods and Protocols*, Methods in Molecular Biology, vol. 1565, DOI 10.1007/978-1-4939-6817-6_17, © Springer Science+Business Media LLC 2017

mutations in the *dystrophin* gene lead to a premature stop codon and result in a lack of dystrophin protein [8]. Becker muscular dystrophy (BMD) is also caused by a mutation in the *dystrophin* gene but results in a milder form of muscular dystrophy, because the mutations are in-frame and lead to producing shorter but functional dystrophin proteins [8]. Several mouse and dog models exist for DMD, but one of the biggest limitations of murine DMD models is the milder phenotype compared to DMD patients [9, 10].

In contrast, the canine X-linked muscular dystrophy (CXMD) is beagle-based and displays body-wide muscle weakness with cardiac symptoms, similar to DMD patients [11–14]. They also develop various human DMD-like phenotypes such as joint contracture and kyphosis [12]. The dog model has a natural point mutation within the acceptor splice site in intron 6 in the *dystrophin* gene, which results in a frameshift and premature termination in exon 8 [15]. To restore the reading frame in the dog model, at least two exons (exons 6 and 8) need to be skipped (removed). The cocktail PMOs were designed to target two sites in exon 6 and a site in exon 8, respectively. A cocktail of these PMOs induced multi-exon skipping (exons 6–9) and resulted in the production of short, but functional dystrophin protein, similar to BMD [3] (Fig. 1). This exon skipping was demonstrated both by local (intramuscular) and systemic (intravenous) injection of PMOs accompanied by improved muscle function

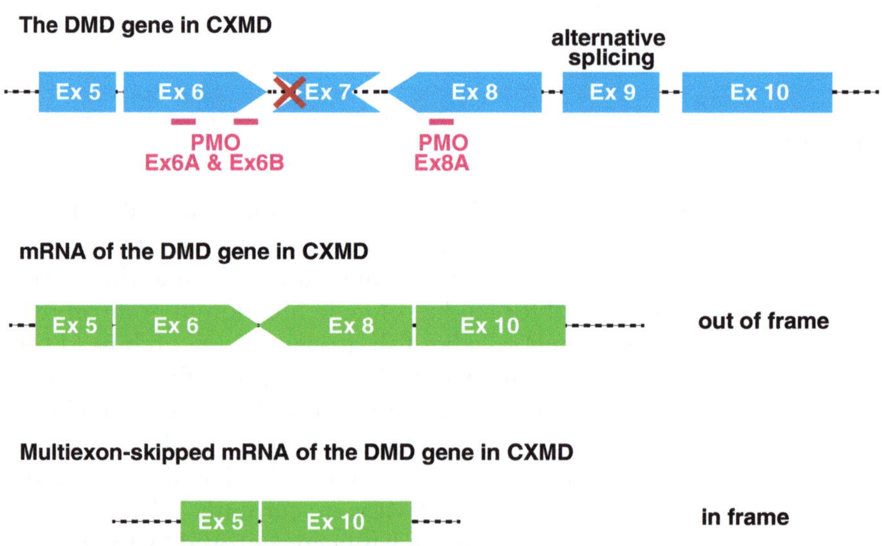

Fig. 1 The dystrophin mutation of the CXMD dog and the exon-skipping strategy. The CXMD dogs harbor a point mutation within the acceptor splice site of exon 7 in the *dystrophin* gene, which results in a frameshift and a lack of dystrophin protein. To restore the open reading frame, exons 6, 7, and 8 need to be skipped. The PMOs designed here target two sites in exon 6 and a site in exon 8, respectively. The injection of a cocktail of these PMOs leads to the multiexon skipping and results in the production of in-frame mRNAs (the exon 5 and the exon 10 are connected), which produce a short but functional dystrophin proteins

[3]. The demonstration of systemic delivery of cocktail PMOs has made a significant impact on the development of therapeutics for DMD, because it is necessary to skip multiple exons to treat most DMD cases [16]. Multiple exon skipping using cocktail PMOs also offers the prospect of selecting in-frame deletions in which the functionality of the truncated dystrophin is optimized [17]. For example, a natural mutation of dystrophin exons 45–55 deletion is associated with very mild or asymptomatic phenotypes in humans [18]. Systemic rescue with exons 45–55 skipping has been demonstrated in a mouse model of DMD using a cocktail of cell-penetrating octa-guanidine-conjugated PMOs (vivo-PMOs) [19, 20]. In this chapter, we summarize the method and protocol of systemic delivery of a cocktail of PMOs to skip multiple exons in dystrophic dogs and the evaluation of the efficacies and toxicity in vivo.

2 Materials

2.1 Design of Antisense Morpholinos

1. Websites to identify exonic splicing enhancers (ESEs), e.g., ESE finder (http://rulai.cshl.edu/cgi-bin/tools/ESE3/esefinder.cgi?process=home) and Rescue-ESE (http://genes.mit.edu/burgelab/rescue-ese/).

2.2 Systemic Injection of Antisense Morpholinos in Dogs

1. CXMD dogs and wild-type littermates.
2. Antisense morpholinos (Gene Tools, LLC). We used a cocktail of three morpholinos (Ex6A: 5′-GTTGATTGTCGGACCCAGCTCAGG-3, Ex6B: 5′-ACCTATGACTGTGGATGAGAGCGTT-3′, Ex8A: 5′-CTTCCTGGATGGCTTCAATGCTCAC-3) to skip exons 6–9 of *dystrophin* (*see* Fig. 1).
3. Saline.
4. Syringe infusion pump (Muromachi, Tokyo, Japan).
5. 22 G Indwelling needles (TERUMO, Tokyo, Japan).
6. 50 ml syringe (TERUMO).
7. Blood test (Veterinary Diagnostic Lab, University of Missouri, MO, USA).

2.3 Clinical Grading of Dogs (Behavior and Gait Test)

1. Video Camera.
2. Stopwatch.

2.4 Clinical Grading of Dogs (Magnetic Resonance Imaging)

1. Thiopental sodium (Mitsubishi Tanabe Pharma, Osaka, Japan).
2. Isoflurane (Abbott Laboratories, Chicago, IL, USA).

2.5 Muscle Sampling (Necropsy)

1. Thiopental sodium (Mitsubishi Tanabe Pharma, Osaka, Japan).
2. Isoflurane (Abbott Laboratories, Chicago, IL, USA).

3. Veterinary surgical instruments: forceps, scalpels, scissors.
4. Tragacanth gum (Sigma-Aldrich, St. Louis, MO, USA).
5. Cork disks (Iwaki-Kagaku, Tokyo, Japan).
6. Liquid nitrogen.
7. Isopentane (Sigma-Aldrich).
8. Metal container.
9. Dry ice.
10. 25 ml syringe.
11. Glass vials.
12. Poly-l-lysine coated slides (Thermo Fisher, Waltham, MA, USA).

2.6 RT-PCR and cDNA Sequencing

1. Trizol (Thermo Fisher).
2. Chloroform (Sigma-Aldrich).
3. Isopropanol (Sigma-Aldrich).
4. 75% ethanol (Sigma-Aldrich).
5. One-Step RT-PCR kit (Qiagen, Hilden, Germany).
6. Forward primer in exon 5: 5′-CTGACTCTTGGTTTGAT TTGGA-3′.
7. Reverse primer in exon 10: 5′-TGCTTCGGTCTCTGTCA ATG-3′.
8. RNase-free water.
9. 2% agarose gel.
10. SYBR Safe DNA gel stain (Invitrogen).
11. Gel extraction kit (Qiagen).

2.7 Western Blotting of Dog Muscles

1. Lysis buffer: 10% SDS, 70 mM Tris–HCl at pH 6.7, 5 mM EDTA-2Na at pH 8.0, Complete Mini proteinase inhibitor cocktail (Roche, Mannheim, Germany), 5% beta-mercaptoethanol at a final concentration.
2. Coomassie (Bradford) Protein Assay Kit (Thermo Fisher).
3. 20% SDS stock solution.
4. 0.5 M Tris–HCl, pH 6.7.
5. 0.5 M EDTA-2Na, pH 8.0.
6. 99% Glycerol (Invitrogen).
7. Beta-Mercaptoethanol (Sigma-Aldrich).
8. 1% bromophenol blue stock solution (BPB) (0.1 g/10 ml distilled water).
9. NuPAGE™ Novex™ 3–8% Tris-Acetate Protein Gels (Thermo Fisher).

10. NuPAGE® Tris-Acetate SDS Running Buffer (20×) (Thermo Fisher).
11. HiMark™ Pre-stained Protein Standard (Thermo Fisher).
12. Polyvinylidene difluoride (PVDF) membrane, pore size 0.45 μm (Millipore, Billerica, MA, USA).
13. Extra thick paper, 2.5 mm thickness, 8.0 cm × 13.5 cm (Thermo Fisher).
14. Methanol.
15. Concentrated Anode Buffer: 0.3 M Tris-HCl, 20% methanol.
16. Anode Buffer: 0.03 M Tris-HCl, 20% methanol.
17. Cathode Buffer: 25 mM Tris-HCl, 20% methanol, 40 mM 6-amino-n-hexanoic acid, 0.01% SDS.
18. Phosphate buffered saline (PBS) containing 0.05% Tween 20 (PBST).
19. Amersham ECL Prime Blocking Reagent (GE Healthcare, Little Chalfont, UK).
20. Antidystrophin C-terminal antibody (Rabbit polyclonal, catalog number 15277, Abcam, Cambridge, UK) or NCL-DYS2 (Mouse monoclonal, Leica Biosystems, Newcastle, UK).
21. HRP-conjugated goat anti-rabbit (or mouse) IgG H + L secondary antibody (Bio-Rad, Hercules, CA, USA).
22. Amersham ECL Select Western Blotting Detection Reagent (GE Healthcare).

2.8 Immunostaining for Dog Muscles

1. Cover glasses (Thermo Fisher).
2. Mouse monoclonal anti-dystrophin antibody NCL-DYS1 against rod domain (Leica Biosystems).
3. Alexa 594 goat anti-mouse IgG_{2a} highly cross-absorbed (Thermo Fisher).
4. ProLong Gold antifade mountant with 4′,6-Diamidino-2-Phenylindole Dihydrochloride (DAPI) (Thermo Fisher).
5. Goat serum (Thermo Fisher).
6. Moisture chamber (Scientific Devise Laboratory, Des Plaines, IL, USA).
7. PBS.
8. Hydrophobic barrier pen (Vector Laboratories, Burlingame, CA, USA).
9. Nail polish.
10. Triton X-100 (Sigma-Aldrich).

3 Methods

3.1 Design of Antisense Morpholinos

1. Identify exonic splicing enhancers (ESEs) in the genome sequence around targeted exons by the programs on the websites.
2. Design 24–30 mer PMOs that are antisense sequence of targeted sites. PMOs can include an ESE or an exon/intron boundary. To design PMOs for cocktails, target the sequences from several exons within the region you aim to skip (*see* **Note 1**).

3.2 Systemic Injection of Antisense Morpholinos in Dogs

1. Restrain dogs manually in accordance with an animal use guideline and set an indwelling needle into a limb vein (a cephalic or saphenous vein).
2. Inject PMOs 120–240 mg/kg (40–80 mg/kg each PMO, 50 ml in total, dissolved in saline) with an infusion pump or syringe driver (2.5 ml/min rate) (*see* **Note 2**).
3. Perform blood tests weekly for toxicity tests. Collect 3 ml blood from a subcutaneous vein of the fore or hind limb (e.g., a cephalic vein). Include complete blood count (CBC), γ-glutamyl transpeptidase (γGTP), aspartate aminotransferase (AST), blood urea nitrogen (BUN), alanine aminotransferase (ALT), creatine kinase (CK), and creatinine.

3.3 Clinical Grading of Dogs (Behavior and Gait Test)

1. Record the behavior and the gait of the dogs by a video camera.
2. Grade gait and movement disturbance by the following categories:
   - Gait and movement disturbance; grade 1 = none, grade 2 = sitting with hind limb extended, grade 3 = bunny hops with hind limbs, grade 4 = shuffling walk, and grade 5 = unable to walk.
   - Mobility disturbance; grade 1 = none, grade 2 = lying down more than normal, grade 3 = cannot jump on hind limbs, grade 4 = increasing difficulty moving around, and grade 5 = unable to get up and move around.
   - Muscle atrophy on the limb; grade 1 = none, grade 2 = suspect hardness, grade 3 = can feel hardness or apparently thin, grade 4 = between grades 3 and 5, and grade 5 = extremely thin or hard.
   - Drooling; grade 1 = none, grade 2 = occasionally dribbles saliva when sitting, grade 3 = some drool when eating and drinking, grade 4 = strings of drool when eating or drinking, and grade 5 = continuous drool.

- Hypertrophy of the tongue; grade 1 = none, grade 2 = slightly enlarged, grade 3 = extended outside dentition, grade 4 = enlarged and slightly thickened, and grade 5 = enlarged and thickened.
- Ability to swallow; grade 1 = none, grade 2 = takes time and effort in eating food, grade 3 = difficulty in eating food from a plate, grade 4 = difficulty in chewing, swallowing, or drinking, and grade 5 = unable to eat.

3. Calculate total grade by adding the scores of each category.
4. Measure how long it takes for the dog to run 15 m.

3.4 Clinical Grading of Dogs (Magnetic Resonance Imaging, MRI)

1. Inject 20 mg/kg thiopental sodium intravenously for general anesthesia, and maintain it by 2–3% isoflurane inhalation. Monitor muscle reflexes, heart, and breathing rate. The normal respiration rate (RR), heart rate (HR), and SpO_2 under general anesthesia are as follows: RR: 10–20 breath/min, SpO_2: 95–100%, HR: 80–120 beats/min.
2. Use 3 Tesla (3 T) Magnetic resonance imaging (MRI) with 18 cm diameter/18 cm length human extremity coil. The setting to obtain T2-weighted images of hind limbs is as follows: TR/TE = 4000/85 ms, slice thickness = 6 mm, slice gap = 0 mm, field of view = 18 cm × 18 cm, matrix size = 256 × 256, and number of acquisition = 3 during fast spin echo.

3.5 Muscle Sampling (Necropsy)

1. Mix 20 g tragacanth gum well with the equal amount of water until it becomes soft and sticky. Put them into 25 ml syringes. The gum in the syringes can be stored in −20 °C freezer.
2. Inject 20 mg/kg thiopental sodium intravenously for general anesthesia, and maintain it by 2–3% isoflurane inhalation. Monitor muscle reflexes, heart, and breathing rate. The normal respiration rate (RR), heart rate (HR), and SpO_2 under general anesthesia are as follows: RR: 10–20 breath/min, SpO_2: 95–100%, HR: 80–120 beats/min.
3. Euthanize dogs by exsanguination from the carotid artery under deep general anesthesia, to avoid the effects caused by bleeding derived factors.
4. Put a metal container of the isopentane into liquid nitrogen to make it cold enough for freezing.
5. Dissect the following muscles: tibialis anterior (TA), extensor digitorum longus (EDL), gastrocnemius, soleus, biceps femoris, rectus femoris, biceps brachii, triceps brachii, deltoid, extensor carpi ulnaris (ECU), extensor carpi radialis (ECR), flexor carpi ulnaris (FCU), flexor carpi radialis (FCR), gracilis, intercostal, abdominal muscles, diaphragm, lateral dorsi,

esophagus, sternocleidomastoid, and the heart. To examine toxicity, collect kidney and liver samples.

6. Cut the samples into small sections (approximately 1–1.5 cm length).
7. Label cork discs and place tragacanth gum on the opposite side of the label on the cork (approximately 0.5–1 cm height).
8. Embed the bottom one-third of the dissected muscles in the tragacanth gum. The longitudinal axis of the muscles should be perpendicular to the cork. Putting additional gum around the bottom of muscles helps to do that.
9. Soak the muscles with the cork into the cold isopentane by tweezers to freeze. Shake the muscles in it constantly until completely frozen, and then keep them on dry ice temporarily.
10. Put the samples in glass vials, and store at −80 °C.
11. Set up cryostat for sectioning. The working temperature is −25 °C. The section thickness is 8 μm for immunohistochemistry, 10–12 μm for hematoxylin and eosin (HE) staining, and 15–20 μm for RT-PCR and Western blotting samples.
12. Mount the cork with muscle block on a disc of the cryostat and trim the muscle to obtain flat sections.
13. For RT-PCR and Western blotting, put 20–40 sections in a frozen 1.5 ml tube and store at −80 °C. For immunohistochemistry and HE staining, transfer every section onto a slide glass and dry at room temperature for at least 30 min (*see* **Note 3**). The slides can be stored in a tightly closed box at −80 °C.

3.6 RT-PCR and cDNA Sequencing

1. Add 1 ml cold Trizol and vortex for 30 s. Incubate for 10 min at room temperature.
2. Add 200 μl chloroform and vortex. Incubate for 2 min at room temperature.
3. Centrifuge the sample at 12,000 × *g* for 15 min at 4 °C.
4. Take only the top layer carefully and transfer to a new tube (*see* **Note 4**). Add 500 μl isopropanol and vortex. Incubate for 10 min at room temperature.
5. Centrifuge at 12,000 × *g* for 10 min at 4 °C. A RNA pellet will be on the bottom of the tube.
6. Remove all supernatant and add 1 ml of cold 75% ethanol.
7. Centrifuge at 7500 × *g* for 5 min at 4 °C.
8. Remove the ethanol as much as possible, dry up for 5–10 min at room temperature. (Open the tube and cover it with a Kimwipe.)
9. Add 15–30 μl RNase-free water and warm at 65 °C for 10 min.

10. Quantify total RNA concentration by UV–Vis spectrophotometer at 260 nm and adjust the concentration at 40 ng/μl with RNase-free water.
11. Set up RT-PCR reaction (total: 25 μl). Mix 1.5 μl 10 μM forward primer, 1.5 μl 10 μM reverse primer, 1 μl dNTP, 5 μl one-step PCR kit buffer, 0.7 μl RNase inhibitor, 1 μl enzyme mixture from one-step PCR kit, 5 μl (200 ng) of RNA, and 9.3 μl RNase-free water.
12. Set parameters for the thermal cycler as follows. 1 cycle of 50 °C for 30 min, 1 cycle of 95 °C for 15 min, then 35 cycles of 94 °C for 1 min, 60 °C for 1 min, 72 °C for 1 min; lastly, 1 cycle of 72 °C for 10 min.
13. Run 5–10 μl of the PCR product on the 1–2% agarose gel and stain the gel by SYBR safe solution for 30 min.
14. Cut the band of interest and collect it to a 1.5 ml tube (*see* **Note 5**). Use the Qiagen gel extraction kit to purify the PCR product. Store the PCR product at 4 °C or −20 °C
15. Sequence the purified PCR product to confirm the exon skipping. The sequence primer is the same primer as for the RT-PCR.

3.7 Western Blotting of Dog Muscles

1. Add 150 μl lysis buffer and vortex for 30 s.
2. Incubate at 37 °C for 5 min and centrifuge at 16,500 × *g* for 15 min.
3. Collect the supernatant carefully in a fresh tube.
4. Dilute the samples to 100-fold with water. Measure the protein concentration by Bradford Protein Assay Kit and adjust the protein concentration to 5 μg/μl with the lysis buffer. Store at −80 °C.
5. Make 2× SDS sample buffer on the same day it will be used. For 2.5 ml of 2× SDS sample buffer, mix 625 μl of 20% SDS, 350 μl of 0.5 M Tris–HCl (pH 6.7), 25 μl of 0.5 M EDTA-2Na (pH 8.0), 500 μl of 99% glycerol, 10 μl of 1% BPB, 125 μl of beta-mercaptoethanol, and 865 μl of distilled water.
6. Mix 12 μl of the 5 μg/μl protein solution and 12 μl of 2× SDS sample buffer. Heat the mixed samples at 70 °C for 10 min before loading.
7. Load 20 μl of the sample (50 μg protein) per lane of the gel and do SDS-PAGE at 150 V for 75 min.
8. Soak PVDF membrane for 20 s in methanol, and then in the anode buffer until before use (at least for 10 min).
9. Soak blotting papers in each transfer buffer at least for 30 min.
10. Soak the gel after the electrophoresis in cathode buffer and rock gently for 5–10 min.

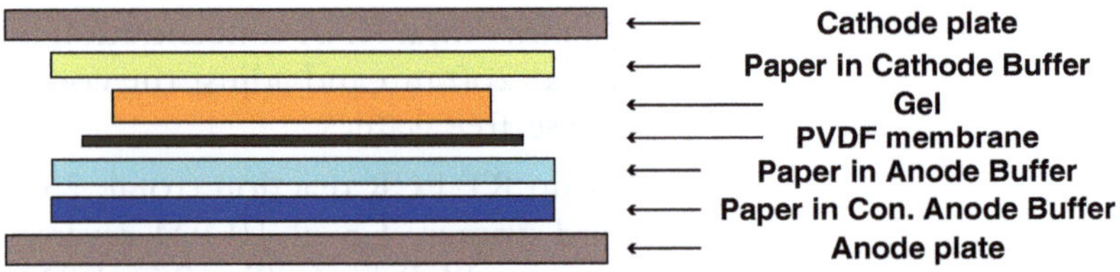

Fig. 2 Semi-dry transfer method for Western blotting. An extra thick paper (2.5 mm thickness) soaked in Conc. Anode buffer; an extra thick paper soaked in Anode buffer; the PVDF membrane, the gel, and an extra thick paper soaked in Cathode buffer are placed between the two plates of a semi-dry blotting machine. The order is from the *bottom* to the *top*

11. Place the blotting papers, the PVDF membrane, and the gel on a semi-dry transfer machine (*see* Fig. 2).
12. Transfer at 20 V for 70 min.
13. Rinse the membrane by PBST twice.
14. Soak the membrane in 2% blocking solution and rock gently for 1 h at room temperature or overnight at 4 °C for blocking.
15. Incubate the membrane with anti-dystrophin antibody in blocking solution (the dilution is 1:2500 for rabbit anti-dystrophin C-terminal antibody or 1:150 for NCL-DYS2) for 1 h at room temperature or overnight at 4 °C.
16. Rinse the membrane twice and wash three times for 10 min by PBST at room temperature.
17. Incubate the membrane with the horseradish peroxidase (HRP)-conjugated secondary antibody in PBST (1:10,000) for 1 h at room temperature.
18. Rinse the membrane twice and wash three times for 10 min by PBST at room temperature.
19. Detect the signal by the ECL select kit and a chemiluminescent imager (Fig. 3).

3.8 Immunostaining for Dog Muscles

1. Dry sections for 1 h.
2. Enclose all the muscle sections in a circle made by a hydrophobic barrier pen.
3. Put the slides in a moisture chamber containing water (approximately 1 mm).
4. Incubate in 15% goat serum in PBS for 1 h at room temperature for blocking.
5. Add the DYS-1 anti-mouse monoclonal antibody diluted by 15% goat serum in PBS (1:150 dilution). 1.25 ml of the diluted

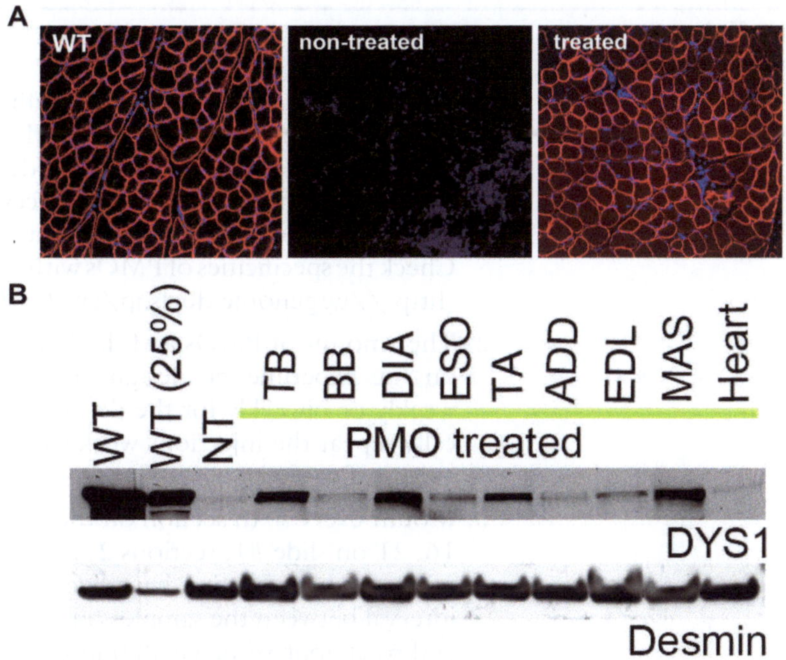

Fig. 3 Rescue of dystrophin expression after cocktail morpholino treatment in CXMD dogs. (**a**) Immunohistochemistry (DYS-1) of the tibialis anterior muscles to compare the expression of Dystrophin in wild-type (WT) (positive control), non-treated CXMD dogs (negative control) and CXMD dogs treated with 120 mg/kg morpholino cocktail. The morpholino cocktail (Ex6A, Ex6B, and Ex8A) was injected five times weekly from saphenous vein. (**b**) Immunoblot for dystrophin with desmin as a loading control for WT, non-treated (NT) and cocktail morpholino-treated dogs. *TB* triceps brachii, *BB* biceps brachii, *DIA* diaphragm, *ESO* esophagus, *ADD* adductor, *EDL* extensor digitorum longus, *MAS* masseter. Images were reproduced from Annals of Neurology, volume 65, issue 6—"Efficacy of systemic morpholino exon-skipping in Duchenne dystrophy dogs" with permission

antibody is required per a slide. Incubate overnight at 4 °C in moisture chambers.

6. Wash by PBS for 5 min three times.
7. Add Alexa 594 goat antibody diluted by 15% goat serum in PBS containing 0.1% Triton X-100 (1:2500 dilution). 1.25 ml of the diluted antibody is required per a slide. Incubate for half an hour at room temperature.
8. Wash by PBS for 5 min five times.
9. Wipe off excess liquid. Drop ProLong Gold antifade mountant with DAPI to a cover glass and mount it. Seal by nail polish.

4 Notes

1. Design several PMOs because the efficacies of PMOs are highly unpredictable. GC contents of PMOs should be between 40 and 65%, but approximately 60% is ideal. Avoid four consecutive "G"s, self-complementary sequences, and self-dimers. When PMOs are injected as a cocktail, avoid heterodimers as well. Check the specificities of PMOs with NCBI Blast or GGGenome (http://gggenome.dbcls.jp/en/hg19/1/+/).
2. The amount of PMOs and the frequency of injections depend on the experimental design. The injection can be repeated weekly or biweekly for the desired number of weeks. We typically repeat the injections weekly or biweekly (every 2 weeks), at least, five times.
3. Mount every sixth section on the same slide (sections 1, 6, 11, 16, 21 on slide #1, sections 2, 7, 12, 17, 22 on slide #2). You will get five slides that have five sections per slide. Adjust the interval between the samples on the same slide (e.g., sections 1 and 6) at approximately 200 μm.
4. Do not include any middle or bottom layer. It causes contamination of protein and DNA.
5. The size of exon 6–9 skipped band is 101 bp.

Acknowledgements

This work was supported by The Friends of Garrett Cumming Research Chair Fund, HM Toupin Neurological Science Research Chair Fund, Muscular Dystrophy Canada, Canada Foundation for Innovation (CFI), Alberta Advanced Education and Technology (AET), Canadian Institutes of Health Research (CIHR), Jesse's Journey—The Foundation for Gene and Cell Therapy, the University of Alberta Faculty of Medicine and Dentistry, Japan Society for the Promotion of Science (JSPS) Postdoctoral Fellowships for Research Abroad, and the Women and Children's Health Research Institute (WCHRI).

References

1. Arora V, Devi GR, Iversen PL (2004) Neutrally charged phosphorodiamidate morpholino antisense oligomers: uptake, efficacy and pharmacokinetics. Curr Pharm Biotechnol 5(5):431–439
2. Lu QL, Rabinowitz A, Chen YC, Yokota T, Yin H, Alter J, Jadoon A, Bou-Gharios G, Partridge T (2005) Systemic delivery of antisense oligoribonucleotide restores dystrophin expression in body-wide skeletal muscles. Proc Natl Acad Sci U S A 102(1):198–203. doi:10.1073/pnas.0406700102
3. Yokota T, Lu QL, Partridge T, Kobayashi M, Nakamura A, Takeda S, Hoffman E (2009) Efficacy of systemic morpholino exon-skipping in Duchenne dystrophy dogs. Ann Neurol 65(6):667–676. doi:10.1002/ana.21627

4. Guncay A, Yokota T (2015) Antisense oligonucleotide drugs for Duchenne muscular dystrophy: how far have we come and what does the future hold? Future Med Chem 7(13):1631–1635. doi:10.4155/fmc.15.116
5. Touznik A, Lee JJ, Yokota T (2014) New developments in exon skipping and splice modulation therapies for neuromuscular diseases. Expert Opin Biol Ther 14(6):809–819. doi:10.1517/14712598.2014.896335
6. Mendell JR, Rodino-Klapac LR, Sahenk Z, Roush K, Bird L, Lowes LP, Alfano L, Gomez AM, Lewis S, Kota J, Malik V, Shontz K, Walker CM, Flanigan KM, Corridore M, Kean JR, Allen HD, Shilling C, Melia KR, Sazani P, Saoud JB, Kaye EM, Eteplirsen Study G (2013) Eteplirsen for the treatment of Duchenne muscular dystrophy. Ann Neurol 74(5):637–647. doi:10.1002/ana.23982
7. Hoffman EP, Brown RH Jr, Kunkel LM (1987) Dystrophin: the protein product of the Duchenne muscular dystrophy locus. Cell 51(6):919–928
8. Flanigan KM (2014) Duchenne and Becker muscular dystrophies. Neurol Clin 32(3):671–688 . doi:10.1016/j.ncl.2014.05.002viii
9. Yu X, Bao B, Echigoya Y, Yokota T (2015) Dystrophin-deficient large animal models: translational research and exon skipping. Am J Transl Res 7(8):1314–1331
10. Rodrigues M, Echigoya Y, Fukada S, Yokota T (2016) Current translational research and murine models for Duchenne muscular dystrophy. J Neuromuscul Dis 3(1):29–48
11. Shimatsu Y, Katagiri K, Furuta T, Nakura M, Tanioka Y, Yuasa K, Tomohiro M, Kornegay JN, Nonaka I, Takeda S (2003) Canine X-linked muscular dystrophy in Japan (CXMDJ). Exp Anim 52(2):93–97
12. Shimatsu Y, Yoshimura M, Yuasa K, Urasawa N, Tomohiro M, Nakura M, Tanigawa M, Nakamura A, Takeda S (2005) Major clinical and histopathological characteristics of canine X-linked muscular dystrophy in Japan, CXMDJ. Acta Myol 24(2):145–154
13. Urasawa N, Wada MR, Machida N, Yuasa K, Shimatsu Y, Wakao Y, Yuasa S, Sano T, Nonaka I, Nakamura A, Takeda S (2008) Selective vacuolar degeneration in dystrophin-deficient canine Purkinje fibers despite preservation of dystrophin-associated proteins with overexpression of Dp71. Circulation 117(19):2437–2448
14. Yugeta N, Urasawa N, Fujii Y, Yoshimura M, Yuasa K, Wada MR, Nakura M, Shimatsu Y, Tomohiro M, Takahashi A, Machida N, Wakao Y, Nakamura A, Takeda S (2006) Cardiac involvement in Beagle-based canine X-linked muscular dystrophy in Japan (CXMDJ): electrocardiographic, echocardiographic, and morphologic studies. BMC Cardiovasc Disord 6:47. doi:10.1186/1471-2261-6-47
15. Sharp NJ, Kornegay JN, Van Camp SD, Herbstreith MH, Secore SL, Kettle S, Hung WY, Constantinou CD, Dykstra MJ, Roses AD et al (1992) An error in dystrophin mRNA processing in golden retriever muscular dystrophy, an animal homologue of Duchenne muscular dystrophy. Genomics 13(1):115–121
16. Aoki Y, Yokota T, Wood MJ (2013) Development of multiexon skipping antisense oligonucleotide therapy for Duchenne muscular dystrophy. Biomed Res Int 2013:402369. doi:10.1155/2013/402369
17. Echigoya Y, Yokota T (2014) Skipping multiple exons of dystrophin transcripts using cocktail antisense oligonucleotides. Nucleic Acid Ther 24(1):57–68. doi:10.1089/nat.2013.0451
18. Nakamura A, Yoshida K, Fukushima K, Ueda H, Urasawa N, Koyama J, Yazaki Y, Yazaki M, Sakai T, Haruta S, Takeda S, Ikeda S (2008) Follow-up of three patients with a large in-frame deletion of exons 45–55 in the Duchenne muscular dystrophy (DMD) gene. J Clin Neurosci 15(7):757–763
19. Echigoya Y, Aoki Y, Miskew B, Panesar D, Touznik A, Nagata T, Tanihata J, Nakamura A, Nagaraju K, Yokota T (2015) Long-term efficacy of systemic multiexon skipping targeting dystrophin exons 45–55 with a cocktail of vivo-morpholinos in mdx52 mice. Mol Ther Nucleic Acids 4:e225. doi:10.1038/mtna.2014.76
20. Aoki Y, Yokota T, Nagata T, Nakamura A, Tanihata J, Saito T, Duguez SM, Nagaraju K, Hoffman EP, Partridge T, Takeda S (2012) Bodywide skipping of exons 45–55 in dystrophic mdx52 mice by systemic antisense delivery. Proc Natl Acad Sci U S A 109(34):13763–13768. doi:10.1073/pnas.1204638109

Chapter 18

In Vivo and Explant Electroporation of Morpholinos in the Developing Mouse Retina

Devi Krishna Priya Karunakaran and Rahul Kanadia

Abstract

Neonatal in vivo electroporations and retinal explant electroporations have been widely employed in understanding the effects of loss or gain of function of protein-coding genes in retinal development. Here, we describe a rapid and efficient delivery of morpholinos to add another tool to perturb gene expression during mouse retinal development.

Key words In vivo electroporations, Explant electroporation, u12 morpholinos, Green fluorescent protein (GFP), Subretinal space, Retinal development

1 Introduction

It is a great time to study the molecular underpinnings of biological processes as biologists have numerous tools available for perturbing gene expression. These various methods of gene silencing include RNA interference (RNAi) using small interfering RNAs (siRNA), short hairpin RNAs, or bifunctional shRNAs, and other methods including using phosphorothioate-linked DNA (S-DNA), CRISPR, and morpholinos (MOs). Among these tools, MOs are widely used for blocking various processes at the RNA level such as blocking poly-A signal sites [1] and zipcode-binding sequences [2, 3], masking microRNA target sites [4], inducing splicing changes such as exon skipping, etc. Indeed, crucial breakthroughs in developmental biology have arisen from strategic application of morpholinos in developing zebrafish embryos.

Developmental biologists have exploited MOs to specifically inhibit functions of a particular gene of interest by blocking protein production at a specific developmental time point. Since development is a dynamic process, injection of MOs at a particular time point can affect various processes including cell cycle regulation, cell fate determination, and differentiation. MOs have high affinity

Hong M. Moulton and Jon D. Moulton (eds.), *Morpholino Oligomers: Methods and Protocols*, Methods in Molecular Biology, vol. 1565, DOI 10.1007/978-1-4939-6817-6_18, © Springer Science+Business Media LLC 2017

for RNA and are employed to inhibit its processing or translation. MOs are synthetic oligonucleotides of about 25 nucleotides in length. They are similar to DNA or RNA, except that they have an uncharged phosphorous-containing linkage and a morpholine ring rather than a ribose ring and uncharged intersubunit linkages. Since MOs have neutral backbones and are resistant to nucleases, they are more stable. MOs are widely used to block the translation of specific transcripts or to alter splicing of pre-mRNA, promoting exon skipping or retention of introns. First, in inhibition of translation, MOs are targeted to sequences 5′ of the translation start site, thereby inhibiting the progression of the initiation complex to the start. An example of such inhibition by MOs has been shown by Cornell et al., where MOs against Notch signaling were assessed in zebrafish neurogenesis [5, 6]. Second, in splicing, MOs can be designed to prevent correct splicing by designing a MO to cause the spliceosome to splice out an exon of interest or preventing the removal of an intron. An example of the use of MOs to alter splicing has been shown for the gene *Nrhl* [7]. Third, MOs can be designed complementary to a microRNA, which can then block expression of the mature miRNA and help in the understanding of the miRNA targets. In both zebrafish and frog, MOs are routinely used to block miRNAs and understand their targets and pathway regulated by them [8–10]. The goal of a gene knockdown strategy is to produce highest efficiency with no off-target effects. Compared to other knockdown methods, morpholinos have been shown to be highly specific with least or no off-target effects. Since morpholinos have a neutral backbone, they are not efficiently electroporated. Electroporation relies on the net negative charge of DNA or RNA to drive them into replicating cells. It is known that electroporation of nucleic acids mostly works in dividing cells and not postmitotic cells. The best explanation offered thus far is that while electroporation can breach the plasma membrane to deliver the DNA into the cytosol, it is not enough to breach the nuclear membrane. Since majority of the labeling associated with electroporation relies on visualization of electroporated cells through transcription of GFP coding sequence that is coelectroporated, the plasmid DNA has to be inside the nucleus. Before mitosis the nuclear envelope of dividing cells breaks down, which then allows the GFP plasmid DNA to become entrapped in the newly formed daughter cell nuclei thereby expressing GFP. In case of MOs, the neutral charge is not conducive to electroporation, so it is important to add carboxyfluorescein or some other fluorophore that adds a negative charge. Addition of a fluorophore to the MO would also mark the cells that are electroporated. However, a single fluorescein tag is not enough for visualization. Therefore, combining the fluoresceinated MOs with plasmids expressing GFP (pCAG-GFP) is recommended. This strategy has been used in developing chicken embryos and in the developing mouse retina [11–13].

The mouse retina is a thin layer of neural tissue that lines the inner part of the eye, which converts the electromagnetic light signal into a chemical signal which is then relayed to the brain via the optic nerve. The vertebrate retina is a part of the central nervous system and contains six neuronal cell types, namely, rod photoreceptors, cone photoreceptors, amacrine cells, bipolar cells, horizontal cells, and ganglion cells; and one glial cell, namely, Müller glia. The retina has a stereotypic architecture with the nuclei of rod photoreceptors and cone photoreceptors forming the outer nuclear layer (ONL); the nuclei of amacrine cells, bipolar cells, horizontal cells, and Müller glia forming the inner nuclear layer (INL); and the nuclei of ganglion cells forming the ganglion cell layer (GCL) [14]. The precise order in which each retinal cell type is born in the mouse is well characterized [15]. Ganglion cells are the first-born cells, which are born around embryonic day (E) 11.5. While cone photoreceptors and horizontal cells are born between E12 and E16, amacrine cells are born both embryonically and postnatally. Bipolar cells and Müller glia are born postnatally, with latter being born after P4 [15]. Rod photoreceptor cells are born throughout retinal development.

While neonatal in vivo electroporations using MOs were performed to understand the postnatal retinal development, explant electroporation was performed to assess the effects of MOs in early embryonically born cell types such as ganglion cells, cone photoreceptors, and horizontal cells. Explant electroporation is an in vitro technique, where the embryonic retina is cultured and electroporated as an explant. Electroporated retinae can stay in culture for about 3–4 days and develop as they would in vivo, producing the aforementioned cell types. Explant culture methods are more generally useful when the knockout of a gene of interest is embryonically lethal [16, 17].

In this chapter, we describe in vivo delivery of MOs into P0 followed by in vitro explant electroporation to deliver MOs into developing embryonic retinae. For P0 in vivo electroporation, the protocol from Matsuda et al. is modified to employ glass needles using FemtoJet instead of Hamilton needles so as to minimize the footprint of the injection site and control the delivery volume [16]. Here, we coelectroporated pCAG-GFP plasmid with U12 or scrambled MOs instead of fluorescein-labeled MOs since the visualization of the electroporated patch has been better with coelectroporation. Besides, coelectroporation of multiple plasmids has been shown to be ~99% efficient [18]. Therefore, coelectroporation of MOs with pCAG-GFP allows us to visualize the effect of gene knockdown by MOs in terms of changes in the number or structure of each postnatally born cell type. In this method, MOs are injected into the subretinal space (between the retinal pigmented epithelium and the retina) for in vivo electroporation. This is then followed by short pulses of electric current, which cause the plasma membranes to become transiently leaky. This allows the DNA + MO to be taken up by cycling progenitor cells. Therefore, any neuron born from these progenitor cells contains MOs. Thus, the effect of downregulating a gene or an isoform can be assessed in postmitotic neurons. Since there is a lack of significant lateral migration of cells in the retina, only a portion of the retina is electroporated while the unelectroporated region of the retina serves as internal control.

2 Materials

2.1 In Vivo Electroporation

Prepare all solutions using ultrapure water (prepared by purifying deionized water to attain a sensitivity of 18 MΩ cm at 25 °C) and analytical grade reagents.

1. 10× PBS stock: 10× phosphate buffered saline (pH 7.4) (working concentration: 1×).
2. 10% Fast Green dye stock, working concentration: 1%.
3. Morpholinos (GeneTools, LLC.): u12 morpholino (u12MO)—TCGTTATTTTCCTTACTCATAAGTT (complementary sequence to U12 nt 9–33) (Fig. 2b′) and control morpholino (cMO)—CCTCTTACCTCAGTTACAATTTAT A. Morpholinos were dissolved in water at a concentration of 0.5 nmol/μl.
4. Plasmids—pCAG-GFP plasmid was obtained from Dr. Constance Cepko's laboratory at the Harvard Medical School, Boston, MA.
5. MO mixture for in vivo electroporation: Mix 10 μg of morpholino and 10 μg of pCAG-GFP along with 10% Fast Green (final concentration should be 1%) and 10× PBS (Final concentration is 1×) and bring it up to 10 μl with ultra-pure water (*see* **Notes 1–3**).
6. Femto jet Injector (Eppendorf, catalog # E5247000013) with foot pedal to aid injection of MO.
7. Capillary injector (Eppendorf, Universal capillary holder, catalog # 920007392).
8. Needle for back-filling micropipettes (28 G, 97 mm long) (Microfil, World Precision instruments Inc., Sarasota, FL, catalog # MF286-5).
9. Glass needles (Origio, Charlottesville, VA, catalog # C060609).
10. Square pulse electroporator.
11. Disposable 30 G ½ needle.
12. Blunt forceps.
13. Tweezer-type electrodes (BTX Harvard apparatus, ECM 830, 7 mm diameter).

14. Cotton swabs.
15. Heating pad.
16. Dissection Microscope.
17. Ice.
18. Loading pipette tips.
19. Petri dish.
20. 4% paraformaldehyde.
21. 5%, 10%, 20%, 30% sucrose in 1× PBS.
22. OCT compound (Tissue-tek).
23. Cryo mold (Tissue-tek, catalog # 4565).

2.2 In Vitro Explant Electroporation

1. Custom explant electroporation chamber.
   - Chamber dimensions: 1/8 × 1/8 × 3/8 in.
   - Dimensions of gold bars (99.97% pure gold): 3 mm × 10 mm × 10 mm.
   - BNC type female connector (catalog # 885-5369, Allied Electronics).
   - Translucent RTV 108 Silicone rubber adhesive seal unit to ensure that the two gold electrodes are locked in position.
   - Plexi glass: 12.5 cm (L) × 12.5 cm (W) × 4 cm (H).
2. MO mixture for explant electroporation: 70 μg of MOs and 70 μg of pCAG-GFP plasmid in 1× PBS (70 μl).
3. Dissecting medium: 50% Dulbecco Modified Eagle's medium + 50% F-12 nutrient mixture.
4. Retinal culture medium (for 100 ml):
   (a) 44 ml of F-12 nutrient mixture.
   (b) 44 ml of Dulbecco Modified Eagle's medium.
   (c) 10 ml of Fetal bovine serum/Fetal calf serum.
   (d) 1 ml of Penicillin/Streptomycin.
   (e) 1 ml of 100× 2 mM l-Glutamine.
5. DPBS: Dulbecco's Phosphate Buffered Saline (1×) (GIBCO/Invitrogen, catalog # 14190-144).
6. Nucleopore membrane filter (Whatman, catalog # 110406, 13 mm, 0.2 μ pore size).
7. Six well plates.
8. 5% CO_2 incubator (37 °C).
9. Heating pad.
10. Embryonic day (E)12.5 pregnant female mouse.

3 Methods

3.1 In Vivo Electroporation

1. Electroporation setup
   - Load 10 μl of morpholino and pCAG-GFP mix into the glass needle using the syringe with needle for backfilling micropipettes and fit it into the capillary injector (*see* **Notes 4** and **5**).
   - Set the pulse frequency to five square pulses of 50 ms duration with 950 ms interval. Voltage is set to 80 V. The measured current of a successful electroporation is 0.08–0.15 A.
2. P0 in vivo electroporation procedure
   - Anesthetize P0 pups by placing them on ice for 2–3 min until they are immobilized.
   - Visualize the suture that seals the eyelids under a stereo-dissecting scope and use 301/2 G needle to separate the eyelid by running the needle tip along the suture (Fig. 1b).
   - The developing eye underneath will be visible and can be made to pop-out by pressing down the two flaps of the eyelids with IRIS dressing forceps.
   - Stabilize the exposed eyeball by the skin and position the globe such that the cornea and lens can be visualized as a clock face (Fig. 1c–e). Then insert the glass needle at 3:00 clock position at a 45° downward angle. The cornea and the lens can be used to gauge the needle insertion site. Once the needle breaks through the sclera, it is crucial to not go too deep as you might pass the retina and deliver the MO in the vitreous (*see* **Note 6**).
   - Inject DNA into the subretinal space by tapping the foot pedal twice. It ensures exact amount of DNA during each delivery. Confirmation that the MO was delivered in the subretinal space can be obtained by visualizing the spread of Fast Green dye along the retina (Fig. 1f). If the spread is not observed, then it can be assumed that the subretinal space was not targeted. Fast Green works well with albino strains such as CD1 or FvB as their RPE is not pigmented. A confounding feature of using Fast Green is that for strains with pigmented RPE, the Fast Green spread is difficult to visualize. Thus, it is recommended that electroporation should be practiced on albino strains (*see* **Notes** 7 and **8**).
   - Close the eyelids and clean the surface with a cotton swab.
   - Dip the tweezer electrodes in 1× PBS buffer placed in the Petri dish to soak the two paddles in PBS.
   - Place the screwed side of the electrode on the DNA injected eye (Fig. 1f, schematic in red).

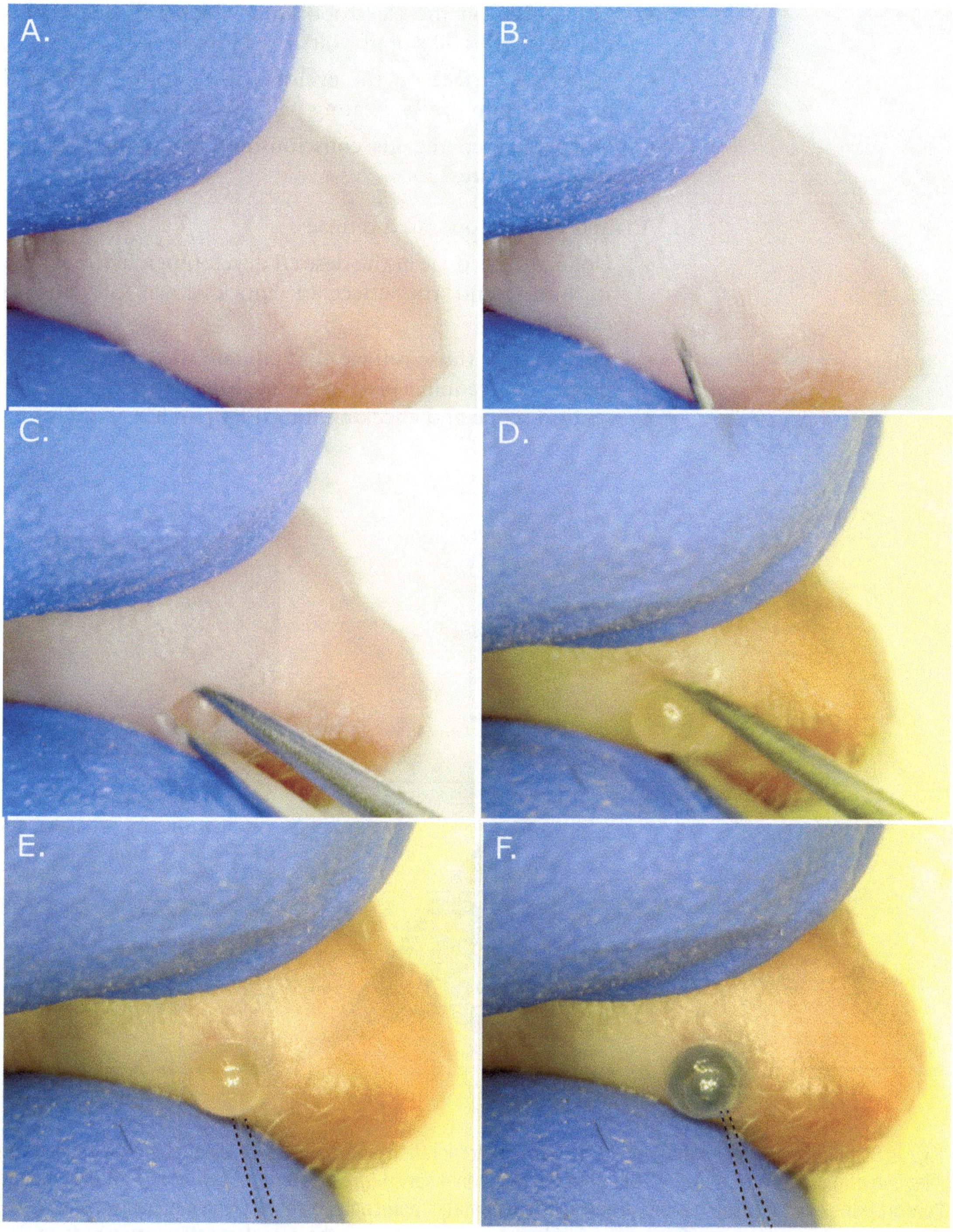

Fig. 1 Steps in P0 in vivo electroporation: (**a–f**) Steps in neonatal in vivo electroporation. (**a**) Anaesthetized pup is held under the stereoscope. (**b**) Using 30 G ½ needle, open the future eyelid mark. (**c–e**) Pop-out the eye ball using the forceps. (**f**) Inject DNA into the subretinal space. The *dotted black lines* indicate the direction of the needle into the eye. If done correctly, the Fastgreen dye is visible

- Hit 'pulse' on the electroporator, which will deliver five pulses of 80–100 square volts (*see* **Note 9**).
- Wipe the surface on the eyelids with a cotton swab and place the pup on the heating pad (*see* **Note 10**).
- Once the pup regains consciousness, put it back in the cage (*see* **Note 11**).

3. Analysis of electroporated retinae
   - Collect the retinae at the desired developmental time point to understand the effect of knock down of gene of interest.
   - Dissect only those retinae that contain the electroporated patch. This requires placing the eyeballs under the UV dissection scope and checking for GFP+ patches (Fig. 2c).

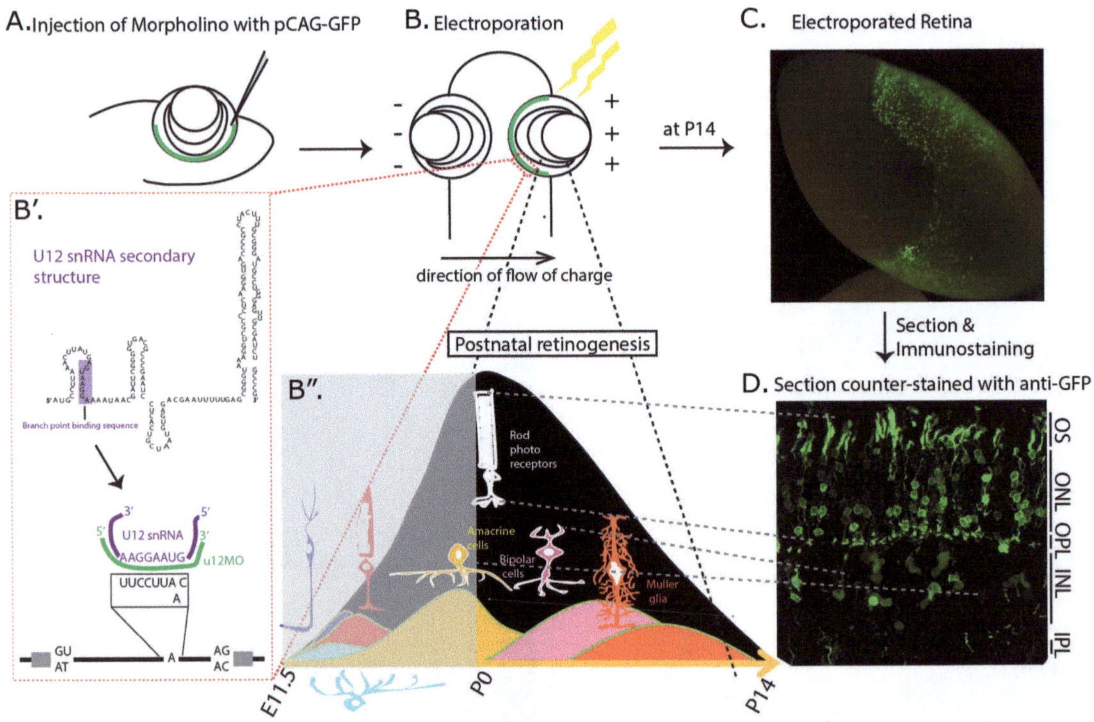

Fig. 2 Schematic representation of P0 in vivo electroporation and analysis of electroporated retinae: (**a**) Injection of morpholino with pCAG-GFP into the subretinal space (*green*). (**b**) Direction of flow of DNA charge flow once the electric pulse (*yellow*) is given. (**b′**) Schematic representation of blocking mechanism of u12 MO. *Gray boxes* represent exons. Sequences inside the boxed region are the complementary sequences to U12 snRNA (*purple*). U12 MO (*green*) prevents the binding of U12 snRNA by binding to the complementary sequence at the branch point. (**b″**) Schematic representation of birth order of retinal cell types. Embryonic development is grayed out since only the cells born postnatally are covered in this technique. All cell types born postnatally from P0 to P14 are schematized. (**c**) At postnatal day (P) 14, retinae are harvested and checked for GFP+ electroporated patch. (**d**) GFP+ retinae are cryopreserved, sectioned, and stained with anti-GFP antibody

- After dissection, fix these retinae in 4% paraformaldehyde.
- After series of dehydration steps with 20% sucrose and 30% sucrose, mount these retinae in OCT such that the patch will be at the bottom of the mold. This will ensure the patch appears first while sectioning.
- Once the sections are obtained, proceed with staining for GFP (since it was coelectroporated with morpholinos) and counter stain for postnatally born retinal cell type specific markers (Fig. 2d).

3.2 In Vitro Explant Electroporation

1. Custom-built explant electroporation chamber

This chamber is custom built with solid gold electrodes and specifications are as described as follows (total volume is ~70 μl) (Fig. 3a, b):

At the center of the plexiglass block, a milling bit (7/64th inch) was used to carve H-shape with the dimensions shown in the Fig. 3a. The black circles represent holes going into the plexiglass (depth of 18 mm) block that were drilled with tap

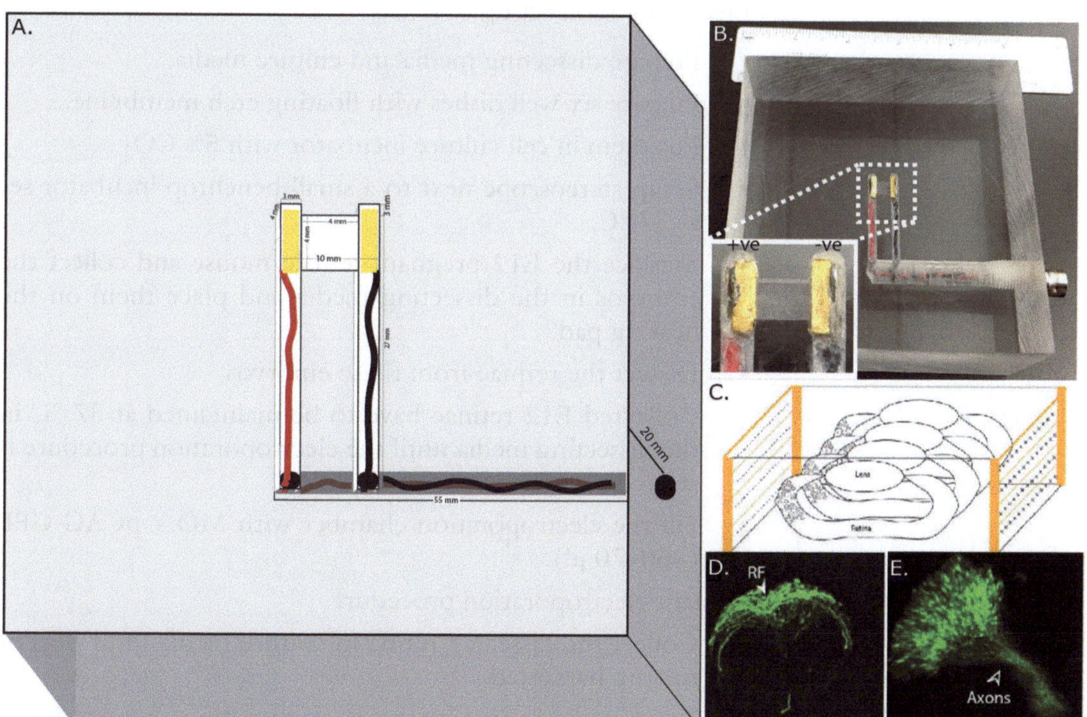

Fig. 3 Custom-built explant electroporation chamber: (**a**) Schematic representation of the explant electroporation chamber made by Edward J. Lechowicz, Jr. (**b**) Custom-built chamber for explant electroporation chamber. (*Insert*) *Boxed region* is the magnified image of the chamber area. (**c**) Arrangement of the retinae and the direction of DNA charge flow through the retina in the chamber. (**d**, **e**) Ganglion cells extending axons in an E12.5 explant retina. *RF* retinal fissure

drill bit (7/64th inch). The 99.99% gold slugs (3.175 mm × 6.35 mm) were hammered to rectangular shape (8 × 1 mm) and were positioned as shown the Fig. 3a. The leads were soldered on the backside with care taken to prevent soldering material from seeping on the side facing the electroporation chamber. Use of solder masking paint can minimize the spread of solder to the side facing the electroporation chamber. The wires soldered to the gold bars were placed at the bottom of the canal (27 mm long). The open areas were made waterproof with room temperature volcanizing (RTV) 108 translucent adhesive. The end of the canal was the site where holes (black circles) were drilled (7/64th drill bit) into the plexiglass (depth of 18 mm). The wires from the leads were threaded into the holes that were connected to a canal that was drilled from the side at the depth of 20 mm from the top with a tap drill bit (21/64 in.) followed by threading with a tap bit 3/8-24. The two wires that were threaded through the holes were drawn through the side canal to be connected to the Female BNC connector (50 Ω, bulkhead Jack solder cup 3/8-24 UNF thread).

2. Electroporation set up
   - Prepare dissecting media and culture media.
   - Prepare six-well dishes with floating etch membrane.
   - Put them in cell culture incubator with 5% CO_2.
   - Setup stereoscope next to a small benchtop incubator set at 37 °C.
   - Sacrifice the E12 pregnant female mouse and collect the embryos in the dissecting media and place them on the heating pad.
   - Dissect the retinae from these embryos.
   - Collected E12 retinae have to be maintained at 37 °C in the dissecting media until the electroporation procedure is started.
   - Fill the electroporation chamber with MO + pCAG-GFP (~60–70 μl).
3. Explant electroporation procedure
   - Collect the dissected retina in culture media until all retinae are harvested.
   - Transfer the retinae to the electroporation chamber using P1000 pipet with a tip that is cut at the end so as to increase the bore size (*see* **Note 12**).
   - With a 55 forceps, gently push the retinae down by gently poking the lens and let the retinae settle to the bottom. Arrange the retinae as shown in the Fig. 3c between the electrodes (*see* **Note 13**).

- Proceed by placing up to four retinae in the chamber filled with MOs as shown in Fig. 3c.
- Pulse the chamber with 50 V.
- Next use the P1000 pipet to aspirate the electroporated retinae into a dish with culture media to transfer them to the filters floating on 2 ml retinal culture media of six-well dish (*see* **Note 14**).
- After 2 days of incubation, you can visualize GFP+ ganglion cells sending out axons (Fig. 3d, e).

4 Notes

1. DNA mix can be stored at −20 °C for future use.
2. Make sure the plasmid preparation is of high concentration to avoid further precipitation of the plasmid. Sometimes, the precipitation introduces salts in the plasmid preparation that can block the tip of the glass needle.
3. Coelectroporation with pCAG-GFP ensures visibility of the electroporated patch clearly under the UV scope for dissection.
4. Glass needles can be cleaned thoroughly using ultrapure water and can be reused. If any particle is stuck at the tip of the needle even after cleaning, avoid reusing it.
5. Remove the loading syringe slowly out of the glass needle as you are pipetting out DNA into the needle to ensure no gap is introduced inside the needle.
6. Occasionally, the opening of the future eyelid causes bleeding. (If this occurs, the suture was not separated properly and care should be used for the subsequent steps.) Make sure the blood is wiped out before injection, which could interfere with injection. Do not cut open the eyelid forcefully as this might puncture the eyeball underneath and damage the eye. Also, do not extend the cut beyond the eyelid mark as this might cause bleeding of the tissue surrounding the eye.
7. Although the technique can be performed on any background strain of mice, CD1 mice are best suited as they exhibit strong maternal behavior as the electroporated pups need to be cared for by the mother until the day of their harvest. Also, albino strains enable easier visualization of the dye when DNA is injected into the subretinal space.
8. If DNA is injected correctly, the dye spreads within the area around the point of injection. If DNA is injected deep inside the eye (incorrect), the dye will be invisible.

9. Holding the pup and placing the electrodes on the eyelids might keep both hands occupied, so adjust the distance between the electroporator and you such that you can hit the "pulse" button on it without losing the position of electrodes.
10. After electroporation, make sure to wipe out any blood or dye with a cotton swab. Make sure to close the future eyelid so that there is no damage when the pups grow.
11. Make sure to finish electroporation of all pups and then place them back in the cage with their mother. To ensure that the mother identifies its pups, collect some bedding at the beginning while taking out pups out of the cage. If the mother does not identify its pups when they are put back in the cage, she might not care for them.
12. Before adding the retina to the chamber, passage the retina through three Petri dishes filled with PBS. This will ensure removal of culture media.
13. Retinae must be oriented the same way during electroporation such that retinal cup faces the bottom of the chamber when electric pulse is passed.
14. Since retinae are very tiny at E12, care must be taken when washing the retinae after electroporation so as not to lose them.

Acknowledgements

This work was supported by P30 from National Institute of Neurological Disorders and Stroke (5P30NS069266) and K99-R00 from National Eye Institute (4R00EY019547) to R.K. We would like to thank Edward J. Lechowicz, Jr. at the University of Connecticut for building the explant electroporation chamber.

References

1. Kaida D, Berg MG, Younis I, Kasim M, Singh LN, Wan L, Dreyfuss G (2010) U1 snRNP protects pre-mRNAs from premature cleavage and polyadenylation. Nature 468(7324):664–668. doi:10.1038/nature09479
2. Sasaki Y, Welshhans K, Wen Z, Yao J, Xu M, Goshima Y, Zheng JQ, Bassell GJ (2010) Phosphorylation of zipcode binding protein is required for BDNF signaling of local β-actin synthesis and growth cone turning. J Neurosci 30(28):9349–9358
3. Eom T, Antar LN, Singer RH, Bassell GJ (2003) Localization of a beta-actin messenger ribonucleoprotein complex with zipcode-binding protein modulates the density of dendritic filopodia and filopodial synapses. J Neurosci 23(32):10433–10444
4. Staton AA, Giraldez AJ (2011) Use of target protector morpholinos to analyze the physiological roles of specific miRNA-mRNA pairs in vivo. Nat Protoc 6(12):2035–2049
5. Cornell RA, Eisen JS (2000) Delta signaling mediates segregation of neural crest and spinal sensory neurons from zebrafish lateral neural plate. Development 127(13):2873–2882
6. Cornell RA, Eisen JS (2002) Delta/Notch signaling promotes formation of zebrafish neural

crest by repressing Neurogenin 1 function. Development 129(11):2639–2648

7. Knapp D, Messenger N, Ahmed Rana A, Smith JC (2006) Neurotrophin receptor homolog (NRH1) proteins regulate mesoderm formation and apoptosis during early Xenopus development. Dev Biol 300(2):554–569. doi:10.1016/j.ydbio.2006.09.038
8. Flynt AS, Li N, Thatcher EJ, Solnica-Krezel L, Patton JG (2007) Zebrafish miR-214 modulates Hedgehog signaling to specify muscle cell fate. Nat Genet 39(2):259–263. doi:10.1038/ng1953
9. Eberhart JK, He X, Swartz ME, Yan YL, Song H, Boling TC, Kunerth AK, Walker MB, Kimmel CB, Postlethwait JH (2008) MicroRNA Mirn140 modulates Pdgf signaling during palatogenesis. Nat Genet 40(3):290–298. doi:10.1038/ng.82
10. Martello G, Zacchigna L, Inui M, Montagner M, Adorno M, Mamidi A, Morsut L, Soligo S, Tran U, Dupont S, Cordenonsi M, Wessely O, Piccolo S (2007) MicroRNA control of Nodal signalling. Nature 449(7159):183–188. doi:10.1038/nature06100
11. Pekarik V, Bourikas D, Miglino N, Joset P, Preiswerk S, Stoeckli ET (2003) Screening for gene function in chicken embryo using RNAi and electroporation. Nat Biotechnol 21(1):93–96. doi:10.1038/nbt770
12. Muramatsu T (2000) In ovo gene electroporation into early chicken embryos. Methods Mol Med 37:369–376. doi:10.1385/1-59259-080-2:369
13. Saito T, Nakatsuji N (2001) Efficient gene transfer into the embryonic mouse brain using in vivo electroporation. Dev Biol 240(1):237–246. doi:10.1006/dbio.2001.0439
14. Livesey FJ, Cepko CL (2001) Vertebrate neural cell-fate determination: lessons from the retina. Nat Rev Neurosci 2(2):109–118. doi:10.1038/35053522
15. Cepko CL, Austin CP, Yang X, Alexiades M, Ezzeddine D (1996) Cell fate determination in the vertebrate retina. Proc Natl Acad Sci U S A 93(2):589–595
16. Matsuda T, Cepko CL (2004) Electroporation and RNA interference in the rodent retina in vivo and in vitro. Proc Natl Acad Sci U S A 101(1):16–22. doi:10.1073/pnas.2235688100
17. Matsuda T, Cepko CL (2008) Analysis of gene function in the retina. Methods Mol Biol423:259–278.doi:10.1007/978-1-59745-194-9_19
18. Barnabe-Heider F, Meletis K, Eriksson M, Bergmann O, Sabelstrom H, Harvey MA, Mikkers H, Frisen J (2008) Genetic manipulation of adult mouse neurogenic niches by in vivo electroporation. Nat Methods 5(2):189–196. doi:10.1038/nmeth.1174

Chapter 19

Intracerebroventricular Delivery in Mice for Motor Neuron Diseases

M. Nizzardo and M. Rizzuti

Abstract

The use of antisense oligonucleotides to target specific mRNA sequences represents a promising therapeutic strategy for neurological disorders. Recent advances in antisense technology enclose the development of phosphorodiamidate morpholino oligomers (MO), which is one of the best candidates for molecular therapies due to MO's excellent pharmacological profile.

Nevertheless, the route of administration of antisense compounds represents a critical issue in the neurological field. Particularly, as regards motor neuron diseases, intracerebroventricular (ICV) injection is undoubtedly the most efficient procedure to directly deliver therapeutic molecules in the central nervous system (CNS). Indeed, we recently demonstrated the outstanding efficacy of the MO antisense approach by its direct administration to CNS of the transgenic mouse models of Spinal Muscular Atrophy (SMA) and Amyotrophic Lateral Sclerosis (ALS).

Here, we describe methods to perform the ICV delivery of MO in neonatal SMA mice and in adult ALS mice.

Key words Intracerebroventricular injection, Motor neuron diseases, Morpholino, Spinal muscular atrophy, Amyotrophic lateral sclerosis, Neonatal mice, Adult mice

1 Introduction

Amyotrophic lateral sclerosis (ALS) and spinal muscular atrophy (SMA) are fatal motor neuron diseases (MNDs) that affect adults or children, respectively. In both cases, diseases can be extremely heterogeneous both from a clinical and a genetic point of view taking into account the age of onset and the motor functions achieved [1, 2]. Currently, no effective therapies are available for these disorders. However, many preclinical therapeutic strategies have been developed demonstrating the potential of the antisense oligonucleotide approach [3, 4]. In particular, recent evidences showed, for both ALS and SMA, the efficacy of phosphorodiamidate morpholino oligomers (MO) in the rescue of the pathogenic mechanisms underlying these disorders [5, 6]. Indeed, the ability of MO to

Hong M. Moulton and Jon D. Moulton (eds.), *Morpholino Oligomers: Methods and Protocols*, Methods in Molecular Biology, vol. 1565, DOI 10.1007/978-1-4939-6817-6_19, © Springer Science+Business Media LLC 2017

modulate the splicing and interfere with gene expression is particularly suitable for SMA and ALS. Moreover, due to its excellent safety and efficacy pharmacological profile, MO represents one of the most promising candidates for MNDs therapies.

The gene responsible for SMA pathogenesis is the survival motor neuron (*SMN*), which in humans exists in two almost identical copies: the telomeric gene *SMN1*, whose mutations cause the disease, and the centromeric one *SMN2*. The degree of similarity between the two *SMN* copies is very high. Indeed, *SMN2* differs from *SMN1* in only five nucleotides, among which only one is placed in the coding region (c.840C>T) [7, 8]. Despite being a silent mutation, this variant modifies a splicing modulator and often results in the alternative splicing of *SMN2*, which accounts for the exon 7 exclusion and the subsequent production of 10% of the full-length isoform [9, 10].

Since in SMA-affected subjects *SMN1* is not functional and the SMN production is completely attributable to *SMN2*, its upregulation or aberrant splicing modification represents a promising therapeutic strategy to increase the full-length isoform production. Numerous regions are involved in *SMN2* splicing regulation such as the intronic splicing silencer ISS-N1 [11]. Several MO sequences have been designed to target ISS-N1 and prevent its repressor ability. In particular, the MO(-10-34), which matches the region enclosed among the nucleotides 10-34, was found to be the most efficacious to target ISS-N1 and promote exon 7 inclusion [12–15] (Fig. 1a). The resulting production of the functional protein can balance the *SMN1* deletion or mutations associated with SMA. In our lab we have observed that direct CNS administration of MO(-10-34) in transgenic SMA mice effectively increased SMN full-length expression leading to robust neuromuscular improvement and survival rescue, with phenotype in some cases essentially indistinguishable from that of healthy heterozygous mice [5].

ALS is the most common adult-onset MND; different genes have been identified as responsible for the disease etiopathogenesis [16]. The vast majority of cases are sporadic, but 10% are familial ALS (fALS). Of these, ~20% of patients are linked to mutations in the superoxide dismutase 1 (*SOD1*) gene causing an overexpression of the protein. Recently, SOD1 protein has also been demonstrated to be involved in sporadic ALS pathogenesis [17, 18]. Indeed, wild-type or mutated SOD1 protein can exist in an aberrant conformation whose accumulation is toxic [19–22]. Since the resulting progressive motor neurons death is likely associated with SOD1-related toxicities, the ability of MO to interfere with the production of SOD1 may represent a promising therapeutic strategy for ALS treatment (Fig. 1b) [23, 24].

We recently tested the MO approach to silence mutant SOD1 in transgenic ALS rodents. We showed that a single

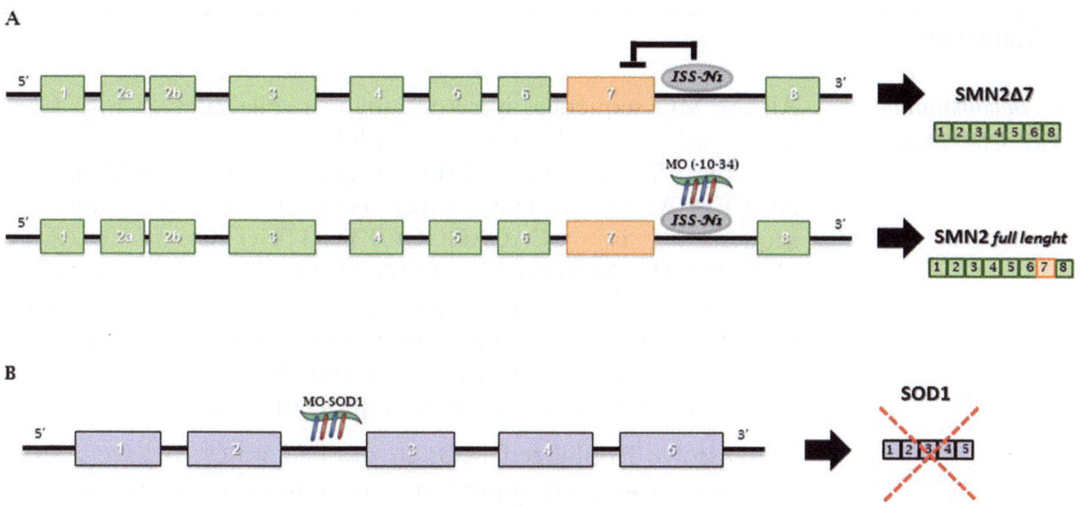

Fig. 1 MO therapeutic strategies for SMA and ALS. (**a**) In order to develop an antisense approach for SMA, MO(-10-34) is designed to target the ISS-N1 region in the *SMN2* gene and promote exon 7 inclusion. MO(-10-34) action results in the production of a functional SMN protein and in the consequent rescue of the SMA pathological phenotype. (**b**) For ALS, MO is designed to silence the production of the overexpressed SOD1 protein reducing the related protein toxicity and ameliorating the ALS pathological phenotype

intracerebroventricular (ICV) injection of MO, designed to reduce the synthesis of human SOD1 in SOD1G93A mice, slowed disease progression, improved neuromuscular function, and increased survival [6].

For neurological disorders, research can rely on different murine models, which have been developed in order to reproduce the human disease in vivo. These animal models are extremely useful to validate different therapeutic strategies at a preclinical level. Specifically, we tested our MO sequences in SMAdelta7 mouse model, which develops a SMA Type 1 phenotype, and in the SOD1G93A mouse model, which mimics the familial ALS associated with SOD1 gene mutation [5, 6].

However, the route of administration of MOs represents a controversial issue. Systemic and local injections display pros and cons associated with the dosage, invasiveness, biodistribution, and with a view to future clinical applications. Indeed, MNDs require the delivery of a consistent drug amount throughout the whole CNS and an efficient biodistribution to spinal cord motor neurons. To achieve these goals the direct ICV administration is undoubtedly the most efficacious procedure.

Here, we describe the procedures of ICV delivery of MOs in SMA and ALS mouse models with a detailed description of the technique and focusing on the differences between the injections in neonatal versus adult mice.

2 Materials

2.1 Reconstitution of Powdered MOs

All the MO sequences were designed in collaboration with Gene Tools and are usually provided lyophilized.

For SMAl, the MO sequence is GTAAGATTC ACTTTCATAATGCTGG (MO(-10-34)). For ALS, the MO sequence is CACAGGCCTTCGTCGTCGCCATAACTC (MO-SOD1). As controls, scrambled MO sequences were designed based on the best control sequence predicted using bioinformatic tool (Gene Tools, www.gene-tools.com). MO sequences are synthesized as custom MO without any modifications.

To reconstitute and prepare aliquots of MOs:

1. Dissolve lyophilized MOs in distilled water to have a stock solution with the appropriate concentration for injection. For MO(-10-34) the final concentration is 42.3 μg/μL, while for MO-SOD1 the final concentration is 66.6 μg/μL.
2. Make sure that the oligo is completely dissolved when making stock solution. If solid remains, heat the vial containing the stock solution at 65 °C for 5–10 min and briefly vortex.
3. After MO reconstitution, we recommend avoiding repeated freezing and thawing that may decrease its activity (*see* **Note 1**). Prepare aliquots of the stock solution in 0.2 mL individual tubes in order to have single dose injection amount for each tube. Prepare tubes with 2.4 μL of stock solution for MO(-10-34) (12 nmol, 101.52 μg) and with 5 μL for MO-SOD1 (40 nmol, 333 μg) (*see* **Note 2**).
4. Store aliquots at +4 °C.

2.2 Mice

To test the efficacy of MOs in vivo we used two transgenic murine models of motor neuron disorders available at the Jackson Laboratories as follows:

1. SMA: We used the triple-mutant SMA mouse that harbors two transgenic alleles and a targeted mutant (Jackson Laboratories, Stock number: 007952; FVB. Cg-Tg(SMN2*delta7)4299Ahmb Tg(SMN2)89Ahmb Smn1tm1Msd), which develops a SMA type 1. The Tg(SMN2*delta7)4299Ahmb allele is an SMA cDNA lacking exon 7 while the Tg(SMN2)89Ahmb allele is the entire human SMN2 gene. Heterozygous Smn knockout mice with SMN2 transgenes were bred to obtain homozygous mice for the knockout Smn alleles (SMA mice, SMN2+/+;SmnD7+/+;mSmn−/−). Transgenic mice were identified by polymerase chain reaction [25].
2. ALS: We used transgenic mice of the strain B6. Cg-Tg(SOD1-G93A)1Gur/J, which carries a high copy number of the mutant

human SOD1 allele containing a Gly93Ala (G93A) substitution (Jackson Laboratories, stock number: 004435). Progeny for experimental analyses was obtained by breeding SOD1G93A transgenics with C57BL/6 wild-type mice. Transgenic mice were identified by polymerase chain reaction [26].

2.3 MO Injection into SMA Neonatal Mice

1. Evans Blue (powder, Sigma-Aldrich). Dissolve in distilled water to have a 0.04% final solution (*see* **Note 3**).
2. Ice.
3. Transilluminator.
4. Nontoxic laboratory pen (Sharpie pen).
5. Micro-Fine insulin syringe, 0.5 mL, 30 G × 8 mm (Becton Dickinson).
6. Warming pad (2biological instruments).

2.4 MO Injection into ALS Adult Mice

1. Oxygen/isofluorane system (2biological Instruments).
2. Digital Lab Stereotaxic unit (2biological Instruments).
3. 2BTOS rodent shaver (Moser).
4. 70% ethanol, 95% ethanol.
5. Iodine.
6. Hydrogen peroxide.
7. A 10 μL Hamilton syringe, 33 G.
8. Nylon sutures, size 5.0.
9. Warming pad (2biological Instruments).
10. Ceftriaxone.

3 Methods

To achieve widespread drug delivery in the CNS, MOs can be administered through a single bolus ICV injection into the mouse cerebral ventricular system. The ICV injection ensures a direct delivery into cerebrospinal fluid (CSF) via the lateral ventricles to the entire brain and spinal cord. Since the two mouse models used have a different mean survival, 13 days for SMA mice (*see* **Note 4**) and 140 days for SOD1, it is necessary to perform MO injection at two different time point according to the strain, at neonatal stage for SMA mice and in adulthood for SOD1 mice. The procedure is different depending on the age of the injected animals; here we describe in details the two type of injection. The first one, performed in neonatal mice, is a rather simple procedure, because the cerebral ventricles can be easily targeted, while the second one, in adult mice, requires an accurate surgical practice due to the definitively developed murine skull cap.

To determine the real efficacy of MO, the procedure has to be performed exactly and accurately (*see* **Notes 3** and **5**).

3.1 MO Injection into SMA Neonatal Mice

1. Start with the procedure when the litter is at postnatal day 1 (P1, *see* **Note 6**).
2. Prepare the equipment for injection and remove the aliquots of MO you need from +4 °C (*see* **Note 1**). The number of aliquots depends on the numbers of neonatal mice to be injected.
3. Load 2.4 μL of diluted MO into a Micro-Fine insulin syringe.
4. Cryo-anesthetize pup: transfer the pup from the cage and place it on ice to induce hypothermia anesthesia. Wait 2–3 min for the pup to become fully anesthetized. Confirm anesthesia by very gently squeezing a paw and monitor for lack of movement or respiration.
5. Hand-mount the pup over a backlight or under the guidance of a transilluminator to visualize the intersection of the coronal and sagittal cranial sutures (Bregma).
6. Identify the injection sites 1 mm lateral and 1 mm anterior to bregma. These landmarks are visible through the skin at P1 (Fig. 2a). Mark the injection site with a nontoxic laboratory pen.
7. Lay the pup on its side with its head directly under the syringe. Turn the pup's head so that the marked injection site is facing up, and gently but firmly hold this position with an open hand. Hold the syringe perpendicular to the surface of the skull and

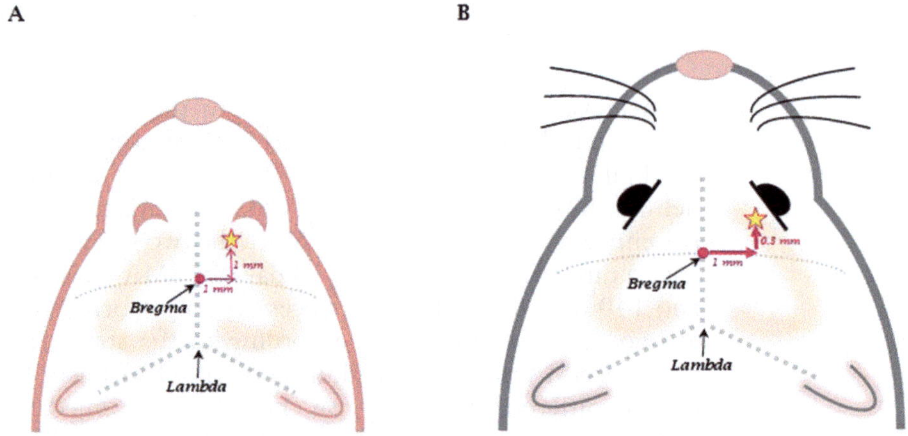

Fig. 2 ICV injection coordinates in neonatal and adult mice. (**a**) Stereotactic coordinates for ICV bolus injection in neonatal mice: 1 mm lateral and 1 mm anterior to bregma. Once identified the correct injection site, the depth of the needle injection through the skull is 2 mm ventral. (**b**) Stereotactic coordinates for ICV MO delivery in adult mice from Bregma: 1 mm lateral, 0.3 mm anterior, and 3.0 mm ventral

insert the needle 2 mm deep to the skin at the marked injection site (*see* **Note 7**).

8. Hold the syringe rigidly so that the plunger can be depressed without moving the needle farther into the brain.
9. Begin slowly injecting the predetermined volume and concentration of MO while monitoring the volume dispensed from the syringe.
10. Slowly withdraw the needle.
11. Following the injections, toe clip the pups for the identification and collect the tail for the genotype [25].
12. Place treated pup in a warming pad for recovery (5–10 min) until their body temperature and skin color return to normal and check that they recover normal movement.
13. Place the pup back into the cage with the biological mother.
14. Repeat from the procedure from **steps 3** to **13** for all the mice to be injected.

3.2 MO Injection into ALS Adult Mice

1. Perform the injection in early symptomatic phase SOD1G93A mice at P85.
2. Prepare the equipment for injection and remove the aliquots of MO you need from +4 °C (*see* **Note 1**). The number of aliquots depends on the numbers of adult mice to be injected (*see* Subheading 2.1).
3. Cleanse all the surgery equipment and the stereotaxic area with 70% ethanol solution in order to create a sterile field, which needs to be maintained throughout the entire surgery procedure.
4. Clean the Hamilton syringe with sterilized water, load it with 5 μL MO-SOD1 and attach it to the stereotaxic apparatus.
5. Turn on the isoflurane/oxygen system and direct gas flow to the anesthesia chamber.
6. Place the mouse in the chamber and anesthetize through isofluorane inhalational (4% induction, and oxygen to 0.4 L/min). Wait until breathing is significantly slowed and toe pinch to ensure the mouse is completely unconscious.
7. Place the mouse into the cranial stereotactic frame and arrange the anesthesia nose cone over the mouse nose. Secure the animal head with the ear bars.
8. Direct the gas flow to the stereotaxic system and subsequently calibrate the isoflurane level down to 2% for the anesthesia maintenance (*see* **Note 5**).
9. At 5 min intervals make sure the mouse is completely unconscious with tail and toe pinch, otherwise place the mouse back into the anesthesia chamber (*see* **Note 8**).

10. Shave the scalp with the rodent shaver and sterilize with a cotton wad soaked in 95% ethanol and then with a cotton wad dipped in iodine.
11. Make in the scalp a 1–1.5 cm midline incision to visualize Bregma and Lambda (*see* **Note 9**).
12. Secure the syringe to the stereotaxic apparatus and guide the needle to preselected stereotaxic coordinates to punch through the skull at 1.0 mm lateral and 0.3 mm anterior to the Bregma (Fig. 2b).
13. Lower the needle ventral to a depth of 3 mm and wait 2–3 min for the brain to seal around the needle.
14. Deliver 5 μL of MO-SOD1 solution through a single bolus injection into the right lateral ventricle at a rate of 1 μL/s.
15. After 5 min, slowly withdraw the needle. Hold a cotton wad over the needle hole to prevent any MO leakage.
16. Suture the incision.
17. Place injected mouse on a warming pad for recovery (10–15 min).
18. Place the mouse back in the home cage.
19. Repeat from the procedure from **steps 4** to **16** for all the mice to be injected.
20. Monitor the injected mice daily after the surgery to check for any infections. If necessary, administer antibiotic systemically through an intraperitoneal injection during postoperative care (Ceftriaxone, 200 mg/kg).
21. Remove the suture thread a week following surgery.

4 Notes

1. When stored cold, MO can come out of solution or can decrease in activity. If it occurs with your aliquots, heat them 10 min at 65 °C and cool to room temperature before the use.
2. At dosage higher than 24 nmol of MO(-10-34) we observed toxicity in neonatal mice [5], therefore we suggest not to further increase the dosage.
3. Precise targeting of the lateral ventricle is critical to optimize the treatment. Inaccurate injections can displace MO in the parenchyma and into the thalamus, reducing its efficacy and variability among injection output could alter the results. Therefore, we strongly recommend repeatedly practicing the injection with an opaque tracer (Evans Blue, 0.04%) before treating with MO until targeting the lateral ventricles is trustworthy and reproducible. Use the same procedure described

earlier for MO injecting only the dye: if the injection has been successful, the dye will spread throughout the ventricular chambers and the borders of the lateral ventricle will be visible in blue. The injection with Evans Blue can be performed both for neonatal and adult mice following point-by-point instruction earlier substituting MO for the tracer.

4. MO ICV injection in adult mice has been reported also for SMA [27]. The transgenic model used reproduces SMA type III (FVB. Cg-Tg(SMN2)2HungSMN1tm1Hung/J, stock number 005058, Jackson Laboratory) displaying a milder pathological phenotype reaching adulthood.
5. The procedure in neonatal mice is usually safe; if you observe mortality, the injections are probably not performed correctly. Cases of death due to anesthesia in adult mice are unusual because the depth of anesthesia can be easily monitored during the procedure.
6. We recommend performing MO injection at P1 or at the latest at P3 in neonatal mice. At this age, the cerebrospinal fluid–brain barrier has not matured yet and MO will diffuse throughout the ventricular system and then follow the flow of cerebrospinal fluid into the brain and spinal cord, where motor neurons reside. A narrow MO therapeutic window has been described for SMA mice [5, 14].
7. To have a reference for the injection depth for neonatal injection we suggest marking 2 mm from the tip of needle with nontoxic maker.
8. Since hypothermia during the surgery may extend the postsurgery recovery, if the mouse has to be placed back to anesthesia chamber before the injection, we recommend laying the mouse on the warming pad for 3–4 min.
9. To enhance Bregma visualization in adult mice, we suggest wiping the skull with a cotton wad dipped in hydrogen peroxide after sterilization.
10. In adult mice, the single ICV bolus injection can limit the possibility both to increase the amount of injected MO and to deliver continuously the compound for a long period whose therapeutic effect might diminish in time due to its half-life. It would be crucial to increase the biodistribution of the compound and the efficacy of the treatment. An alternative route of administration can be the infusion into the cerebrospinal fluid via the cerebral ventricles through an osmotic pump that allows for a slow, continuous delivery of MO. However, Rigo and colleagues have demonstrated, in adult homozygous SMA mice (*Smn−/−*; *SMN2+/+*), that ICV bolus injection is a more efficient method of delivering antisense oligonucleotide (ASO) targeted to the *SMN2* pre-mRNA compared to ICV

infusion, based on the half-maximal effective dose, on the half-maximal, effective concentration and on prolonged pharmacological activity [27]. The same comparison needs to be performed for MO to identify the most efficient delivery route in adult mice.

11. Even if the majority of the studies demonstrated the efficacy of MO treatment following ICV injection, it remains possible that MO action outside of the blood–brain barrier can be required to fully correct motor neuron disorders phenotype. Moreover, ICV injection in particular in adult mice can be stressful requiring anesthesia and a surgical procedure. Other minimally invasive delivery methods can be considered. Intravenous injection (facial vein in neonatal mice and tail vein in adult mice) can be an alternative route of administration and has been already described for MO in SMA [5, 13–15].

Acknowledgement

This work was supported by AriSLA, Pilot Grant ALSsiMO to M.N. and Telethon, GGP14025 to M.N.

References

1. Ludolph AC, Brettschneider J, Weishaupt JH (2012) Amyotrophic lateral sclerosis. Curr Opin Neurol 25:530–535
2. D'Amico A, Mercuri E, Tiziano FD, Bertini E (2011) Spinal muscular atrophy. Orphanet J Rare Dis 2:6–71
3. Hua Y, Vickers TA, Okunola HL, Bennett CF, Krainer AR (2008) Antisense masking of an hnRNPA1/A2 intronic splicing silencer corrects SMN2 splicing in transgenic mice. Am J Hum Genet 82:834–848
4. Passini MA, Bu J, Richards AM, Kinnecom C, Sardi SP, Stanek LM, Hua Y, Rigo F, Matson J, Hung G, Kaye EM, Shihabuddin LS, Krainer AR, Bennett CF, Cheng SH (2011) Antisense oligonucleotides delivered to the mouse CNS ameliorate symptoms of severe spinal muscular atrophy. Sci Transl Med 3(72):72ra18
5. Nizzardo M, Simone C, Salani S, Ruepp MD, Rizzo F, Ruggieri M, Zanetta C, Brajkovic S, Moulton HM, Müehlemann O, Bresolin N, Comi GP, Corti S (2014) Effect of combined systemic and local morpholino treatment on the spinal muscular atrophy Δ7 mouse model phenotype. Clin Ther 36(3):340–356
6. Nizzardo M, Simone C, Rizzo F, Ulzi G, Ramirez A, Rizzuti M, Bordoni A, Bucchia M, Gatti S, Bresolin N, Comi GP, Corti S (2016) Morpholino-mediated SOD1 reduction ameliorates an amyotrophic lateral sclerosis disease phenotype. Sci Rep 6:21301
7. Burglen L, Lefebvre S, Clermont O, Burlet P, Viollet L, Cruaud C, Munnich A, Melki J (1996) Structure and organization of the human survival motor neuron (SMN) gene. Genomics 32:479–482
8. Monani UR, Lorson CL, Parsons DW, Prior TW, Androphy EJ, Burghes AH, McPherson JD (1999) A single nucleotide difference that alters splicing patterns distinguishes the SMA gene SMN1 from the copy gene SMN2. Hum Mol Genet 8:1177–1183
9. Vitte J, Fassier C, Tiziano FD, Dalard C, Soave S, Roblot N, Brahe C, Saugier-Veber P, Bonnefont JP, Melki J (2007) Refined characterization of the expression and stability of the SMN gene products. Am J Pathol 171:1269–1280
10. Sendtner M (2010) Therapy development in spinal muscular atrophy. Nat Neurosci 7:795–799
11. Singh NK, Singh NN, Androphy EJ, Singh RN (2006) Splicing of a critical exon of human Survival Motor Neuron is regulated by a unique silencer element located in the last intron. Mol Cell Biol 26(4):1333–1346

12. Mitrpant C, Porensky P, Zhou H, Price L, Muntoni F, Fletcher S, Wilton SD, Burghes AHM (2013) Improved antisense oligonucleotide design to suppress aberrant SMN2 gene transcript processing: towards a treatment for spinal muscular atrophy. PLoS One 8(4):e62114
13. Zhou H, Janghra N, Mitrpant C, Dickinson R, Anthony K, Price L, Eperon I, Wilton S, Morgan J, Muntoni F (2013) A novel morpholino oligomer targeting ISSN1 improves rescue of severe SMA transgenic mice. Hum Gene Ther 24(3):331–342
14. Porensky PN, Mitrpant C, McGovern VL, Bevan AK, Foust KD, Kaspar BK, Wilton SD, Burghes AHM (2012) A single administration of morpholino antisense oligomer rescues spinal muscular atrophy in mouse. Hum Mol Genet 21:1625–1638
15. Zhou H, Meng J, Marrosu E, Janghra N, Morgan J, Muntoni F (2015) Repeated low doses of morpholino antisense oligomer: an intermediate mouse model of spinal muscular atrophy to explore the window of therapeutic response. Hum Mol Genet 24(22):6265–6277
16. Renton AE, Chio A, Traynor BJ (2014) State of play in amyotrophic lateral sclerosis genetics. Nat Neurosci 17(1):17–23
17. Rosen DR, Siddique T, Patterson D, Figlewicz DA, Sapp P, Hentati A, Donaldson D, Goto J, O'Regan JP, Deng HX (1993) Mutations in Cu/Zn superoxide dismutase gene are associated with familial amyotrophic lateral sclerosis. Nature 362(6415):59–62
18. Rotunno MS, Bosco DA (2013) An emerging role for misfolded wild-type SOD1 in sporadic ALS pathogenesis. Front Cell Neurosci 7:253
19. Bruijn LI, Houseweart MK, Kato S, Anderson KL, Anderson SD, Ohama E, Reaume AG, Scott RW, Cleveland DW (1998) Aggregation and motor neuron toxicity of an ALS-linked SOD1 mutant independent from wild-type SOD1. Science 281:1851–1854
20. Ezzi SA, Urushitani M, Julien JP (2007) Wild-type superoxide dismutase acquires binding and toxic properties of ALS-linked mutant forms through oxidation. J Neurochem 102:170–178
21. Bosco DA, Morfini G, Karabacak NM, Song Y, Gros-Louis F, Pasinelli P, Goolsby H, Fontaine BA, Lemay N, McKenna-Yasek D, Frosch MP, Agar JN, Julien JP, Brady ST, Brown RH Jr (2010) Wild-type and mutant SOD1 share an aberrant conformation and a common pathogenic pathway in ALS. Nat Neurosci 13:1396–1403
22. Guareschi S, Cova E, Cereda C, Ceroni M, Donetti E, Bosco DA, Trotti D, Pasinelli P (2012) An over-oxidized form of superoxide dismutase found in sporadic amyotrophic lateral sclerosis with bulbar onset shares a toxic mechanism with mutant SOD1. Proc Natl Acad Sci U S A 109(13):5074–5079
23. Ralph GS, Radcliffe PA, Day DM, Carthy JM, Leroux MA, Lee DC, Wong LF, Bilsland LG, Greensmith L, Kingsman SM, Mitrophanous KA, Mazarakis ND, Azzouz M (2005) Silencing mutant SOD1 using RNAi protects against neurodegeneration and extends survival in an ALS model. Nat Med 11:429–433
24. Smith RA, Miller TM, Yamanaka K, Monia BP, Condon TP, Hung G, Lobsiger CS, Ward CM, McAlonis-Downes M, Wei H, Wancewicz EV, Bennett CF, Cleveland DW (2006) Antisense oligonucleotide therapy for neurodegenerative disease. J Clin Invest 116:2290–2296
25. Le TT, Pham LT, Butchbach ME, Zhang HL (2005) SMNDelta7, the major product of the centromeric survival motor neuron (SMN2) gene, extends survival in mice with spinal muscular atrophy and associates with full-length SMN. Hum Mol Genet 14:845–857
26. Gurney ME, Pu H, Chiu AY, Dal Canto MC, Polchow CY, Alexander DD, Caliendo J, Hentati A, Kwon YW, Deng HX et al (1994) Motor neuron degeneration in mice that express a human Cu,Zn superoxide dismutase mutation. Science 264:1772–1775
27. Rigo F, Chun SJ, Norris DA, Hung G, Lee S, Matson J, Fey RA, Gaus H, Hua Y, Grundy JS, Krainer AR, Henry SP, Bennett CF (2014) Pharmacology of a central nervous system delivered 2′-O-methoxyethyl-modified survival of motor neuron splicing oligonucleotide in mice and nonhuman primates. J Pharmacol Exp Ther 350(1):46–55

Chapter 20

Delivery of Morpholino Antisense Oligonucleotides to a Developing Ovine Conceptus via Luminal Injection into a Ligated Uterine Horn

Xiaoqui Wang and Kathrin A. Dunlap

Abstract

In vivo delivery of morpholino antisense oligonucleotides (MAO) directly into the uterine lumen of a peri-implantation period pregnant sheep is an effective technique for evaluation of gene products for conceptus development. The highly phagocytic conceptus is undergoing rapid morphological change, thereby the available MAO are readily consumed and delivered to developing cells. Here, we describe the method for preparation and surgical delivery of MAO—Endo-Porter complex to developing ovine conceptus on day 8 postmating. Also outlined are methods for posttreatment sample recovery on day 16 postmating.

Key words Ovine, Conceptus, Peri-implantation, Uterine ligation, Morpholino antisense oligonucleotide, Endo-Porter

1 Introduction

In all species, implantation of the blastocyst precedes and is required for successful placentation. It is a highly coordinated process that involves blastocyst hatching and shedding of the zona pellucid; precontact blastocyst orientation, apposition, and attachment; and adhesion of the trophoblast to the endometrial luminal epithelium [1, 2]. In sheep, the peri-implantation time period is particularly sensitive, as a large percentage of early pregnancy loss occurs during this time [3, 4]. The blastocyst is undergoing a rapid morphological transition from a spherical to a tubular and ultimately filamentous form between days 8 and 16 of gestation, thereby this is an exciting time period for investigation of gene products that may be necessary for conceptus development [1, 3, 5–9]. While there are emerging technologies to facilitate domestic animal gene targeting, there are still limited options to perform in vivo studies to determine the biological role of genes expressed in the conceptus of domestic animals using loss of function. More

Hong M. Moulton and Jon D. Moulton (eds.), *Morpholino Oligomers: Methods and Protocols*, Methods in Molecular Biology, vol. 1565, DOI 10.1007/978-1-4939-6817-6_20, © Springer Science+Business Media LLC 2017

than a decade ago, Luu et al. [10] demonstrated the usefulness of morpholino antisense oligonucleotides in the mouse uterus to knockdown expression of calbindin-9 K protein in endometrial epithelia of pregnant mice, which prevented implantation. This morpholino approach to in vivo loss-of-function studies has been adapted by our laboratory for use in sheep and is the focus of this manuscript [6–9, 11].

2 Materials

Prepare and store all reagents at room temperature (unless otherwise indicated). All water used in solution preparation is double-distilled water (unless otherwise indicated). All materials should be handled and disposed of following University regulations.

2.1 Morpholino Antisense Oligonucleotide (MAO) and Endo-Porter Ordering

1. 3′-lissamine modified targeted MAO and mismatch control MAO (Gene Tools, Philomath, OR).
2. Endo-Porter Aqueous Delivery Reagent (Gene Tools, Philomath, OR).

2.2 MAO Solubilization and Preparation

1. Eppendorf Snap-Cap Safe-lock™ conical tube, 1.5 mL (product number 022600028, Fisher Scientific, Sugarland, TX).
2. OPTI-MEM medium (product number 11058-021, Invitrogen, Carlsbad, CA).

2.3 MAO In Vivo Delivery

1. Animals: Female sheep, day 8 postmating.
2. Ovine surgical instruments.
3. Umbilical tape, 3/8″ (product number 17439, Jorgenson Labs, Loveland, CO).
4. Precision Glide hypodermic needle, 18 gauge × 1″ (product number 305195, Becton-Dickinson, Franklin Lakes, NJ).
5. Insulin syringe with slip tip, 1 mL (product number 329654, Becton-Dickinson, Franklin Lakes, NJ).
6. Sovereign Tom Cat catheter with nonslip suturing adaptor, 3.5 French × 4.5″L (product number 008187, MWI Animal Health, Boise, ID).
7. Chromic gut suture, 3-0 (product number U204H, Ethicon, West Somerville, NJ).
8. Chromic gut suture, 0 (product number U246H, Ethicon, West Somerville, NJ).
9. Disposable skin staples, 35 regular (product number 618029, MWI Animal Health, Boise, ID).

10. 5% Glycerol–0.9%Saline solution: Add about 100 mL water to a 1 L graduated cylinder. Weigh 9.0 g NaCl and transfer to the cylinder. Add water to a volume of 900 mL. Add 50 mL Glycerol. Mix. Make up to 1 L with water. Transfer to autoclavable safe bottle. Autoclave for 20 min to sterilize.

2.4 Tissue Recovery and Fixation

1. Phosphate-buffered saline (PBS) solution: Add about 300 mL water to a 1 L graduated cylinder. Weigh 16.0 g NaCl, 0.4 g KH_2PO_4 (monobasic), 2.28 g NA_2HPO_4 (dibasic), and transfer to cylinder. Add water to a volume of 900 mL. Mix and adjust pH to 7.3. Make up to 1 L with water.
2. Precision Glide hypodermic needle, 16 gauge × 1″ (product number 305197, Becton-Dickinson, Franklin Lakes, NJ).
3. Luer-lok™ syringe, 10 mL (product number, 309604, Franklin Lakes, NJ).
4. Square dish with grid, Fisherbrand™ Petri dish (product number, FB0875711A, Fisher Scientific, Sugarland, TX).
5. OCT Compound (product number, 4583, Tissue-Tek®, Electron Microscopy Sciences, Hatfield, PA).
6. Tissue Path™ Disposable Base Molds (product number, 22-363-553, Fisher Scientific, Sugarland, TX).

3 Methods

All materials should be handled and disposed of following University regulations. All animal handling and surgical procedures should be conducted in congruence with the Institutional Animal Care and Use Committee (IACUC) guidelines.

3.1 Morpholino Antisense Oligonucleotide (MAO) and Endo-Porter Ordering

MAO Design: Compile sequence information for gene of interest including Genebank name and mRNA sequence with marked start codon (ATG). Order both gene of interest and mismatch control (*see* **Note 1**). Order all MAO with 3′-lissamine modification for epifluorescent delivery verification postdelivery (*see* **Note 2**).

3.2 MAO and Endo-Porter Preparation

1. Fluorescent MAO is light sensitive, thereby handle under low light conditions and keep samples wrapped in foil to protect against light. Store lyophilized MAO in cool dark location at room temperature.
2. MAO stock solution: Dissolve 1000 nmol MAO into 1 mL of sterile double-distilled water (*see* **Note 3**). Aliquot as 0.1 mL into foil wrapped 1.5 mL Eppendorf Snap-Cap Safe-lock™ conical tubes (product number 022600028, Fisher Scientific). Store at −20 °C for up to 5 days, −80 °C for up to 12 months.

3. MAO working solution: Add 0.1 mL of MAO stock solution (100 nmol) to 0.1 mL Endo-Porter aqueous delivery reagent into 0.8 mL OPTI-MEM medium (product number 11058-021, Invitrogen). Protect working solution in foil and place on ice until use (*see* **Note 4**).

3.3 MAO In Vivo Delivery

1. Animals: Female sheep (*Ovis aries*) are evaluated for estrous cycle activity and at the time of estrus are mated. Surgical delivery of the MAO to the conceptus will occur on day 8 postmating. Ewes are deprived of feed and water for 24 h prior to surgery.
2. Anesthesia: Ewes are maintained on a surgical plane of anesthesia via administration of Isoflurane and oxygen delivered via inhalation masks.
3. Surgical field preparation: The anesthetized ewe will be placed in dorsal recumbency in a surgical cradle. The abdomen and inguinal area pertinent to the area for performing a midventral laparotomy will be shaved to the skin and then scrubbed twice with Betadine solution to achieve a sterile field for the subsequent surgical procedure.
4. Midventral laparotomy and ovarian evaluation: Under sterile operating conditions, the ovaries and uterine horns are exteriorized following a midventral incision. The ovaries are assessed for presence of a functional corpus luteum (*see* **Note 5**). The uterine horn ipsilateral to a viable corpus luteum is selected for uterine horn ligation and subsequent MAO injection.
5. Uterine horn ligation: The base of the uterine horn ipsilateral to the corpus luteum will be double ligated using nonabsorbable umbilical tape to prevent migration and growth of the conceptus through the uterine body into the contralateral uterine horn. The ligature consists of two ties using square knots (Fig. 1). This surgical procedure does not affect conceptus implantation or fetal development in sheep [6–9, 11–13].
6. MAO Injection (*see* **Note 6**): The MAO complex will be introduced into the lumen of the uterus via the uterotubal junction. An 18 gauge blunt tipped surgical needle will be used to make an entry point into the oviduct just anterior to the uterotubal junction to allow for insertion of the Tom Cat catheter. The MAO complex will be drawn up into a 1 mL slip tip syringe and fitted to the end of the catheter for discharge into the uterine lumen. The catheter is inserted into the oviduct and threaded approximately 2 cm or just into the uterine lumen (Fig. 2). After discharging the solution the catheter is withdrawn and the uterine horn gently massaged to distribute the solution throughout the uterine horn. The exterior of the uterus is rinsed with the glycerol–saline solution to prevent

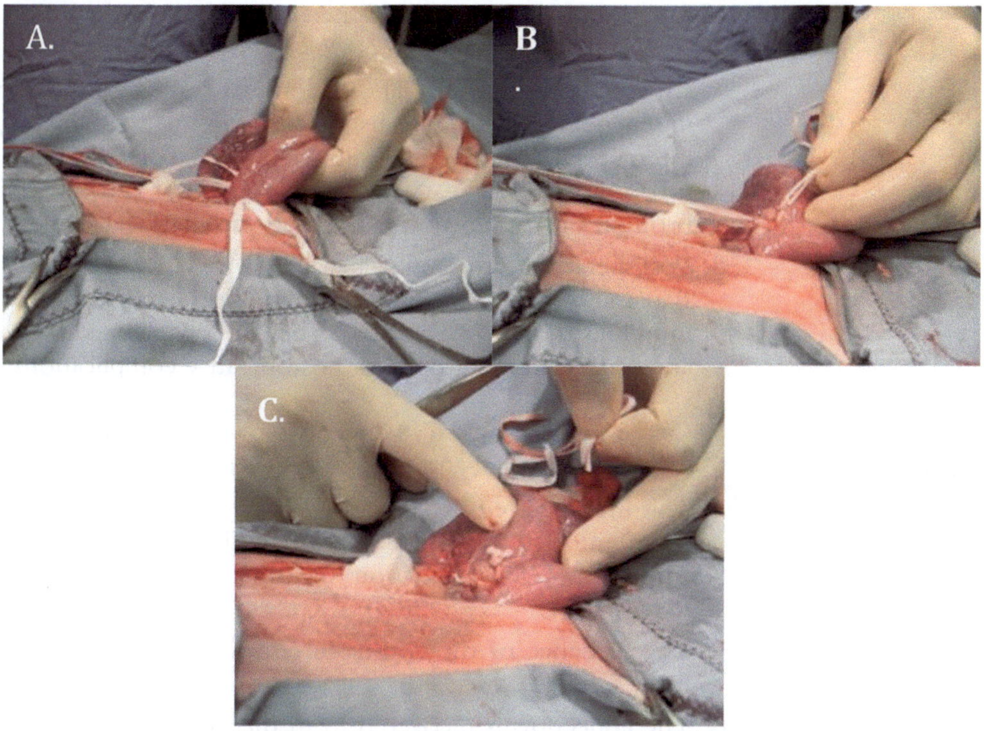

Fig. 1 Placement of double ligature at base of uterine horn ipsilateral to corpus luteum. (**a**) Two sections of umbilical tape are drawn around the base of the uterine horn and through the bifurcation to serve as an anchor. (**b**) Using a repeated square knot the ligature is tightened at the base of the uterine horn. (**c**) The ends of the umbilical tape are clipped revealing the presence of the two, adjacent ligatures at the base of the uterine horn

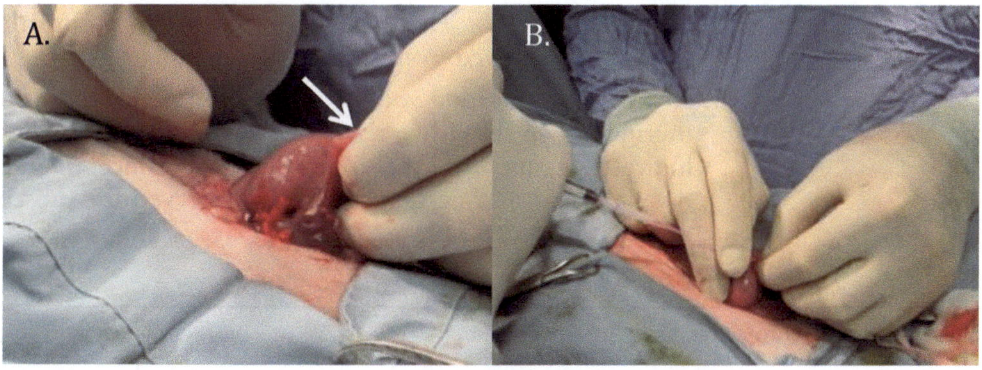

Fig. 2 Placement and infusion of Tom Cat catheter for delivery of MAO complex into the uterine lumen. (**a**) The white arrow denotes the section of oviduct just apical to the uterotubal juncture. This is the site of entry by 18-gauge blunt surgical needle to create an opening for threading of the catheter. (**b**) The Tom Cat catheter has been threaded into the opening created in the oviduct, past the uterotubal juncture. The MAO complex (*pink colored* solution) is deposited from the 1 mL slip-tip syringe, through the catheter and into the uterine lumen

formation of adhesions and then placed back in the body cavity (*see* **Note 7**).

7. Closure of body wall: The abdominal wall is closed with an independent suture line using chromic gut (#3) in a simple interrupted pattern. The subcutaneous tissue in closed in a simple continuous pattern with chromic gut (#0). The skin is brought into apposition and closed with surgical staples.
8. Postoperative care: Animals will receive an i.m. injection of Banamine (analgesic) and placed in an isolated recovered area until fully mobile (*see* **Note 8**). They will be returned to full feed and water and monitored for postoperative complications.

3.4 Tissue Recovery and Fixation

1. On gestational day 16 (8 days post in vivo MAO delivery) ewes will be necropsied and uterine tissues removed.
2. The uterus is removed from the body cavity by dissection at the point of the cervix and at the individual ovarian pedicles. This allows for evaluation of ovary, oviduct, and uterine tissues.
3. The presence of the corpus luteum ipsilateral to the ligated horn is noted.
4. The conceptus is recovered from the uterine horn via uterine lavage with PBS. A 10 mL syringe fitted with a 16-gauge needle is filled with PBS solution. This solution is infused into the uterine lumen via injection into the uterine lumen adjacent to the ligatures at the base of the uterine horn (*see* **Note 9**).
5. The tip of the uterine horn is opened and the respective sides of the opening secured with hemostats so that solution contained within the uterine lumen will drain into a collecting vessel (*see* **Note 10**). The uterus is massaged to ensure that all solution is removed from the uterine lumen.
6. At this time the total volume of solution recovered from the uterine lumen should be recorded, as should a gross morphological assessment of the conceptus. The conceptus, uterine flushing, and uterine tissues can be preserved for subsequent analysis (*see* **Note 11**).

4 Notes

1. When ordering, ask for associated MAO design to block the translation initiation of mRNA instead of alternative splicing site [6]. It is necessary to evaluate both the gene target of interest as well as a mismatch or scrambled control MAO [12, 14]. Placing the order for both MAO at the same time will expedite the design process. After receive the design from Gene Tools, BLAST the sequence to check whether the

designed MAO sequence: (1) targets the gene of interest; (2) picks up any off-target gene. Redesign if needed, otherwise place the order from Gene Tools.

2. Use of the 3′-lissamine modification will allow for rapid visualization of efficacy of MAO delivery to the conceptus tissue. Samples recovered at day 16 (8 days postdelivery) will appear pink in color to the naked eye (Fig. 3). This uptake can be validated via use of epifluorescent evaluation of frozen sections of conceptus tissues.
3. It is easiest to add 1 mL of sterile water directly to the stock vial of lyophilized MAO. If it does not go into solution, it can be heated to 37 °C. Care must be taken to cool to room temperature prior to creating aliquots. All activities should be conducted in a laminar flow hood to maintain sterility.
4. The amount of MAO required for 8 day in vivo conceptus treatment is 100 nmol + 0.1 mL Endo-Porter aqueous delivery reagent per sample (i.e., uterine horn). This working stock solution is added to the OPTI-MEM and mixed with intent of in vivo delivery within 4 h. Solution should be protected from light and maintained on ice until injection into the uterine lumen.
5. If a viable corpus luteum is not visualized at the time of surgery it is advisable not to treat the animal as she is likely not pregnant and/or undergoing pregnancy loss. The horn selected for treatment should be ipsilateral to the corpus luteum.

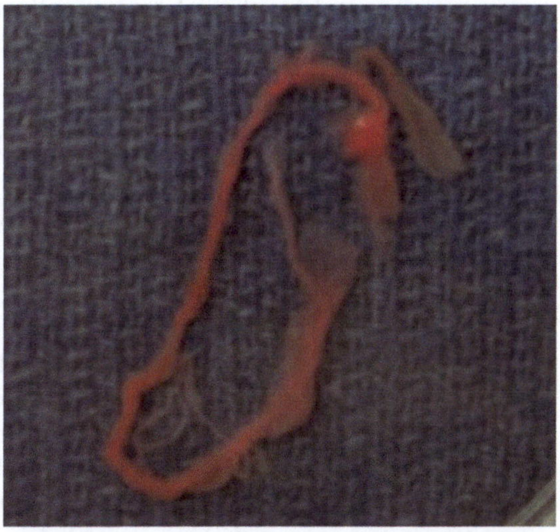

Fig. 3 Visual assessment of conceptus uptake of 3′lissamine labeled MAO. A typical elongated, filamentous conceptus at day 16 of gestation (8 days postinfusion with 3′lissamine labeled control MAO). Pink coloration is expected and denotes uptake of MAO complex by conceptus

6. The laboratory of Dr. Thomas Spencer has recently utilized an alternative method for in vivo delivery of MAO to the sheep uterus. As opposed to a single infusion of MAO solution, there is a timed release of MAO via the use of Alzet 2ML1 Osmotic Pump [12, 15, 16]. Further, the Spencer laboratory has shown that delivery of a standard control MAO via a bolus injection into the lumen of a ligated uterine horn on day 8 postmating does reduce conceptus length when evaluated at day 14 postmating [12]. However, despite this difference, similar studies have reported morphologically normal and elongated conceptuses by day 16 postmating thereby the decreased length on day 14 does not appear to affect continued elongation or establishment of pregnancy [6–9, 11, 12].
7. The glycerol–saline solution should be warmed to animal's body temperature prior to delivery into the abdominal cavity. This can be achieved via placement in a warm water bath until use in surgery.
8. As sheep are flock species it is helpful to minimize time that they are isolated. Once they are physically stable postsurgery it is beneficial to move them to an area where they are in contact with other sheep to minimize stress.
9. It is necessary to take care that PBS is infused into the uterine lumen and not into the space between the endometrium and myometrium. This will be apparent as there should be essentially no resistance to delivery of solution if the lumen has been penetrated. If the PBS is delivered into the uterine wall a balloon-shaped projection will form in the wall of the uterus where the solution is collecting. Another measure of efficacy of delivery into the lumen is the steady flow of solution out of the opened tip adjacent to the uterotubal junction. If there is a lack of flow exiting the uterus it is advisable to reposition the needle prior to proceeding with PBS delivery.
10. Use of a sterile collecting dish capable of containing greater than 10 mL of liquid is required. A grid dish is helpful as it allows for measurement of relative conceptus size upon collection. An untreated ovine conceptus at gestational day 16 is an elongated filamentous structure that is easily visible with the naked eye (approximately 190 mm in length [5]). Dependent upon treatment, the conceptus may appear stunted, fragmented, or even undergoing apoptosis. These conditions can make collection more challenging. Thereby, it is advisable to document via digital imaging at the time of collection.
11. For assessment of MAO delivery it is advisable to freeze the conceptus in OCT compound within a disposable base mold. To do so place a thin layer of OCT in the base of the mold then add the conceptus tissue to the center and surround and cover

with additional OCT. Take care to avoid the formation of bubbles. Freeze in liquid nitrogen vapor and store at −80 °C for subsequent frozen sectioning [6]. If tissue is available, additional sections can be snap frozen in liquid nitrogen for subsequent RNA and protein extraction and evaluation. All samples should be preserved at −80 °C. The flush should be clarified by centrifugation, spun at 3000 × *g* at 4 °C for 15 min, and the supernatant removed and aliquoted for storage at −80 °C.

Acknowledgement

The authors would like to thank Fuller W. Bazer, M. Carey Satterfield, and Thomas E. Spencer, as well as former graduate students and postdoctoral fellows in the Bazer, Spencer and Satterfield laboratories for their participation in development of the in vivo MAO delivery method for use in pregnant sheep. This research was supported by the Agriculture and Food Research Initiative competitive grants 2011-6705-20028 and 2011–67015-20067 from the U.S. Department of Agriculture National Institute of Food and Agriculture, as well as support via National Institutes of Health grant HD052745.

References

1. Guillomot M, Flechon JE, Wintenberger-Torres S (1981) Conceptus attachment in the ewe: an ultrastructural study. Placenta 2:169–182
2. Guillomot M (1995) Cellular interactions during implantation in domestic ruminants. J Reprod Fertil Suppl 49:39–51
3. Spencer TE, Johnson GA, Bazer FW, Burghardt RC (2007) Fetal-maternal interactions during the establishment of pregnancy in ruminants. Soc Reprod Fertil Suppl 64:379–396
4. Wu G, Bazer FW, Satterfield MC, Li X, Wang X, Johnson GA, Burghardt RC, Dai Z, Wang J, Wu Z (2013) Impacts of arginine nutrition on embryonic and fetal development in mammals. Amino Acids 45:241
5. Spencer TE, Johnson GA, Bazer FW, Burghardt RC (2004) Implantation mechanisms: insights from the sheep. Reproduction 128:657–668
6. Dunlap KA, Palmarini M, Varela M, Burghardt RC, Hayashi K, Farmer JL, Spencer TE (2006) Endogenous retroviruses regulate periimplantation growth and differentiation. Proc Natl Acad Sci 103:14390–14395
7. Wang W, Ying W, Dunlap KA, Lin G, Satterfield MC, Burghardt RC, Wu G, Bazer FW (2014) Arginine decarboxylase and agmatinase: an alternative pathway for de novo biosynthesis of polyamines for development to mammalian conceptuses. Biol Reprod 90:84
8. Wang W, Frank JW, Little DR, Dunlap KA, Satterfield MC, Burghardt RC, Hansen TR, Wu G, Bazer FW (2014) Functional role of arginine during the peri-implantation period of pregnancy I. Consequences of loss of function of arginine transporter SLC7A1 mRNA in ovine conceptus trophectoderm. FASEB J 28:2852–2863
9. Wang W, Frank JW, Wu J, Dunlap KA, Satterfield MC, Burghardt RC, Romero JJ, Hansen TR, Wu G, Bazer FW (2014) Functional role of arginine during the peri-implantation period of pregnancy II. Consequences of loss of function of nitric oxide synthase NOS3 mRNA in ovine conceptus trophectoderm. Biol Reprod 91:59
10. Luu KC, Nie GY, Salamonsen LA (2004) Endometrial calbindins are critical for embryo implantation: evidence from in vivo use of morpholino antisense oligonucleotides. Proc Natl Acad Sci 101:8028–8033
11. Ruiz-Gonzalez I, Minten M, Wang X, Dunlap KA, Bazer FW (2015) Involvement of TLR7 and TLR8 in conceptus development and establishment of pregnancy in sheep. Reproduction 149:305–316

12. Brooks K, Spencer TE (2015) Biological roles of interferon tau (IFNT) and type 1 IFN receptors in elongation of the ovine conceptus. Biol Reprod 92:1–10
13. Dorniak P, Bazer FW, Spencer TE (2011) Prostaglandins regulate conceptus elongation and mediate effects of interferon tau on the ovine uterine endometrium. Biol Reprod 84:119–1127
14. Schulte-Merker S, Stainer DYR (2014) Out with the old, in with the new: reassessing morpholino knockdowns in light of genome editing technology. Development 141:3103–3104
15. Brooks K, Burns G, Spencer TE (2015) Biological roles of hydroxysteroid (11-Beta) dehydrogenase 1(HSD11B1), HSD11B2, and glucocorticoid receptor (NR3C1) in sheep conceptus elongation. Biol Reprod 93:38
16. Brooks K, Burns G, Spencer TE (2015) Peroxisome proliferator activator receptor gamma (PPARG) regulates conceptus elongation in sheep. Biol Reprod 92:42

Chapter 21

Surface Plasmon Resonance-Based Concentration Determination Assay: Label-Free and Antibody-Free Quantification of Morpholinos

Jordan Boutilier and Hong M. Moulton

Abstract

Surface plasmon resonance (SPR) is a physical process that allows label-free and real-time detection of biomolecular interactions. SPR provides a rapid and quantitative method for studying interactions of macromolecules such as proteins and nucleic acids. Antisense Morpholino oligomers are widely used to regulate gene expression and the US FDA has approved a Morpholino drug for treatment of Duchenne muscular dystrophy. Here, we describe an antibody-free, label-free, high-throughput, and walk-away SPR method for quantification of Morpholino compounds extracted from biological specimens. This provides a valuable way for determining pharmacokinetics and pharmacodynamics of Morpholino oligomers in biological matrices for research and therapeutic applications.

Key words Surface plasmon resonance, SPR, Concentration analysis, Bioavailability, Morpholino oligomers, PMO, PPMO, Label-free, Antibody-free

1 Introduction

Surface plasmon resonance (SPR) is a phenomenon that allows real-time, label-free detection of biomolecular interactions. The phenomenon occurs when polarized light hits an electrically conducting sensor surface at the interface between two media and generates plasmons, electron charge density waves. This leads to reduction of the intensity of reflected light at a specific angle. This angle, in proportion to the mass on the sensor surface, changes as molecules associate and dissociate and the interaction profile is recorded in real time (Fig. 1). SPR detection provides a high-throughput, walk-away method for characterizing binding events with size ranging from hundreds of Daltons [1] to whole cell [2]. More recently, there has been an interest in using SPR for the screening and design of nanotherapeutics [3]. A number of instruments are available for SPR-based applications [4], although

Hong M. Moulton and Jon D. Moulton (eds.), *Morpholino Oligomers: Methods and Protocols*, Methods in Molecular Biology, vol. 1565, DOI 10.1007/978-1-4939-6817-6_21, © Springer Science+Business Media LLC 2017

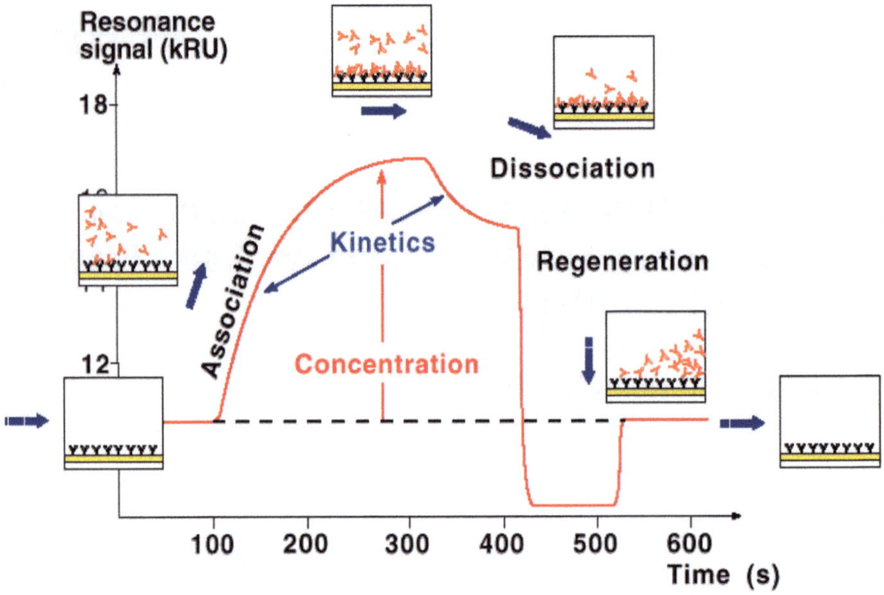

Fig. 1 The sensorgram is a rich source of information. As the mobile-phase analyte is injected over the sensor chip surface it interacts with immobilized ligand. The progress of the interaction is monitored as a change in refractive index and plotted as response units (RU) (Y-axis) against time (X-axis) generating a sensorgram in the control software

Biacore has been a dominant name in the market since the 1980s [5] and has been used in many peer-reviewed scientific publications. The remainder of this section will introduce some basic principles for defining an SPR concentration assay, and then focus more specifically on the application of SPR in quantifying Morpholino oligomers (henceforth referred to as Morpholinos) isolated from biological matrices.

There are three major steps in a SPR-based assay (Fig. 1): immobilization, interaction (association/dissociation), and regeneration [6]. The information obtained from SPR analysis will only be useful if the assay is carefully designed with a good understanding of each of these three steps.

1.1 Immobilization: The Process Where Ligand Is Immobilized to a Sensor Chip Surface

Choosing a right sensor chip is an integral part of SPR method development. Depending on the nature of molecule to be coupled and the type of assay being performed, SPR users can choose from a range of sensor chips currently on the market, offering flexibility to create biospecific surfaces suitable for many of the common ligand/analyte interactions [5], including: Streptavidin/Biotin, NTA-Ni_2^+/His, RaMFc/Mab, and Anti-GST/GST.

Alternatively, it is possible to covalently couple proteins, nucleic acids, carbohydrates, or small molecules directly onto the CM sensor surface via an amine ($-NH_2$), thiol ($-SH$), aldehyde ($-CHO$), carboxyl ($-COOH$), or alcohol ($-OH$) functional group [6].

1.2 Interaction: The Process for Monitoring the Binding Interaction Between an Analyte and the Immobilized Ligand in a Mobile Phase

After choosing a chip that is biocompatible with the analyte of interest, a number of parameters must be defined by the user. Depending on how each of the parameters below is defined, they will influence the sensitivity (magnitude of the binding response over time) and the quality (reproducibility) of a concentration assay. These include (1) *Calibration*: the use of standards and controls in an appropriate matrix, with stringent and repeated calibration to achieve accurate and precise concentration measurements, (2) *Nonspecific binding*: formulating a running buffer that reduces nonspecific interactions, (3) *Dilutions*: determining a dilution scheme for all samples such that a signal is generated for each sample at one or more dilutions falling within the dynamic range of the assay.

1.3 Regeneration: The Process of Reestablishing Baseline by Removing Any Bound-Specific Analyte and Other Nonspecific Agents from the Immobilized Ligand and the Chip Matrix

The goal of this step is to find a condition (i.e., regeneration solution) that will completely remove the analyte, but not harm immobilized ligand and the matrix on the chip surface. Incomplete regeneration may cause a decrease in signal-to-noise ratio over multiple cycles due to a decrease in the available ligand-binding sites and concomitant increasing baseline. Also keep in mind that hundreds of sample analyses (and therefore hundreds of regeneration cycles) are usually performed on the same chip surface, necessitating a regeneration condition that allows for complete baseline regeneration but also preserves the integrity of the chip surface.

A number of components (acid/base, solvents, detergents) can be used for making regeneration solutions, including ethanol, TFE, guanidine, formamide, urea, NaN_3, NaCl, and EDTA [5, 7]. However, in our experience none of these conditions are optimal for removing a Morpholinos from its cDNA ligand, whereas NaOH damages Biotin/SA surface.

1.4 SPR Applied to Quantification of Morpholinos

Morpholinos have been broadly used for various research applications and recently there has been an increased interest in Morpholino-based therapeutics [8–14], due to their unique chemical properties and safety profile in patients with muscular dystrophy [9]. Morpholinos have unique chemical properties that profoundly distinguish them from their negatively charged nucleic acid counterparts, most notably the neutral (nonionic) charge conferred by the phosphorodiamidate backbone structure [15, 16]. The SPR method described herein has been optimized specifically for determining the concentration of Morpholino compounds isolated from biological matrices—a critical step in determining tissue biodistribution [17] and pharmacokinetic [18, 19] properties of Morpholino compounds in vivo.

The basic idea of the method is to apply the SPR principle for quantifying interactions between a Morpholino (the analyte) extracted from biological matrices and the biotin-modified cDNA (the ligand) over the surface of a sensor chip. This interaction

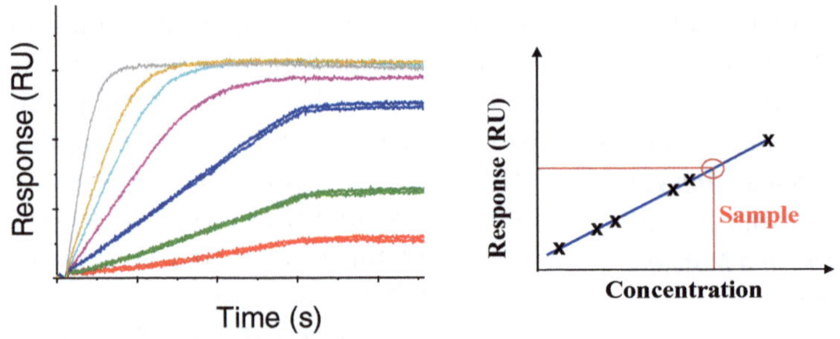

Fig. 2 Sensorgram and calibration curve. A example of sensorgram of a Morpholino at several difference concentrations (*Left*). The response (RU) of a set of standard samples from a sensorgram and the concentrations of sample can be used to build a calibration curve (*right*)

causes a change in the refractive index, which is proportional to the amount (concentration) of the Morpholino bound to the cDNA attached to the sensor chip. The concentration of Morpholino in the sample can be calculated from a calibration curve using a set of standard samples (Fig. 2). The method described later has a dynamic range of quantification of a Morpholino between 2–7 nM and 200 nM in tissue, plasma, or urine samples.

2 Materials

2.1 Extraction of Morpholinos from Tissues (See Note 1)

1. Tissue collected from mock-treated and Morpholino-treated animals.
2. Tissue Digestion Buffer: 20 units/mL proteinase K (Sigma) in tissue PE/LB lysis buffer (Genotech) (*see* **Note 2**).
3. A MagNA Lyser tissue homogenizer (Roche Life Science).
4. MagNA Lyser Green Bead (Roche): 2-mL screw-capped tubes, prefilled with 1.4-mm (diameter) ceramic beads (*see* **Note 3**).
5. A heating block shaker or shaking incubator.
6. A centrifuge.
7. A lyophilizer.
8. Ice cold acetonitrile (*see* **Note 4**).
9. Razor blades.
10. Dry ice or a −80 °C freezer.

2.2 Plasma or Urine Samples

1. Plasma or urine samples collected from mock-treated and Morpholino-treated animals.
2. Running Buffer: *see* Subheading 2.3.
3. A centrifuge.

2.3 Buffers and Morpholino Stock Solutions for SPR Analysis

1. Running Buffer: 1 L of 1× HBS-P with additional 150 mM NaCl. Prepare by diluting 100 mL 10× HBS-P into 900 mL sterile filtered water and adding 8.76 g NaCl. The 10× concentrated stock HBS-P solution contains 0.1 M HEPES, 1.5 M NaCl, and 0.5% v/v Surfactant P20 (GE Healthcare Life Sciences).
2. Regeneration Solution: It can be prepared in a laboratory as a 10 mM glycine HCl solution (pH1.75) by mixing equal volume of the pH 1.5 and pH 2.0 of the 20 mM Gly-HCl solutions (GE Healthcare Life Sciences).
3. Immobilization Solution: 1 μM biotin-cDNA in the Running Buffer.
4. Wash Solution: the Running Buffer with 50 mM NaOH.
5. Biotin-cDNA ligand: synthetic cDNA ligand (complementary to the Morpholino target sequence) must be ordered with a 5′ biotin-TEG modification (*see* **Note 5**).
6. Morpholino Stock Solution: 1 mM either in ultrapure water or in filtered PBS, used for making standards to build a calibration curve and for making control samples.

2.4 SPR Components

1. Series S Sensor Chip SA for immobilization of biotinylated nucleic acids.
2. 15 mm 4.0 mL polypropylene vials.
3. 7 mm 0.8 mL rounded polypropylene micro vials.
4. Biacore T100 instrument (GE Healthcare Life Sciences).
5. Series S Sensor Chip SA (streptavidin chip, GE Healthcare Life Sciences).
6. Microplate 96-well.
7. Microplate foil.
8. Type 3 Penetrable caps made of Kraton G (SEBS), ventilated (GE Healthcare Life Sciences).
9. Type 2 Penetrable caps made of Kraton G (SEBS), ventilated (GE Healthcare Life Sciences).

3 Methods

3.1 Extraction of Morpholinos from Tissues

1. Thaw frozen tissue samples at room temperature.
2. Cut and weigh ~60 mg of tissue and place into prelabeled MagNA Lyser Green Bead tubes (*see* **Note 6**).
3. Add 0.3 mL Tissue Digestion Buffer to each tissue sample (*see* **Note 7**).
4. Homogenize samples in the tissue homogenizer to make tissue lysates at the maximum speed of the instrument for 30 s. Tissue samples should be completely pulverized before proceeding to the next step, repeat this step if necessary.

5. Place tissue lysates into a shaking incubator with temperature controlled at 60 °C, set at 320 rpm overnight (*see* **Note 8**).
6. Following the overnight incubation step, pipette 600 μL of ice cold acetonitrile to each tissue lysate at 2:1 of acetonitrile:lysate ratio and vortex the tubes rigorously.
7. Centrifuge the tubes at 10,000 × *g* for ten minutes to pellet insoluble debris and transfer the supernatants to fresh vials. The supernatants should contain Morpholinos.
8. Freeze the supernatants on dry ice or in −80 °C and lyophilize overnight (*see* **Note 9**).
9. Rehydrate the lyophilized samples in 400 μL of Running Buffer.
10. Centrifuge samples at 10,000 × *g* to pellet any residual debris carried over from the extraction process.
11. Using a pipette, transfer 100 μL of rehydrated lysate samples to a 96-well plate, this is the 1× plate.
12. If necessary, use the same rehydrated lysate sample in **step 9** to make additional dilution plates with the Running Buffer to ensure the concentration measurement for each test sample will fall within the dynamic range of the calibration cure. For example, prepare dilution plates at 1/4×, 1/10×, and 1/50×.

3.2 Prepare Plasma and Urine Samples for SPR Analysis

1. Dilute serum or urine samples starting with 1/100X dilution in the Running Buffer (*see* **Note 10**).
2. Centrifuge samples at 10,000 × *g* to remove insoluble materials. The supernatants should contain Morpholinos.
3. Transfer 100 μL of the supernatant to a 96-well plate.

3.3 Prepare Standard and Control Samples for SPR Analysis

Standards and controls should be prepared at the same time and under the same conditions as the unknowns.

1. Prepare standard samples for a calibration curve by titrating a Morpholino Stock Solution at known concentrations ranging from 0 nM to 400 nM in 300 μL Tissue Digestion Buffer, in blank plasma, or in blank urine. (*see* **Notes 11** and **12**).
2. In a similar manner, prepare a set of control samples at 1, 50, and 100 nM. These samples will be run at regular intervals over the course of the assay to assess quality and reproducibility of calibration over a plate or between plates.
3. Add 600 μL cold acetonitrile to the standard and control samples at 2:1 acetonitrile to the Tissue Digestion Buffer ratio (v:v). Vortex the tubes rigorously.
4. Centrifuge the tubes at 10,000 × *g* for 10 min to pellet insoluble debris and transfer the supernatants to fresh vials.
5. Freeze and lyophilize the samples.

6. Reconstitute lyophilized standard or control sample each in 300 µL running buffer.
7. For analysis of Morpholinos in plasma with low dilutions (<1/100×), prepare a set of standard samples by titrating a Morpholino Stock Solution at known concentrations into blank plasma and then add 20% (v/v) 4 M ammonium sulfate to each of the standards followed by centrifuge (*see* **Note 12**).

3.4 Immobilization of Biotin-cDNA Ligand to the Sensor Chip SA (See Note 13)

1. Open the instrument control software.
2. Insert blank SA chip to be immobilized.
3. Select the flow cell(s) to immobilize with the corresponding cDNA ligand (*see* **Note 14**).
4. Run the immobilization program with the Immobilization Solution as the running buffer at 10 µL/min until it reaches to the maximum immobilization level (typically ~2200 RU) (*see* **Note 15**).
5. After the last ligand injection, inject a *short pulse* of the Wash Solution to wash the surface of the sensor chip and remove any unbound ligand molecules.

3.5 Prepare Instrument for Concentration Analysis

1. Fill the two glass bottles with the Running Buffer and Regeneration Buffer.
2. Open the instrument control software.
3. Insert the sensor chip prepared in Subheading 3.4 (*see* **Note 16**).
4. Use the Wizard/Template to set up a new concentration assay.
5. Set the number of regeneration cycles to one.
6. Designate flow cells. Select the flow cell(s) you desire for analysis.
7. Prior to initiating the concentration assay, perform a start-up step by running three cycles of the running buffer. The running solution is the Running Buffer.

3.6 Injection Parameters

1. The contact time during the concentration measurements should be set to 120 s at a flow of 30 µL/min followed by a 'dissociation time' of 15 s.
2. For regeneration of the sensor chip surface, inject a single pulse of the Regeneration Solution. Set the contact time to 5 s at a flow rate 50 uL/min with a 10 s stabilization period at the beginning of the sensorgram to allow baseline stabilization.

3.7 Specify the Variables

1. Enter analyte name in top entry.
2. Enter concentrations of calibration points and standards.
3. Run standard curve samples at the beginning, middle, and at the end of the assay. Control samples should be run at regular

intervals (about every 20 cycles) throughout the assay (*see* **Note 17**).

4. Fill out the Sample Table by entering in all of samples at their respective dilutions.
5. Prime the system with the Running Buffer after inserting the sensor chip.
6. Define auto-sampler rack types and sample positions using 'Rack Positions Screen.'
7. Select the *pooling* option for samples intended for repeated testing during the assay (e.g., calibration and control samples), this option enables resampling from a single vial. Under *Menu* select '*Automatic Positioning*' and choose the '*Pooling*' option. Select YES for any calibration points or control samples that require pooling.
8. Save the Method file in the appropriate folder and begin concentration assay.

3.8 Evaluation of the Results

1. Biacore software automatically generates a results tab for each flow cell selected for use during the concentration assay. Flow cells from a single assay can be analyzed individually or two at time (e.g., Referenced-subtracted). Biacore software automatically reports the following values for each sample: cycle number, sample ID, dilution factor, response (RU), calculated concentration (nM), and confidence value (% CV) if applicable.
2. The linear range of the assay is determined by plotting the standard sample data with RU on the Y-axis and concentration of Morpholino standards (nM) on the X-axis. A calibration curve generated using lyophilized samples typically yields linear regression value with $R^2 > 0.96$ over 2–200 nM concentration range (Fig. 3).

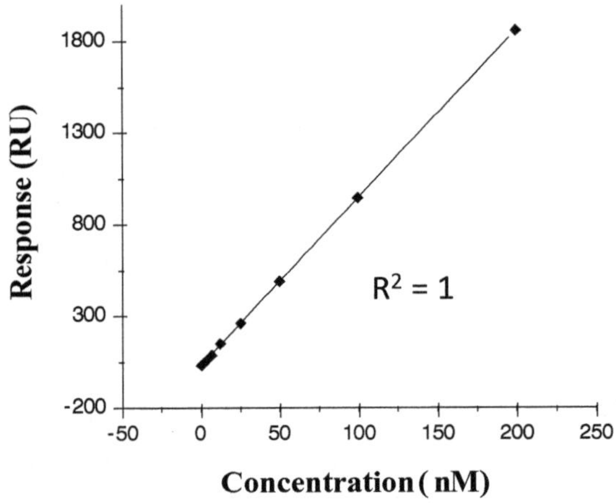

Fig. 3 A calibration curve of a Morpholino with a linear response between 2 and 200 nM

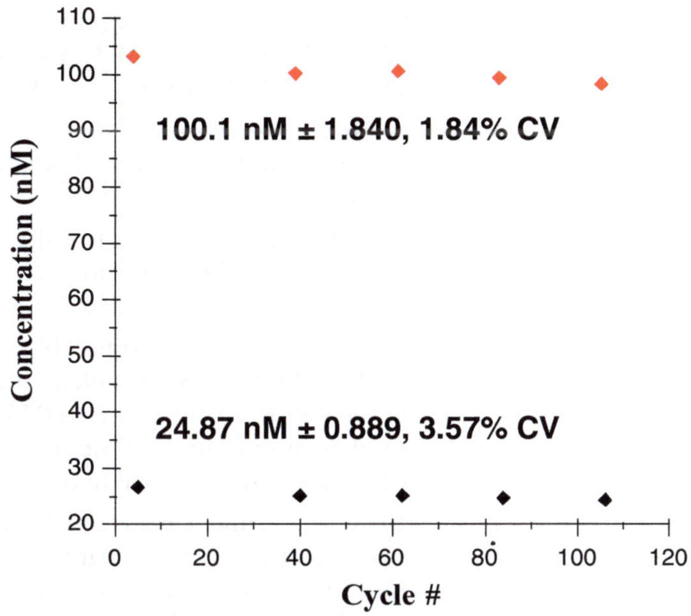

Fig. 4 The SA/biotin-cDNA sensor chip was robust. Two control samples at 25 nM and 100 nM were injected over a flow cell immobilized with the biotin-cDNA ligand at cycle # 4, 40, 62, 84, and 106 to determine the robustness of the sensor chip and effectiveness of the Regeneration Solution. There was very little change in response from the cycle #4 to cycle #106 as shown by the little change in concentrations calculated from a standard curve

3. A set of control samples at two to three concentrations are injected at the beginning, middle, and end of a plate to confirm the robustness of the sensor chip and reliability of the calibration curve (Fig. 4).
4. The lower limits of detection and quantitation (LLOD/LLOQ) can be determined using 2- and 10-fold above the standard deviation of the background signal, respectively (*see* **Note 18**).
5. Results from the concentration analysis can then be used to calculate pharmacokinetic properties of a target Morpholino.

4 Notes

1. All extraction conditions need to be compatible with the integrated fluidics cartridge (IFC). Leftover debris from the homogenization/extraction steps can clog the IFC and/or degrade the sensor chip surface.
2. If the Morpholino is modified with a delivery peptide, e.g. an arginine- and/or lysine-rich cell-penetrating peptide, trypsin should be added to the Tissue Digestion Buffer to degrade the

peptide. Prepare a 50 mg/mL trypsin solution 1:1 in cofactors (250 mM Tris, 50 mM $CaCl_2$) and dilute it with the Tissue Digestion Buffer to achieve a final concentration of trypsin at 5 mg/mL.

3. 2 mL 'tough' tubes and ceramic beads can be ordered separately.
4. Acetonitrile has modest toxicity in small doses. Use personal protective equipment and refer to the appropriate material safety data sheet before handling.
5. We ordered the 5′ modified biotin-TEG-modified oligonucleotides (DNA ligand complementary to the target Morpholino sequence) from Integrated DNA Technologies. Adding TEG modification reduces steric hindrance and allows easier access of the Morpholino analyte to the cDNA ligand.
6. Include specimens from mock-treated controls, i.e., blank tissue or plasma, for measuring background. All control samples must go through the same sample preparation steps as the test samples, including homogenization, extraction, lyophilization, and all dilution steps.
7. If there is a large discrepancy in the weight of the tissue samples, it might be necessary to preweigh each tissue and lyse it with the Tissue Digestion Buffer at 200 mg tissue per 1 mL of the buffer.
8. We find that incubating for 8–12 h completely digests most tissues, but depending on the size of the tissue, digestion could last up to 16 h (especially for more fibrous tissues like the diaphragm). Let samples go until the tissue is completely digested. Place Parafilm around the tops of the sample tubes prior to the overnight incubation step to avoid evaporation of lysate. **Caution**, inconsistent homogenization and extraction efficiencies will compromise the quality (precision/accuracy) of the assay and can result in serious error in data interpretation.
9. If a screw cap microtube is used, turn the cap loose (about one turn) to allow vapor to escape during the lyophilization process. If a snap cap eppendorf tube is used, use a thumb tack to poke a hole in the top of the frozen tube prior to the lyophilization step.
10. Plasma samples that are diluted > 100-fold do not require extraction. In our experience plasma diluted at least 100-fold (in running buffer) can be injected directly over the flow cell. For plasma samples at lower dilutions, we found that adding ~20% (v/v) of 4 M $(NH_4)_2SO_4$ was useful for reducing background; for example, dilute 10 uL of plasma in 90 uL of running buffer, plus 20 μL of 4 M $(NH_4)_2SO_4$. If it is necessary to pretreat your diluted unknown samples with $(NH_4)_2SO_4$,

remember to also use it to prepare your standard curve and control samples at the same concentration (**step 7** in Subheading 3.3). For urine samples, the 1/10 dilution in the Running Buffer has a low background signal (Table 1) and the concentration of a Morpholino in urine is normally fairly high for easy detection.

11. Great care should be given when preparing the standards. This will help assess the quality and the reproducibility of the assay over time. 'Good' calibration is achieved by 'spiking' blank/ untreated matrix with known concentrations of the target Morpholino compound. We find it best to use an 11-point curve that has been calibrated at the following concentrations: 400, 200, 100, 50, 25, 12.5, 6.25, 3.125, 1.56, 0.78, 0 nM. **All of the standard curve and control samples should be subjected to the exact same sample preparations as the unknowns including extraction and lyophilization**.
12. For Morpholino analysis using plasma samples, prepare the calibration curve and control samples in blank plasma. The % blank plasma in each standard and control sample should be same as the % in the unknown samples.
13. For concentration analysis, biotin-cDNA ligand complementary to the target Morpholino sequence must be immobilized onto a SA chip. The most suitable chip for concentration determination of Morpholino oligomers is a CM dextran matrix sensor chip preimmobilized with streptavidin (SA). The SA senor chip achieves the greatest (ligand) binding density compared to other chips.
14. Up to four ligands can be immobilized onto the sensor chip (one per each flow cell).
15. New chips must be preconditioned with NaOH to ensure the removal of unbound SA from the chip surface. Then a biotin-modified cDNA will only bind to the chip-bounded

Table 1
Background signal (relative unit (RU)), lower limit of quantification (LLOQ), and upper limit of quantification (ULOQ) of a Morpholino in various sample matrices

Sample matrix	Background (RU)	LLOQ	ULOQ
Running buffer	0–3	1.5	100–200
Tissues(extracted and lyophilized	0–3	2	100–200
Plasma (0.1X)	20–30	7	100–200
Urine (0.1X)	6–8	3	100–200

SA. Complete saturation of ligand can be achieved using a single 10 min pulse of 1 μM biotin-cDNA. The binding density of a flow cell completely saturated with SA on the surface should yield about 2200 RU.

16. Great care should be taken when storing and managing sensor chips—each chip should be labeled with appropriate ligand/s to the corresponding flow cell/s. We find it is best to store individual sensor chips in a plastic 45 mL conical vial. Each sensor chip should be assigned a unique identifier code that includes the date (in the event that multiple chips are immobilized with the same ligand at different times).
17. We find it best to run control samples several times throughout the course of each concentration assay as a quality control measure (Fig. 4).
18. Only samples within the linear calibration range are reliable. We found that the linear range for a Morpholino sample is between 2 and 200 nM. Samples having signals outside this range should not be used for concentration determination. If a sample is below the limit of detection, the sensitivity can be increased by using a flow cell with higher ligand density on the chip surface and/or by optimizing the sample preparation conditions in order to reduce the level of background "noise" caused by nonspecific binding to the sensor chip surface. If a sample generates a signal greater than the upper limit of detection, try diluting and reanalyzing the sample.

References

1. Davis TM, Wilson WD (2000) Determination of the refractive index increments of small molecules for correction of surface plasmon resonance data. Anal Biochem 284(2):348–353. doi:10.1006/abio.2000.4726
2. Quinn JG, O'Neill S, Doyle A, McAtamney C, Diamond D, MacCraith BD, O'Kennedy R (2000) Development and application of surface plasmon resonance-based biosensors for the detection of cell-ligand interactions. Anal Biochem 281(2):135–143. doi:10.1006/abio.2000.4564
3. Schneider CS, Bhargav AG, Perez JG, Wadajkar AS, Winkles JA, Woodworth GF, Kim AJ (2015) Surface plasmon resonance as a high throughput method to evaluate specific and non-specific binding of nanotherapeutics. J Control Release 219:331–344. doi:10.1016/j.jconrel.2015.09.048
4. Mullett WM, Lai EP, Yeung JM (2000) Surface plasmon resonance-based immunoassays. Methods 22(1):77–91. doi:10.1006/meth.2000.1039
5. Rich RL, Myszka DG (2000) Survey of the 1999 surface plasmon resonance biosensor literature. J Mol Recognit 13(6):388–407. doi:10.1002/1099-1352(200011/12)13:6<388::AID-JMR516>3.0.CO;2-#
6. BIACORE Application handbook. (BIAcore, Inc., Piscataway, N.J.).
7. Andersson K, Areskoug D, Hardenborg E (1999) Exploring buffer space for molecular interactions. J Mol Recognit 12(5):310–315. doi:10.1002/(SICI)1099-1352(199909/10)12:5<310::AID-JMR470>3.0.CO;2-5
8. Cirak S, Arechavala-Gomeza V, Guglieri M, Feng L, Torelli S, Anthony K, Abbs S, Garralda ME, Bourke J, Wells DJ, Dickson G, Wood MJ, Wilton SD, Straub V, Kole R, Shrewsbury SB, Sewry C, Morgan JE, Bushby K, Muntoni F (2011) Exon skipping and dystrophin restoration in patients with Duchenne muscular dystrophy after systemic phosphorodiamidate morpholino oligomer treatment: an open-label, phase 2, dose-escalation study. Lancet 378(9791):595–605. doi:10.1016/S0140-6736(11)60756-3

9. Nakano S, Ozasa S, Yoshioka K, Fujii I, Mitsui K, Nomura K, Kosuge H, Endo F, Matsukura M, Kimura S (2011) Exon-skipping events in candidates for clinical trials of morpholino. Pediatr Int 53(4):524–529. doi:10.1111/j.1442-200X.2011.03330.x
10. Takeda S (2009) Exon skipping therapy for Duchenne muscular dystrophy by using antisense Morpholino. Rinsho shinkeigaku 49(11):856–858
11. Bauer R, Katus HA, Muller OJ (2010) Exon skipping with morpholino oligomers: new treatment option for cardiomyopathy in Duchenne muscular dystrophy? Cardiovasc Res 85(3):409–410. doi:10.1093/cvr/cvp397
12. Goyenvalle A, Babbs A, Powell D, Kole R, Fletcher S, Wilton SD, Davies KE (2010) Prevention of dystrophic pathology in severely affected dystrophin/utrophin-deficient mice by morpholino-oligomer-mediated exon-skipping. Mol Ther 18(1):198–205. doi:10.1038/mt.2009.248
13. Yokota T, Lu QL, Partridge T, Kobayashi M, Nakamura A, Takeda S, Hoffman E (2009) Efficacy of systemic morpholino exon-skipping in Duchenne dystrophy dogs. Ann Neurol 65(6):667–676. doi:10.1002/ana.21627
14. Gebski BL, Mann CJ, Fletcher S, Wilton SD (2003) Morpholino antisense oligonucleotide induced dystrophin exon 23 skipping in mdx mouse muscle. Hum Mol Genet 12(15):1801–1811
15. Summerton J, Stein D, Huang SB, Matthews P, Weller D, Partridge M (1997) Morpholino and phosphorothioate antisense oligomers compared in cell-free and in-cell systems. Antisense Nucleic Acid Drug Dev 7(2):63–70
16. Summerton J (1999) Morpholino antisense oligomers: the case for an RNase H-independent structural type. Biochim Biophys Acta 1489(1):141–158
17. Yin H, Moulton HM, Betts C, Seow Y, Boutilier J, Iverson PL, Wood MJ (2009) A fusion peptide directs enhanced systemic dystrophin exon skipping and functional restoration in dystrophin-deficient mdx mice. Hum Mol Genet 18(22):4405–4414. doi:10.1093/hmg/ddp395
18. Moulton HM (2008, 29 May) RNA binding, pharmacokinetics and tissue distribution of antisense morpholino oligomers targeted to Ebola virus. In: Northwest Biacore user group meeting, Seattle, WA
19. Boutilier J, Wu B, Lu QL, Kole R, Moulton HM (2009) PPMO–promising therapeutics for Duchene muscular dystrophy. In: 12th annual meeting American society of gene therapy, San Diego, CA

Chapter 22

Ultrasensitive Hybridization-Based ELISA Method for the Determination of Phosphorodiamidate Morpholino Oligonucleotides in Biological samples

Umar Burki and Volker Straub

Abstract

Determining the concentration of oligonucleotide in biological samples such as tissue lysate and serum is essential for determining the biodistribution and pharmacokinetic profile, respectively. ELISA-based assays have shown far greater sensitivities compared to other methods such as HPLC and LC/MS. Here, we describe a novel ultrasensitive hybridization-based ELISA method for quantitating morpholino oligonucleotides in mouse tissue lysate and serum samples. The assay has a linear detection range of 5–250 pM (R2 > 0.99).

Key words Phosphorodiamidate morpholino oligonucleotides, ELISA, Hybridization, MNase, Pharmacokinetic, Biodistribution

1 Introduction

In addition to determining the efficacy, detecting the levels of oligonucleotides in biological samples such as blood and tissues is essential for determining the pharmacokinetic and pharmacodynamic (PK/PD) relationship. This, in turn, enables optimization of efficacy, while minimizing toxicity of the drug, thus significantly increasing the chances of successful development to a therapeutic drug.

An enzyme-linked immunosorbent assay (ELISA)-based approach for detecting phosphorothioate (PTO) oligonucleotides has demonstrated far greater sensitivity compared to traditional HPLC and LC/MS method, with detection levels in the picomolar range [1, 2]. The PTO ELISA was successfully adapted for detection of 2′-O-methyl phosphorothioate (2′OMe) oligonucleotides, and as a result, this ELISA is currently the method of choice for 2′OMe detection in preclinical and clinical studies [3–6]. The 2′OMe ELISA is based on a hybridization principle first described

Hong M. Moulton and Jon D. Moulton (eds.), *Morpholino Oligomers: Methods and Protocols*, Methods in Molecular Biology, vol. 1565, DOI 10.1007/978-1-4939-6817-6_22, © Springer Science+Business Media LLC 2017

by Yu et al., where a complementary capture probe binds to the 2′OMe AON and then a detection probe is ligated to the oligo [2]. In contrast, PMOs lack the OH group at the terminal end, which is required for ligating the detection probe, and therefore, this approach cannot be used directly for PMOs. This may explain why a similar ELISA method has not been developed for PMO detection, and thus, LC/MS is currently the method of choice for PMO studies.

We recently developed and published the first ELISA method for detecting PMO and peptide conjugated PMOs (P-PMOs) in biological samples [7]. Here, we describe the method which has been adapted from a nuclease protection design first reported by Wei et al. in 2006 [1]. The method presented here has been significantly modified from the original design, and as a result, the assay has an improved detection limit of 5 pM, which is tenfold more sensitive than the original ELISA design. The assay has been successfully validated in mouse serum and tissue lysate and subsequently tested on in vivo samples to determine the pharmacokinetic and biodistribution (PK/BD) profile of both PMO and P-PMO in mice [7].

Briefly, the principle of the ELISA (shown in Fig. 1) is based on hybridization of PMO with a complementary DNA/PTO probe. The DNA probe contains PTO terminals ends which are nuclease resistant and this stabilizes the PMO/Probe hybrids during the nuclease step. The probe is dual labeled with a biotin and a digoxigenin at each end, where biotin is used to anchor the probe

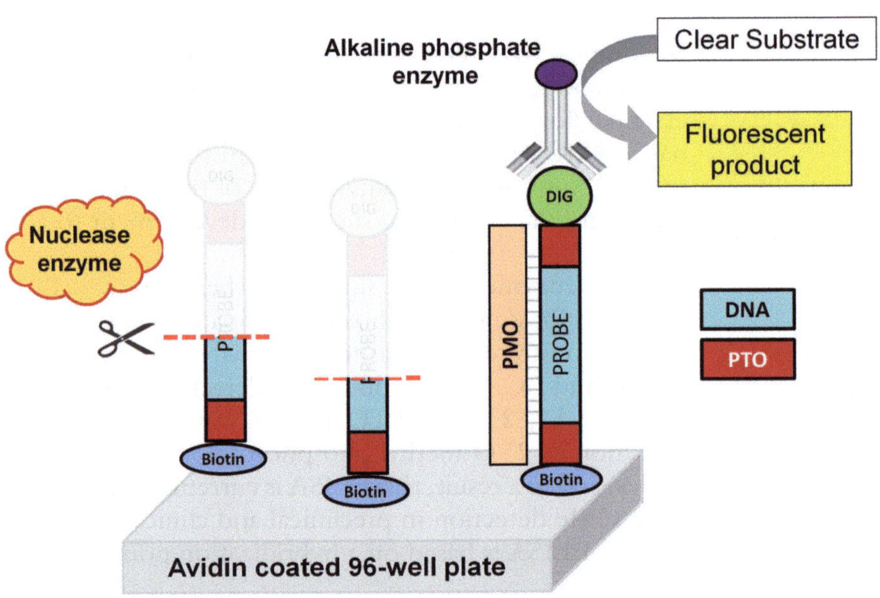

Fig. 1 Overview of the enzyme-linked immunosorbent assay (ELISA) principle based on hybridization of phosphorodiamidate morpholino oligonucleotides (PMO) with a dual-labeled DNA/phosphorothioate (PTO) probe

to NeutrAvidin-coated 96-well plates and digoxigenin is used for subsequent detection. The PMO/probe hybrids are differentiated from the single probe using micrococcal nuclease (MNase), which preferentially cleaves single-stranded DNA. The remaining hybrids are detected using the antidigoxigenin antibody conjugated with an alkaline phosphatase enzyme, which converts a clear substrate into a fluorescent solution. The intensity of the fluorescence is determined using a fluorescence microplate reader, and a standard curve of PMO/P-PMO in serum or tissue lysate is used to determine levels of each oligonucleotide present in samples.

2 Materials

Prepare all solutions using ultrapure water (prepared by purifying deionized water to attain a sensitivity of 18 MΩ cm at 25 °C) and analytical grade reagents. Prepare and store all reagents at room temperature (unless indicated otherwise).

2.1 Oligonucleotides

1. **PMO**: 25 mer PMO is purchased from Gene Tools, LLC. (Philomath, OR, USA). Prepare 1 mM stock solution by adding 1 mL of water to 1000 nmoles of lyophilized PMO in a glass vial. Vortex thoroughly for 1 min and store at 4 °C (*see* **Note 1**).
2. **P-PMO**: P-PMO was synthesized by conjugation the PMO with a cell-penetrating peptide Pip6a with the sequence RXRRBRRXRYQFLIRXRBRXRB (R, arginine; B, β-alanine; X, aminohexanoic acid), as previously described [8]. Prepare 0.1 mM P-PMO stock in water, then vortex for 30 s and freeze at −20 °C as 10 μL aliquots in PCR tubes.
3. **Probe**: Full length PTO/DNA probe complementary for the target PMO sequence is purchased from Eurofins MWG Operon (Ebersberg, Germany). The probe is designed with 7 phosphorothioate bonds at both ends of the probe. The probe is dual labeled with biotin at the 3′-end and digoxigenin at the 5′-end. Prepare 0.1 mM stock solution by suspending the lyophilized probe in 1× Tris–EDTA (TE) buffer. Vortex thoroughly for 1 min and store at 4 °C.

2.2 ELISA Buffers and Stock Solutions

1. **Neutravidin protein stock solution**: Prepare 1 mg/mL stock solution by adding 10 mL water to 10 mg Neutravidin protein. Vortex for 30 s and store at 4 °C (*see* **Notes 2** and **3**).
2. **Neutravidin coating buffer**: Prepare the coating buffer by emptying the contents of one carbonate–bicarbonate buffer capsule (Sigma-Aldrich, UK) in 100 mL water and mix thoroughly. Store at room temperature.

3. **Wash buffer**: Dissolve one Tris-buffered saline tablet (Sigma-Aldrich, UK) in 500 mL of deionized water (50 mM Tris–HCl, 150 mM sodium chloride, pH 7.6) and add 0.5 mL Tween-20 (0.1% v/v). Prepare fresh and store at room temperature.
4. **Dilution buffer**: is the same as wash buffer but with the addition of control muscle lysate or serum as required.
5. **Hybridization buffer**: is the same as wash buffer but with the addition of 1 M NaCl and 0.03 mg/mL BSA (*see* **Note 4**). Dissolve one Tris-buffered saline tablet in 385 mL of deionized water and store at room temperature. On the day of the assay transfer 38.5 mL in a conical tube and add 10 mL of 5 M NaCl stock solution (Sigma-Aldrich, UK) and 1.5 mL of 10× BSA stock (see **step 7** in Subheading 2.2). Finally, add 50 μL Triton™ X-100 (0.1% v/v) and keep at room temperature (*see* **Note 5**).
6. **MNase buffer 10× stock:** Weigh 30.29 g of Trizma base (Sigma-Aldrich, UK) and 2.79 g of $CaCl_2$ and add to a 500 mL glass bottle. Add 200 mL of 5 M NaCl and add water to a volume of 400 mL. Mix and adjust pH with HCl to get pH 8.2 and make up to 500 mL with water. Aliquot minimum 5 mL in Bijous tubes and store at −20 °C. This is a 10× concentrated buffer and when diluted in water will yield 50 mM Tris–HCL (pH 8.5), 200 mM NaCl, 5 mM CaCl.
7. **BSA 10× stock:** prepare 10× concentrated BSA stock solution by dissolving 500 mg BSA (Sigma-Aldrich, UK) in 50 mL water to get 10 mg/mL BSA. Aliquot minimum 1 mL in eppendorf tubes and store at −20 °C.
8. **Trypsin 10× stock (For P-PMO only)**: Supplement trypsin stock solution (Type II-s) 25 g/L in 0.9% NaCl (Sigma-Aldrich, UK) with 100 mM CaCl to prevent autolysis. Prepare 1 mL aliquots in eppendorf tubes and store at −20 °C.

2.3 ELISA Components

1. NeutrAvidin stock solution (Thermo Fisher, UK). Store at 4 °C.
2. Micrococcal nuclease (MNase): 2000,000 gels units/mL concentration (New England BioLabs, UK). Store at −20 °C.
3. Antidigoxigenin-alkaline phosphatase Fab fragments (sheep) (Roche, UK). Store at 4 °C.
4. SuperBlock (TBS) blocking buffer (Thermo Fisher, UK). Store at 4 °C.
5. AttoPhos substrate and buffer (Promega, UK) Store at 4 °C (*see* **Note 6**).
6. Abgene adhesive plate seal (VWR, UK).
7. 50 mL reagent reservoirs (VWR, UK).

8. Greiner Bio-One 96-well black Fluotrac™ 600 plates (VWR, UK).
9. Greiner multiwell plate lids (Sigma-Aldrich, UK).
10. ABgene® natural deep well storage plates, 96-well, 0.8 mL, (VWR, UK).
11. Multichannel pipette.

2.4 Serum Collection Components

1. MiniCollect 0.8-mL serum tubes with clot activator gel (Greiner Bio-One, UK) (*see* **Note 7**).
2. Heparinized capillary tubes calibrated at 15 μL volume and plungers (BHR Pharmaceuticals, UK).
3. 25 gauge hypodermic needle.
4. Liquid paraffin oil.
5. 50 mL conical tube with small 1 cm diameter hole made at the bottom.
6. Hair clippers.

2.5 Tissue Collection and Lysate Preparation Components

1. Phosphate-Buffered Saline (PBS).
2. Small animal dissection instruments, e.g., scissors and forceps.
3. Petri dish.
4. Absorbent paper.
5. Precision scales.
6. 7 mL Bijou tubes.
7. RIPA buffer (Thermo Fisher Scientific, UK). Store at 4 °C.
8. Proteinase-K recombinant, PCR Grade (Roche, UK). Prepare a stock solution of 20 mg/mL in water. Store at 4 °C.
9. TissueRuptor® with probes (Qiagen, UK) (*see* **Note 8**).

3 Methods

3.1 Serum Collection and Preparation

1. For control serum collection, collect blood from multiple untreated mice (i.e., *via* cardiac bleed) in MiniCollect serum tubes. Allow the blood to clot for 30 min at 4 °C before centrifuging at 16,100 × *g* for 10 min to separate the serum from the blood. Pool and aliquot the serum as required. Store at −20 °C.
2. For PK analysis collect 15 μL serial blood samples as required following administration of PMO/P-PMO (*see* **Note 9**).
3. For serial blood collection from saphenous vein in hind limbs, place the mouse in the 50 mL tube with the nose pointing toward the hole at the bottom and hold securely by gripping the tail.

4. Using the forefinger and thumb gently tease out the hind limb and using the clippers shave a small area to expose the saphenous vein.
5. Apply a light coating of liquid paraffin oil to help form a neat blood drop.
6. Using the 25 g needle, carefully pierce the vein and wait for the blood drop to form.
7. Collect the blood in the 15 μL heparinized capillary tube and immediately eject into the MiniCollect serum tube (*see* **Note 10**).
8. Place tube at 4 °C for 30 min to allow the blood to clot before centrifuging at 16.2 k rpm for 5 min to separate the serum from the blood.
9. Extract 5 μL of serum from each sample into a PCR tube. Label and store at −20 °C.

3.2 Tissue Collection and Lysate Preparation

1. Wash freshly excised tissue using PBS in a Petri dish and gently dry on absorbent paper and then place tissues in bijou tubes for weighing and labeling. Store tissues at −20 °C (*see* **Note 11**).
2. For lysate preparation, thaw the frozen tissues at room temperature and prepare fresh tissue lysis buffer using RIPA buffer supplemented with 2 mg/mL Proteinase-K. Then add lysis buffer to each tissue at 100 mg/mL (w/v) (*see* **Note 12**).
3. Homogenize the tissues using TissueRuptor® for a few seconds until the tissues are fully homogenized and then incubate at 55 °C overnight (*see* **Note 13**).
4. Following incubation transfer 1 mL of the homogenate to an eppendorf tube and centrifuge at 16.2 k rpm for 15 min. Extract the supernatant and aliquot 500 μL of the supernatant in PCR tubes. Store samples at −20 °C ready for analysis.

3.3 Additional Steps for Preparing P-PMO Samples and Standards

1. P-PMO samples and standards require overnight incubation with Trypsin to digest the peptide portion of the P-PMO as it interferes with detection. Thaw aliquots of 10× trypsin stock solution at room temperature and add to P-PMO samples and standards as described as follows.
2. For serum samples, after extracting 5 μL of serum from 15 μL blood into a PCR tube, add 5 μL of trypsin stock solution and 40 μL of water. Vortex for 10 s and incubate overnight at 37 °C. Following incubation store the samples at −20 °C ready for analysis.
3. For tissue lysate samples, when extracting supernant following overnight incubation at 55 °C, extract 400 μL of supernatant and add 100 μL of trypsin. Vortex for 10 s and incubate overnight at 37 °C. Following incubation store the samples at −20 °C ready for analysis.

4. For serum standards, add 5 μL of control serum to a PCR tube, add 5 μL of trypsin stock solution, and add 5 μL of 0.1 mM P-PMO and 35 μL of water. Vortex for 10 s and incubate overnight at 37 °C. Following incubation store the samples at −20 °C ready for analysis. (*see* **Note 14**).
5. For tissue lysate standards add 5 μL of 0.1 mM P-PMO and 5 μL of trypsin stock solution, and add 40 μL of control lysate. Vortex for 10 s and incubate overnight at 37 °C. Following incubation store the samples at −20 °C ready for analysis.

3.4 ELISA Procedure

Warm all reagents to room temperature and carry out all procedures at room temperature unless otherwise specified.

1. For coating one 96-well plate: transfer 15 mL of NeutrAvidin coating buffer to a 15 mL conical tube and add 15 μL of NeutrAvidin stock solution to get 1 μg/mL dilution. Vortex for 10 s and using a multichannel pipette and a reservoir add 100 μL to each well in the black 96-well plate (100 ng/well). Seal with a PCR film, place a lid and incubate at 37 °C for 2 h (*see* **Note 15**).
2. Prepare PMO/P-PMO serial dilutions for a standard curve by diluting the PMO stock solution to get 2 nM starting dilution. First of all warm PMO/P-PMO stock solution to room temperature and vortex for 1 min. For a typical 1 mM PMO stock solution this is best done stepwise, so first dilute 5 μL of 1 mM PMO stock solution in 5 mL of water in a bijou tube to get 1 μM concentration and vortex thoroughly for 10 s. Then dilute 1 mL in 4 mL water to get 200 nM concentration and vortex thoroughly for 10 s. Finally add 5 μL in 495 μL of dilution buffer in eppendorf tube to get 2 nM starting dilution. For a typical 0.1 mM P-PMO stock solution, first dilute 5 μL in 5 mL of water in a bijous tube to get 0.1 μM concentration and vortex thoroughly for 10 s. Then add 10 μL in 490 μL of dilution buffer in eppendorf tube to get 2 nM starting dilution. Serially dilute the starting PMO/P-PMO dilution twofold by diluting 250 μL in 250 μL dilution buffer (*see* **Note 16**). Make 12 serial dilutions in total and vortex for 10 s each time. Then transfer 150 μL of each dilution to a deep 96-well plate.
3. Prepare the probe dilution by diluting the stock solution to 0.5 nM final concentration. For a typical 0.1 mM stock solution the dilution is again best done stepwise, so first dilute 5 μL of stock solution in 5 mL water in a bijou tube to get 0.1 μM concentration and vortex thoroughly for 10 s. Then dilute this 1 in 200 by adding 25 μL to 5 mL hybridization buffer to get 0.5 nM concentration and vortex thoroughly for 10 s.
4. Using a multichannel pipette and a reservoir carefully add 150 μL of 0.5 nM probe solution to the PMO standards in the deep 96-well plate.

5. For control wells add 150 μL of dilution buffer in 4 wells. Then add 150 μL of 0.5 nM probe dilution to top 2 wells and hybridization buffer in the other two wells. (*See* Fig. 2 for a suggested plate map.)
6. Seal the hybridization plate with a PCR film, place a lid and incubate at 37 °C for 30 min to allow the probe to hybridize with the PMO.
7. Remove the Neutravidin-coated plate from the incubator after 2 h and discard the coating solution over a sink and dry on absorbent paper.
8. Wash 3× with 100 μL wash buffer per well and dry on absorbent paper.
9. Using a multichannel pipette add 200 μL of SuperBlock (TBS) Blocking Buffer to each well and leave for 5 min, then discard solution and dry on absorbent paper.
10. Remove the hybridization plate from the incubator and carefully remove the PCR film. Then using a multichannel pipette transfer 100 μL from each well to the Neutravidin-coated plate in duplicates using fresh tips each time.
11. Seal the plate with a PCR film, place a lid and incubate at 37 °C for 30 min to allow the biotin-labeled probe to bind to the Neutravidin coating.

	Standard dilutions (pM)		Sample dilutions					
	1	2	3	4	5	6	7	8
1	2		A: 1		A: 2		A: 3	
2	1		B: 1		B: 2		B: 3	
3	0.5		C: 1		C: 2		C: 3	
4	0.25		D: 1		D: 2		D: 3	
5	0.125		E: 1		E: 2		E: 3	
6	0.0625		F: 1		F: 2		F: 3	
7	0.03125		G: 1		G: 2		G: 3	
8	0.015625		H: 1		H: 2		H: 3	
9	0.0078125		I: 1		I: 2		I: 3	
10	0.00390625		J: 1		J: 2		J: 3	
11	0.001953125		K: 1		K: 2		K: 3	
12	MC		CMAX		AB		SUB	

Fig. 2 Suggested plate map showing the standard PMO/P-PMO dilutions alongside samples and controls in duplicate wells. Standard dilutions are best prepared by twofold dilutions starting from 2 nM concentration. Each sample (i.e., A, B, C, etc.) can be tested at three different dilutions (e.g., 1, 2, 3) to ensure at least one dilution is within the linear detection range. There are four control wells where no PMO/P-PMO is added to determine the background signal. *MC* is maximum cleaved well where in absence of PMO the probe is maximally cleaved by MNase to give the lowest possible signal from probe wells. *CMAX* is concentration of probe which gives maximum possible signal in absence of MNase. *AB* is antibody and *SUB* is substrate background

12. Prepare MNase buffer solution by adding 5 mL of MNase stock solution (10×) and 0.5 mL of BSA (10×) in 50 mL conical tube and make up to 50 mL with water.
13. Remove the plate from the incubator after 30 min and discard the coating solution and dry on absorbent paper.
14. Wash 3× with 100 μL wash buffer per well (*see* **Note 17**).
15. Transfer 30 mL of MNase buffer solution to 50 mL conical tube and add 3 μL of MNase enzyme. Vortex for 10 s then using a multichannel pipette and a 50 mL reservoir add 150 μL in all wells (30 U/well) except the control "cmax" wells. Add enzyme-free MNase solution in the "cmax" wells (*see* **Note 18**).
16. Seal the plate with a PCR film, place a lid and incubate at 37 °C for 1 h to allow minimum of 99% single stranded probes to be cleaved by MNase (*see* **Note 18**).
17. Prepare the antibody solution by adding 32 mL SuperBlock (TBS) Blocking Buffer in 50 mL conical tube and add 80 μL of Tween-20 (0.25% v/v). Remove 2 mL into a bijou tube, which is reserved for substrate control wells and add 6 μL of Antidigoxigenin AP-conjugated antibody (1/5000 dilution) to the remaining 30 mL blocking buffer. Vortex for 10 s.
18. Remove the plate from the incubator after 1 h and discard the MNase solution and dry on absorbent paper.
19. Wash 3× with 100 μL wash buffer per well and dry on absorbent paper.
20. Using a multichannel pipette and a reservoir add 150 μL of antibody solution in all wells except control substrate wells. Add antibody-free solution in the substrate control wells.
21. Seal the plate with a PCR film, place a lid and incubate at 37 °C for 30 min to allow the antibody to bind to the remaining digoxigenin-labeled probes.
22. Prepare the substrate solution by adding 30 mL of Attophos buffer in 50 mL conical tube and add 9 mg of Attophos substrate. Vortex for 10 s and protect from light.
23. Remove the plate from the incubator after 1 h and discard the antibody solution and dry on absorbent paper.
24. Wash 3× with 100 μL wash buffer per well and dry on absorbent paper.
25. Using a multichannel pipette and a reservoir add 150 μL of substrate solution in all wells.
26. Seal the plate with a PCR film, place a lid and incubate at 37 °C for 30 min. When transferring to and from the incubator, the plate should be protected from light using an aluminum foil.

27. Remove the plate from the incubator after 30 min and after removing the lid and the PCR film, place the plate in the plate reader and run the program.
28. Plate reader program:
    (a) Briefly shake for 10 s at 2 mm radius/600 rpm.
    (b) Measure fluorescence intensity at 444 nm excitation and 555 nm emission.

3.5 Data Analysis

1. First determine the linear detection range of the assay by plotting the standard dilutions data with fluorescence intensity (FI) on the y-axis and PMO/P-PMO concentration (pM) on the x-axis.
2. Transform the x-axis into log values to generate a typical semi-log standard curve graph (Fig. 3a) and then transform the y-axis to generate a double log-linear graph (Fig. 3b).
3. Display the linear regression values and equation on the graph.
4. The linear detection range is determined by two factors, where the linear regression values must be above $R^2 > 0.99$ and the lower limit of quantitation (LLOQ) must be at least twofold above background control values (i.e., mc wells).
5. Determine the upper and lower limits of the detection range and using the following equation calculate the PMO/P-PMO concentrations for samples which fall within the linear range: Concentration of morpholino = (Fluorescence intensity/ Gradient)^(1/index)
6. Gradient and Index values are obtained from the linear detection graph as shown in Fig. 3b.
7. Finally multiply the calculated concentration values in the well by the dilution factor to determine the concentration of morpholino in the original sample.

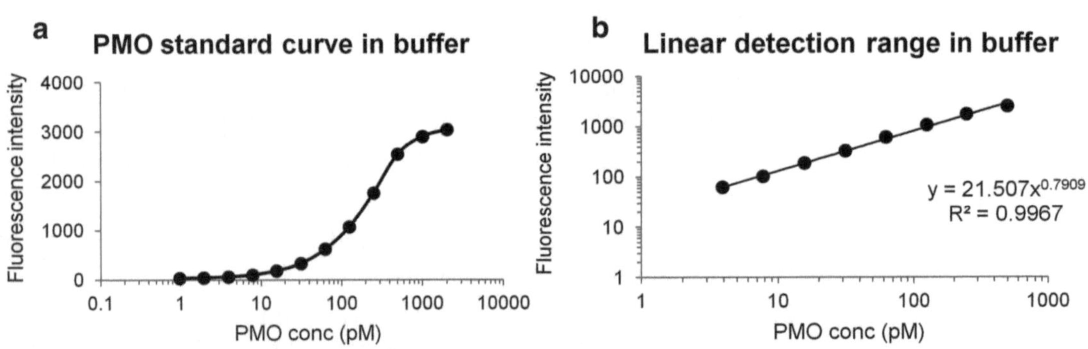

Fig. 3 Typical PMO standard curve in buffer plotted on a semi-log graph (**a**) and the resulting linear detection range on double-log graph (**b**). The latter shows the equation and regression values, which are used to determine the concentration of PMO in samples

4 Notes

1. PMO stock solutions prepared in water are very stable and can be stored at room temperature, 4 °C or −20 °C. It is important to ensure the PMO stock solution is thawed to room temperature and thoroughly vortexed for 1 min before use. If the concentration of stock solution is lower than expected then it may be necessary to heat the stock solution at 65 °C for 10 min before vortexing.
2. The Neutravidin stock solution will turn cloudy over time but we find that it has no effect on performance.
3. You can also buy readymade Neutravidin-coated plates from Thermo Scientific, which use the same components as described in this protocol. The advantage of readymade plates is that it saves time but the downside is that it is more costly than self-coated plates. We find there is no difference in performance between both plates.
4. We find that addition of BSA is essential for getting consistently high sensitivity.
5. Prepare the hybridization buffer fresh each time.
6. Attophos substrate can be purchased in bulk 1 g size, which can then be aliquoted in 9 mg for one plate or 18 mg for two plates in eppendorf tubes. These must be kept at 4 °C protected from light.
7. Both serum and plasma are suitable for detecting PMO/P-PMO in blood.
8. The probes for homogenizing tissues are disposable but can be used multiple times if cleaned in 7% ethanol and autoclaved.
9. 15 μL blood samples consistently yield 5 μL of serum or plasma to be easily extracted.
10. We find it is often unavoidable for serum or plasma samples to become hemolyzed but this has no effect on the detection or sensitivity of the assay.
11. It is important to wash away as much blood as possible to minimize contaminating tissue lysate. This is especially true for harvesting tissues when PMO/P-PMO levels are high. Ensure each tissue is washed separately to prevent cross-contamination between tissues, especially from kidney and liver.
12. Tissue lysate can be made more or less concentrated than 100 mg/mL; however, we find that this concentration works best for all tissues.
13. Muscle lysate can be made from a pool of various muscle tissues such as quadriceps (QD), tibialis anterior (TA), gastrocnemius (GST), and triceps (TCP).

14. 0.1 mM P-PMO is diluted by 1/10 to 0.01 mM, which needs to be diluted to 2 nM starting dilution. Therefore, dilute 0.01 mM P-PMO using 5 μL in 500 μL water and then 10 μL in 490 μL dilution buffer.
15. The Neutravidin coating can be done at 37 °C for up to 4 h or alternatively it can be left at room temperature overnight.
16. Multiple standards and samples can be quickly prepared using the deep well plate and a 200 μL multichannel pipette. Add 200 μL dilution buffer or diluents (serum or tissue lysate) as required in each well and dilute 200 μL from subsequent well to get the required twofold serial dilutions.
17. Ensure the plate is not left to dry by quickly adding the MNase solution after removing the wash buffer for the third time. This ensures the PMO/Probe hybrids are not left in salt-free environments for longer than necessary to minimize the risk of destabilizing the hybrids.
18. The amount of MNase required to cleave single stranded probe by a minimum of 99% is predominantly sequence dependent. MNase preferentially cleaves between AT bases; therefore sequences with low proportion of AT bases in the central DNA region require greater amounts of MNase than sequences with a high proportion of AT bases. Also the latter probes require MNase incubation at room temperature to maintain the integrity of the PMO/probe hybrids. It is therefore recommended that the optimal MNase concentration and incubation temperatures are determined for each probe.

Acknowledgements

We acknowledge the support of Sylvia Aitken Charitable Trust in funding this work and Michael John Gait for kindly providing the P-PMO.

References

1. Wei X, Dai G, Marcucci G, Liu Z, Hoyt D, Blum W, Chan KK (2006) A specific picomolar hybridization-based ELISA assay for the determination of phosphorothioate oligonucleotides in plasma and cellular matrices. Pharm Res 23(6):1251–1264. doi:10.1007/s11095-006-0082-3
2. Yu RZ, Baker B, Chappell A, Geary RS, Cheung E, Levin AA (2002) Development of an ultrasensitive noncompetitive hybridization-ligation enzyme-linked immunosorbent assay for the determination of phosphorothioate oligodeoxynucleotide in plasma. Anal Biochem 304(1):19–25. doi:10.1006/abio.2002.5576 S0003269702955764 [pii]
3. Goemans NM, Tulinius M, van den Akker JT, Burm BE, Ekhart PF, Heuvelmans N, Holling T, Janson AA, Platenburg GJ, Sipkens JA, Sitsen JM, Aartsma-Rus A, van Ommen GJ, Buyse G, Darin N, Verschuuren JJ, Campion GV, de Kimpe SJ, van Deutekom JC (2011) Systemic

administration of PRO051 in Duchenne's muscular dystrophy. N Engl J Med 364(16):1513–1522. doi:10.1056/NEJMoa1011367

4. Heemskerk HA, de Winter CL, de Kimpe SJ, van Kuik-Romeijn P, Heuvelmans N, Platenburg GJ, van Ommen GJ, van Deutekom JC, Aartsma-Rus A (2009) In vivo comparison of 2'-O-methyl phosphorothioate and morpholino antisense oligonucleotides for Duchenne muscular dystrophy exon skipping. J Gene Med 11(3):257–266. doi:10.1002/jgm.1288
5. Verhaart IE, Tanganyika-de Winter CL, Karnaoukh TG, Kolfschoten IG, de Kimpe SJ, van Deutekom JC, Aartsma-Rus A (2013) Dose-dependent pharmacokinetic profiles of 2'-O-methyl phosphorothioate antisense oligonucleotidesin mdx mice. Nucleic Acid Ther 23(3):228–237. doi:10.1089/nat.2012.0398
6. Verhaart IE, van Vliet-van den Dool L, Sipkens JA, de Kimpe SJ, Kolfschoten IG, van Deutekom JC, Liefaard L, Ridings JE, Hood SR, Aartsma-Rus A (2014) The dynamics of compound, transcript, and protein effects after treatment with 2OMePS antisense oligonucleotides in mdx mice. Mol Ther Nucleic Acids 3:e148. doi:10.1038/mtna.2014.1 mtna20141 [pii]
7. Burki U, Keane J, Blain A, O'Donovan L, Gait MJ, Laval SH, Straub V (2015) Development and application of an ultrasensitive hybridization-based ELISA method for the determination of peptide-conjugated phosphorodiamidate morpholino oligonucleotides. Nucleic Acid Ther 25(5):275–284. doi:10.1089/nat.2014.0528
8. Betts C, Saleh AF, Arzumanov AA, Hammond SM, Godfrey C, Coursindel T, Gait MJ, Wood MJ (2012) Pip6-PMO, A new generation of peptide-oligonucleotide conjugates with improved cardiac exon skipping activity for DMD treatment. Mol Ther Nucleic Acids 1:e38. doi:10.1038/mtna.2012.30 mtna201230 [pii]

Index

Hong M. Moulton and Jon D. Moulton (eds.), *Morpholino Oligomers: Methods and Protocols*, Methods in Molecular Biology, vol. 1565, DOI 10.1007/978-1-4939-6817-6, © Springer Science+Business Media LLC 2017

M

N

O

P

MIX
Papier aus verantwortungsvollen Quellen
Paper from responsible sources
FSC® C105338

If you have any concerns about our products,
you can contact us on
ProductSafety@springernature.com

In case Publisher is established outside the EU,
the EU authorized representative is:
Springer Nature Customer Service Center GmbH
Europaplatz 3, 69115 Heidelberg, Germany

Printed by Libri Plureos GmbH
in Hamburg, Germany